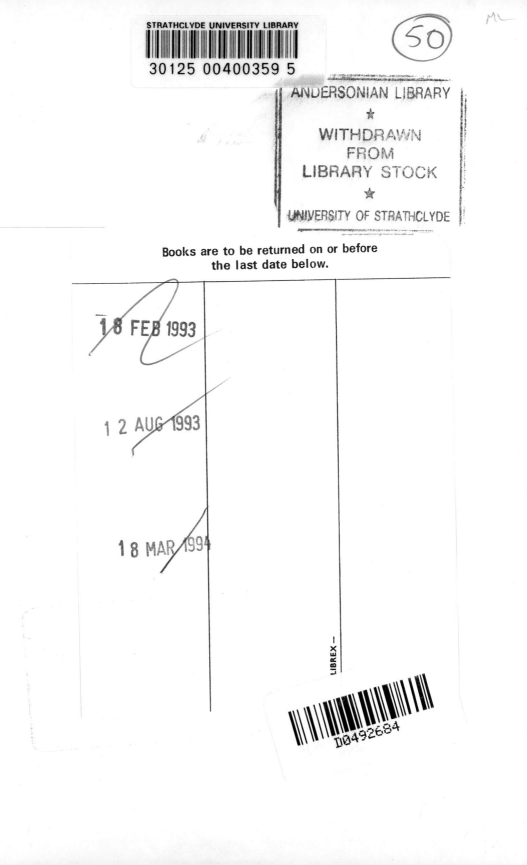

Insect Neurochemistry and Neurophysiology • 1986

Experimental and Clinical Neuroscience

Insect Neurochemistry and Neurophysiology • 1986,
edited by *A. B. Borkovec and Dale B. Gelman,* 1986

Molecular Pathology of Nerve and Muscle: *Noxious Agents
and Genetic Lesions,* edited by *Antony D. Kidman, John
K. Tomkins, Carol A. Morris, and Neil A. Cooper,* 1983

Neural Membranes, edited by *Grace Y. Sun, Nicolas
Bazan, Jang-Yen Wu, Giuseppe Porcellati, and
Albert Y. Sun,* 1983

Insect Neurochemistry and Neurophysiology • 1986

Edited by

A. B. Borkovec and Dale B. Gelman

USDA, Beltsville, Maryland

The Humana Press • Clifton, New Jersey

Library of Congress Cataloging in Publication Data

Insect neurochemistry and neurophysiology, 1986.

(Experimental and clinical neuroscience)
Papers presented at the Second International Conference on Insect
Neurochemistry and Neurophysiology (ICINN—1986) held at the University of Maryland on
Aug. 4–6, 1986.
Includes index.
1. Insects—Physiology—Congresses. 2. Neurochemistry—Congresses.
3. Neurophysiology—Congresses.
I. Borkovec, A. B. (Alexej B.), 1925— . II. Gelman, Dale B. III. International
Conference on Insect Neurochemistry and Neurophysiology (2nd: 1986: University of
Maryland) IV. Series. [DNLM: 1. Insects—physiology—congresses. 2. Neurochemistry
—congresses. 3. Neurophysiology—congresses. WL 104 I59 1986]
QL495.I496 1986   595.7'01'88   86-27753
ISBN 0-89603-119-5

© Copyright 1986 by The Humana Press Inc.
Crescent Manor
PO Box 2148
Clifton, NJ 07015

Preface

The nature and diversity of presentations at the second International Conference on Insect Neurochemistry and Neurophysiology (ICINN–86) held at the University of Maryland on August 4–6, 1986, attest to the vitality and broad scope of research in insect neuroscience. The present volume is a written account of the invited lectures, contributed papers, and posters presented at the conference, and as such, serves as a fair indicator of the trends in current research in this field here and abroad.

The principal portion of this book consists of seven review papers that were presented by invited speakers. Although the topics vary widely, they reflect on and emphasize the main theme of the conference, i.e., the nature and function of molecular messengers that communicate between the central nervous system and organs or tissues involved in the growth, development, reproduction, and behavior of insects. This emphasis is continued in the following three sections on neurochemistry, neurophysiology, and neuroanatomy, although no conscious effort was made by the organizers to highlight these particular fields of neuroscience. It is evident that the recent advances in both physical and chemical analytical techniques have made possible the acquisition of structurally defined probes, the long sought-after tools for unraveling the secrets of endogenous communication. Each section of short papers derived from the oral and poster presentations at the conference is prefaced by an overview that highlights and summarizes the section's content.

ICINN–86, as well as the ICINN–83 that preceded it by three years, and future ICINNs that we hope will follow, could not be organized without the help and financial support of sponsoring organizations. The Agricultural Research Service, US Department of Agriculture, and the National Science Foundation were the principal US Government contributors. The industrial sponsors were the American Cyanamid Company, E. I. DuPont de Nemours & Co., Hoechst-Roussel Agri-Vet Company, Imperial Chemical Industries, Mobay Chemical Corporation, Monsanto Company, Nippon Soda Company, Rohm and Haas Company, Stauffer Chemical Company, Sumitomo Chemical Company, and Zoecon Corporation. To these far-sighted sponsors we give our sincere thanks and appreciation.

Alexej B. Borkovec
Dale B. Gelman

CONTENTS

Contents

Contents

Reviews

INSECT NEUROPEPTIDES - PURE AND APPLIED

Michael O'Shea

University of Geneva

Laboratoire de Neurobiologie, 20 Bd d'Yvoy
CH-1211 Geneva 4, Switzerland

Had this essay been written before the conference, as it should have, it would probably have started with a general statement lamenting how few insect neuropeptides were actually sequenced. Fortunately this sentiment is no longer justified. We are entering a period of rapid proliferation of information on the primary structures of insect neuropeptides. An example of the increased pace of peptide analysis was provided in this meeting by the remarkable presentation of J. Cook and G.M. Holman. In proposing primary structures for several new myotropic peptides, at a single stroke these authors have at least doubled the number of structured insect peptides. Perhaps the single most important factor in facilitating this advance is the application of gas phase sequencing, a highly sensitive modification of Edman sequencing which can yield primary structures from as little as 10 pmoles of pure peptide. In combination with reverse phase HPLC, this method has greatly reduced the problem of obtaining structural information on peptides. It is no longer true to say that the main impedement to rapid growth of this area is the difficulty of obtaining enough pure peptide from small organisms. Moreover it is probably true to say that the rate of progress in the next few years

will depend largely on the availability of gas
phase sequencing. Other methods, including
molecular cloning (see below) are also likely to
contribute to an increased knowledge of insect
neuropeptide structures. Clearly we are entering
a new and exciting era. There seems little doubt
that between now and the next ICINN meeting there
may be twenty to thirty fully sequenced insects
peptides. In parallel with this development in
neurochemistry there will also be a significant
increase in our understanding of the neurobiology
of insect peptidergic systems. Application of
this new knowledge to the development of novel
control methods represents an important
challenge.

        In this short essay I will first briefly
review the insect neuropeptide field
concentrating on the two most studied peptides,
proctolin and the adipokinetic hormones.
Secondly I will describe some strategies which
are likely to be useful in the discovery of new
peptides. Finally I will consider how
fundamental knowledge of neuropeptidergic systems
might provide clues as to how pure research in
this field can be applied.

PROCTOLIN

        The first insect neuropeptide to be
sequenced was proctolin (Arg-Tyr-Leu-Pro-Thr).
Proctolin was first discovered in the American
cockroach, Periplaneta americana, by BROWN
(1967). It was shown to have a potent contractile
effect on the hindgut or proctodeal muscles and
its isolation and purification were guided by the
hindgut muscle bioassay (BROWN and STARRATT,
1975). In 1975, STARRATT and BROWN succeeded in
determining the amino acid sequence of proctolin
purified from an extract of about 125,000
cockroaches.

        Proctolin is not restricted to the cockroach
but appears to be widely distributed among the
arthropods. It is known to occur in at least six
orders of insect (BROWN, 1977) though curiously

it is absent from Lepidoptera. More recently, proctolin has been convincingly demonstrated in crustacea, for example in the lobster Homarus (SCHWARZ et al., 1984) and crayfish (BISHOP et al., 1984). Evidence for the existence of proctolin in non-arthropod invertebrates or in vertebrates is at present incomplete. It seems possible, however, that proctolin itself or closely related peptides exist outside the arthropods. In the leech, for example, proctolin-immunoreactive neurons have been localized in the central ganglia (LI and CALABRESE, 1983). In the rat CNS, proctolin immunoreactivity has been reported in serotonin-containing neurons (HOLETS et al., 1984). It remains to be seen in these examples whether the immunoreactivity is due to the presence of authentic proctolin or to proctolin-related peptides.

Actions of proctolin are diverse and not restricted to the activity that permitted its isolation. No longer should proctolin be considered to be only a "gut peptide". Once synthetic proctolin became available it was possible to examine its bioactivities in a variety of preparations and this was quickly done. For example, PIEK and MANTEL (1977) showed that proctolin causes contraction of insect skeletal muscle and induces a myogenic rhythm of contraction and relaxation. Later MILLER (1979) demonstrated an action of proctolin on the insect heart. In addition, there is now evidence of actions of proctolin on neurons in the insect CNS (WALKER et al., 1980). In crustaceans proctolin has demonstrated activity on skeletal muscle, causing a slow graded contracture of dactyl muscle without depolarization (SCHWARZ et al., 1980) and a potentiation of the action of abdominal flexor motoneurons (BISHOP et al., 1984).

Proctolin-containing neurons were found first by application of HPLC methods to neuronal extracts (O'SHEA and ADAMS, 1981). Subsequently, the availability of synthetic proctolin allowed the development of specific proctolin rabbit antisera (BISHOP et al., 1981; ECKERT et al., 1981; SCHWARZ et al., 1984) and immunochemical methods. A whole-mount immunohistochemical method was developed and used in the mapping and identification of putative proctolinergic neurons in the insect (BISHOP and O'SHEA, 1982) and crustacean CNS (SIWICKI and BISHOP, 1985). The application of immunohistochemistry indicated the presence in the cockroach, and more recently in the locust (Schistocerca sp.) and crustacean CNS, of a subpopulation of proctolinergic skeletal motoneurons.

Several lines of indirect evidence led us initially to consider the possibility that proctolin might be involved in the process of skeletal neuromuscular transmission in insects. For example, it was well established that skeletal muscle in the locust Schistocerca was highly sensitive to proctolin. At subnanomolar concentrations it causes contracture of the extensor tibialis muscle of the locust hindleg (PIEK and MANTEL, 1977). Furthermore, MAY et al., (1979) showed that tonic skeletal muscle fibers in the locust would generate depolarizing potentials to local iontophoretic application of proctolin and that areas of high proctolin sensitivity corresponded to regions where nerve-muscle contact is made.

In spite of the growing evidence suggesting roles for proctolin on skeletal muscle, we were reluctant to accept that it might be a transmitter for skeletal motoneurons. This was due in part to the fact that transmitters had already been assigned to insect skeletal motoneurons. Thus excitatory motoneurons probably employ L-glutamic acid, or aspartic acid, inhibitory motoneurons use GABA and the modulatory dorsal unpaired motoneurons are octopaminergic (EVANS and O'SHEA, 1977). If there were no exceptions to this then proctolin would have to be a co-transmitter. Perhaps it was easier to assume that the sensitivity of skeletal muscle to proctolin was due not to its possible transmitter status but to its role as a circulating myoactive hormone. The problem with this idea was that proctolin seems unsuitable for a hormonal role. It cannot be measured in the hemolymph and is is very rapidly hydrolysed in the circulation (QUISTAD et al., 1984). Questions concerning precisely how proctolin might be delivered to muscles are now being resolved.

We now have evidence in two different insect systems for proctolin as a skeletal neuromuscular transmitter. The first to be characterized involves the identified slow coxal depressor of Ds motoneuron of the cockroach Periplaneta americana and its target muscle, the coxal depressor 177d (ADAMS and O'SHEA, 1983; O'SHEA and BISHOP, 1982). The proctolin content of this cell was established initially by immunocytochemistry and then by performing bioassay and reverse-phase high pressure liquid chromatography (HPLC) on extracts made from the isolated Ds soma. The bioassay used to detect proctolin in fractions from the HPLC was the proctolin-sensitive locust skeletal muscle - the extensor tibialis or ETi muscle of Schistocerca. This muscle was first shown to be proctolin-sensitive by PIEK and MANTEL (1977). At about 10^{-10} M proctolin induces a myogenic rhythm of contraction and relaxation. The frequency of the contractures is increased by higher concentrations and can be used to quantify the levels of proctolin in test samples added to the muscle (KESHISHIAN and O'SHEA, 1985). Using such a bioassay in conjunction with HPLC we have been able to confirm the proctolin-immunocytochemical localization of proctolin to the Ds motoneuron (O'SHEA and BISHOP, 1982).

The Ds motoneuron is one of at least five motoneurons that innervate the coxal depressor muscles (PEARSON and ILES, 1971). By comparing the pattern of proctolin-immunoreactivity with the known pattern of projection by the Ds motoneuron we have established that the Ds neuron is the sole proctolin positive motoneuron of the coxal depressor muscles (WITTEN and O'SHEA, 1985). This evidence suggests that there exists a subpopulation of proctolinergic motoneurons in the skeletal motoneuronal pool of the cockroach and we now believe that a similar situation exists in the locust Schistocerca.

Proctolin immunoreactivity is clearly seen in the Ds nerve terminals on the coxal depressor muscle. This is consistent with a transmitter role for proctolin, a conclusion which is also supported by the results of physiological experiments. Thus it seems that in insect skeletal muscle, sensitivity to proctolin is explained by the local release of the peptide by peptidergic skeletal motoneurons. If this is the case then the locust extensor tibialis (ETi) muscle, used in the proctolin bioassay, may also be innervated by a proctolinergic motoneuron. Indeed our evidence suggests that proctolin is a transmitter of the SETi motoneuron in this system. Physiological experiments on both the identified Ds and SETi motoneurons indicate that proctolin functions as a co-transmitter at the insect neuromuscular junction. Only the Ds example will be described.

Stimulation of the Ds motoneuron results in a biphasic response in the coxal depressor muscle (ADAMS and O'SHEA, 1983). The two types of response can be associated first with the transient actions of a classical transmitter and second with the slower and persistent actions of the peptide proctolin. When stimulated singly the Ds motoneuron produces a brief excitatory junctional potential (EJP) followed by transient increase in force. These transient effects are consistent with the well recognized effects of L-glutamate at the insect neuromuscular junction. When stimulated in a burst, however, the Ds motoneuron produces two effects that are not explained by the release of classical transmitter. These are first a delayed slow catch-like tension which outlasts the duration of the burst and second a slowing of the relaxation phase of the transient contractures. The delayed

slow catch-like tension and the modulation of
twitch-relaxation produced by Ds stimulation is
due to co-release of proctolin.

The discovery of the proctolinergic nature
of the Ds motoneuron suggested that skeletal-
muscle sensitivity to proctolin in insects might
in general be associated with peptidergic
innervation. Several questions arise from this
observation which are of importance to insect
neurobiology in particular, and may also have
general neurobiological significance. For
example what is the functional significance of
the neuropeptide neuromuscular transmitter ? We
are far from being able to provide a general
answer to this question in insects. To date we
have studied only two examples. In both,
proctolin is employed as a co-transmitter in slow
excitatory motoneurons. In contrast to the
effects of glutamate, proctolin produces a
persistent contracture without depolarization.
In the locust this contracture appears to be
independent of extracellular calcium. Proctolin
also has modulatory effects. In the cockroach
preparation it slows the rate of relaxation of
contraction due to the classically acting
transmitter of the Ds motoneuron. In the locust
it modulates the frequency of an intrinsic rhythm
of contraction and relaxation. Clearly there may
be other effects that have escaped our attention
or that will be revealed when other peptidergic
neuromuscular systems are investigated. In
summary, our observations add another layer of
complexity to the already complex insect
neuromuscular system.

ADIPOKINETIC HORMONES

Adipokinetic hormone (AKH) was the name
given by MAYER and CANDY (1969) to a small
peptide located in the corpus cardicum (CC) of
the desert locust Schistocerca gregaria. The
first function assigned to this hormone was the
release of diglycerides from the fat body which
occurs a few minutes after the initiation of
flight. In 1976 the chemical structure of the

peptide was reported (STONE et al., 1976) as
pGlu-Leu-Asn-Phe-Thr-Pro-Asn-Trp-Gly-Thr-NH$_2$.
Complete sequence analysis was achieved by a
combination of amino acid analysis, enzymatic
digestion, dansyl Edman Degradation and low
resolution mass spectroscopy (STONE et al., 1976;
STONE and MORDUE, 1980). Synthetic adipokinetic
hormone and its [Tyr]^1 analogue are currently
available from Peninsula Labs.

Recent progress in research on AKH has
paralleled in many respects recent advances in
our understanding of the neurobiology of
proctolin. For example, like proctolin it
appears to be a peptide with functions in
addition to that for which it was first
recognized. Thus, AKH acts on flight muscle to
suppress carbohydrate oxidation and stimulate
fatty acid oxidation (ROBINSON and GOLDSWORTHY,
1977). It also produces contracture of a locust
skeletal muscle (extensor tibialis) at 10^{-8} M
(O'SHEA, unpublished observations) and has a heart
acceleratory effect, though at rather high
concentration (STONE and MORDUE, 1980). Further
indication that AKH has functions other than the
regulation of lipid metabolism during flight is
its presence in the CC of pre-adult stages of
development, prior to the emergence of flight
behavior and the development of wings. Indeed,
AKH can be detected in late embryonic locusts and
in the first instar.

The major impetus to expanded studies on the
possible transmitter functions of AKH was the
development of an AKH antiserum by Schooneveld
and his associates (SCHOONEVELD et al., 1983).
This has provided the first direct evidence for
the existence of neurons that may use AKH as a
neurotransmitter. Neurons containing AKH-
immunoreactivity have not yet been uniquely
identified, but the fact that AKH can cause
skeletal muscle contraction suggests that
motoneurons may be among the AKH-ergic neuronal
populations. If this is confirmed and if the
motoneurons can be uniquely identified we will
have a second peptidergic subpopulation of

motoneurons to study. This will support the view
that insect motoneurons are functionally and
chemically subdivided according to peptide co-
transmitter type and will provide an additional
stimulus to further physiological studies of
peptidergic systems in insects.

Adipokinetic hormone as identified by STONE
et al. (1976) is known now to be one member of a
family of AKH-like peptides in grasshoppers, two
additional members of which are now sequenced.
In fact, the AKH-peptides of grasshoppers are
themselves a part of a larger family of arthropod
peptides that include prawn red-pigment
concentrating hormone (CARLSEN et al., 1976) and
the MI and MII peptides of the cockroach (see
below) (O'SHEA et al., 1984; WITTEN et al., 1983;
WITTEN, 1984). The adipokinetic hormones of
grasshoppers include the original AKH (or AKH I)
and the so called AKH II S (pGlu-Leu-Asn-Phe-Ser-
Thr-Gly-Trp-NH$_2$) and AKH II L (pGlu-Leu-Asn-Phe-
Ser-Ala-Gly-Trp-NH$_2$). The sequences of these
were recently obtained by the Mordue group using
gas-phase methods. AKH II S is the second
adipokinetic hormone of the CC of Schistocerca.
It was first isolated and partly characterized
(amino acid composition) by CARLSEN et al.
(1979). AKH II L is the homologous hormone found
in Locusta and differs by only one amino acid
(GÄDE et al., 1984). The AKH I compound in
Schistocerca and Locusta is the same in both
genera of grasshopper. The functional reason for
the presence of two AKH peptides in the
grasshopper CC is unknown. There is, however,
some evidence suggesting that the peptides may
play different roles during different stages of
postembryonic development. Thus while in the
adult locust (Schistocerca) there is 5 to 6 times
more AKH I than II, in the early developmental
stages the two peptides are present in
approximately equal amounts (HEKIMI and O'SHEA,
1985).

Recent work (HEKIMI and O'SHEA, 1986) on the
locust corpus cardiacum has resulted in the first
identification of a precursor protein for

adipokinetic hormone. This precursor is
currently being sequenced by the gas phase
method. Work on the process of peptide
biosynthesis in insects is in its infancy. Both
recombinant molecular techniques and direct
biochemical methods are like to be important in
this area. We look forward to rapid progress in
this important field since it is likely to yield
new peptide structures and may provide clues on
how peptide synthesis is regulated in insects.
The AKH peptides provide a convenient model for
the study of insect peptide biosynthesis because
they are made rapidly and in large amounts in the
CC, a gland that can easily be sustained in
vitro.

STRATEGIES FOR DISCOVERY OF NEW INSECT PEPTIDES

Serious technical impedements to the
isolation, purification and sequencing of insect
peptides no longer exist. Direct biochemical
methods include HPLC for purification and FAB-
Mass Spectrometry and gas phase sequencing for
determination of primary structure. In addition
there are presently available techniques for
peptide characterization which circumvent the
need for peptide isolation (recombinant DNA
methods). Here I will briefly outline some
experimental strategies that may be helpful in
expanding the list of isolated and sequenced
insect peptides.

The first strategy can be called "activity-
directed". The activity-directed strategy is to
date the only one to produce fully characterized
neuropeptides in insects. The existing insect
neuropeptides therefore provide the best examples
of how this strategy works. Proctolin for
example was purified from the cockroach (BROWN
and STARRATT, 1975) using its physiological
activity (contraction of the hindgut) to develop
and monitor the purification procedures. An
attractive feature of the activity-directed
approach is that prior to purification there is
already strong evidence for a functional role of
the active compound. In order to perform well in

the isolation of neuropeptides, bioassays ought
to be sensitive and specific. These features
combine to permit the purification of rare
bioactive compounds from complex mixtures
represented by a crude extract. It is important
to realize that while the bioassay used in the
purification of an active peptide might provide a
clue or hint to its physiological role, the
bioactivity need not be "physiological" and it
will almost certainly not represent the only
bioactivity of the compound. An example in point
is represented by the MI and MII peptides. These
were purified from the cockroach (P. americana)
CC but the bioassay used to detect them initially
was developed in the grasshopper (S. nitens).
The MI and MII peptides are not present in the
grasshopper CC and their activity in this species
is therefore not strictly "physiological". The
assay in the grasshopper was the heart-like
pulsatile skeletal muscle associated with the
hindleg tibial extensor muscle. The MI and MII
peptides cause an acceleration of the beating and
a sustained contracture of this muscle, and they
are also metabolically active causing an increase
in blood lipid. These activities are shared by
the AKH peptide of the grasshopper CC. The
cockroach peptides appear to have multiple
actions and could have been purified by a
different assay, as indeed was AKH.

The purification of MI and MII using a
skeletal muscle/heart beat preparation perhaps
indicates a useful general guideline for peptide
purification. It is well known in vertebrates
that cardiac muscle is responsive to a large
number of neuroeffector compounds, including
peptides. The same may be true in insects. In
the insects we have also argued that a variety of
peptides are likely to be involved in the control
of skeletal muscle. Preparations based on
skeletal and/or cardiac muscle may therefore be
useful as screening assays in the purification of
a number of insect neuropeptides.

Having established a screening bioassay
there are a few useful guidelines that can help

in the initial preparation of tissue extracts for assay. The first, which was adopted in the purification of AKH, MI and MII and could have simplified the initial purification of proctolin, is to take advantage of the solubility of small peptides in methanol. Significant selection for peptides and rejection of protein can be achieved at the first step of extraction by homogenizing tissue in a high concentration of methanol (we used 90:9:1, methanol, water, acetic acid in the extraction of MI and MII). Another idea, which we have found useful is to take advantage of hydrophobic interactions and pass extracts through a Waters C_{18} reverse phase SepPak prior to bioassay. The SepPak de-salts the extract, removes protein and concentrates a large-volume of aqueous extract into a small volume of easily evaporated methanol.

The second major category of strategy can be called "hunch directed". This strategy depends initially on developing methods that, based hopefully on a good idea, are highly likely to indicate new bioactive neuropeptides. An example of a hunch is that a peptide precursor will contain sequences of as yet unknown neuropeptides which could be recognized in the sequence of the precursor. It is sometimes possible to isolate and sequence the peptide precursor directly as is the case with the AKH precursor. It is more likely however that their amino acid sequences will be determined by inference from the nucleotide sequence of the precursor's mRNA. This is because the precursors are synthetic intermediates and likely to be present in vanishingly small amounts, also their relatively large size may present a formidable problem for amino acid sequencing. That the precursor structure can lead to the discovery of new peptides is supported by the following. First, different neuropeptides may occur in a common precursor and second neuropeptides can be recognized in a protein because they are typically flanked by recognizable "processing sequences". Processing sequences are frequently two or more basic amino acids, or if the peptide

is amidated the sequence gly-x-x(x=lysine or
arginine). Likely neuropeptide sequences can
then be synthesized and evidence can be
accumulated for their status as neurotransmitters
or hormones. For example, antibodies can be
raised, bioactivities can be tested, presence in
neurons and calcium-dependent release
investigated. An example of this approach is
provided by the work of Scheller and his
associates on the characterization of cDNA
(SCHELLER et al., 1983; NAMBU et al., 1983;
KREINER et al., 1984). The recent revealing of a
network of peptidergic neurons in the abdominal
ganglion of Aplysia using antibodies to amino acid
sequences contained in a presumed neuropeptide
precursor illustrates its potential usefulness
(KREINER et al., 1984). Another example is
provided by the cloning of the Aplysia egg laying
hormone (ELH) gene (SCHELLER et al., 1983). The
gene sequence strongly suggested the existence of
other important peptides. The inferred protein
sequence of the ELH precursor contains a number
of potential processing sites, suggesting the
production of several peptides. One of these,
α bag cell peptide (α BCP) has been synthesized
and is known to be both bioactive and released
from bag cells (ROTHMAN et al., 1983; SIGVARDT
et al., 1983).

Molecular biological studies of insect
neuropeptides are in their infancy. There is
little doubt however that studies analogous to
those carried out in molluscs can be carried out
in insects. In fact work is currently in
progress to isolate AKH and proctolin genes and
we hope in the near future to have their
nucleotide sequences. There is of course no
guarantee that these sequences will encode new
neuropeptides but there is certainly sufficient
evidence from other studies to suggest this as a
high probability. Recombinant DNA studies provide
the possibility of complete structural
characterization of neuropeptide candidates prior
to the assigning of any neurobiological function,
but they are not the only possibility in the
hunch-directed strategy. Among the many others, I

have selected the following to offer as examples
of the way specific approaches may be devised.

The detection of second messenger activation
might provide a useful strategy. The assumption
embodied in the approach is that many
biologically active peptides will act via a
second messenger. If we therefore assay not for
a specific biological action but for the ability
of compounds to elevate levels of second
messenger (cAMP, for example), we can exploit a
more general characteristic of active peptides.
Such an approach, since it depends on a process
common to many peptides with different specific
bioactivities, could be applied to the isolation
and characterization of many biologically active
peptides. This imaginative approach was
described by MORDUE and MORGAN (1984) and has
been applied in the isolation of diuretic
hormones in locusts. It is a particularly
attractive idea in this case, and may prove to be
so in many others, because isolation of diuretic
hormone by more traditional methods (activity-
directed) has been hampered by a poor and
inconvenient assay of diuretic activity. In this
example, Mordue and Morgan measured the elevation
of cAMP in Malpighian tubules to detect diuretic
hormone. The same assay in different target
tissues (muscle, for example) could of course be
used to detect and purify novel peptides of as
yet unknown biological function. In the future,
particularly as new second messengers are
discovered and as assays for them are developed,
this approach may become more generally
important. It may for example be useful not only
in the isolation of naturally occurring
compounds, but also in screening synthetic
ligands.

Another useful hunch might be that peptides
released from neural or endocrine sources upon
depolarization are likely to have biological
activity. Suppose, for example, that we wish to
search for and isolate new biologically active
peptides from the insect gut. Certainly there
are good reasons to think that many active

peptides exist in the gut's innervation. But the nerves on the gut represent a very small fraction of the gut mass, which is composed mostly of muscle. If we pursue the conventional activity-directed strategy we must start with a crude, complex extract of gut tissue. The bioactive peptides will represent a very small part of the total mass. The chromatographic profile of such an extract will be extremely complex too. If, however, we were able to chromatograph only those peptides that are released from the gut when the tissue is exposed to saline containing a high concentration of potassium, a selection for biologically active peptides is achieved. The "releasable cocktail" will be far less complex than the total extract but more importantly it will be highly biased towards peptides likely to be of biological interest. The simplified chromatographic profile of the released compounds can then be compared to the much more complex profile of the total extract to identify the compounds of potential interest. A technical difficulty with this experiment is that a very small proportion of the total store of peptides are released into a large volume of perfusing saline from which they must be recovered for subsequent chromatography. The use of C_{18} Sep Paks as described by ADAMS and O'SHEA (1983) might help to overcome this problem in many cases.

Exploitation of structural homology represents another useful approach. In this strategy the observation that neuropeptides in different organisms show varying degrees of structural similarity is exploited. In the extreme case, they may be identical. Thus for instance proctolin is found in a wide variety of different insects and crustaceans. There are several ways to exploit expected homology. For example, the isolation of a closely related but different compounds in different species can be facilitated by using a "non physiological" bioassay. Thus a substance P-like compound was isolated from octopus using a mammalian substance P hypotension assay. This compound, eledoisin,

was in fact purified and sequenced before
substance P (ERSPAMER and ANASTASI, 1962).
Unfortunately its function in octopus remains
obscure. Several other methods of recognizing
new peptides with structural similarity to known
peptides can and have been devised. For example,
LEUNG and STEFANO (1984) found enkephalins in the
mussel (Mytilus edulis) using a mammalian
receptor binding assay to recognize enkephalin-
like compounds in an organism very distantly
related to mammals. In addition a wide variety
of approaches exploiting immunological
recognition can be used. For example, DUVE et
al. (1982) have isolated a pancreatic
polypeptide-like activity from the blowfly.
Certainly there is a wealth of immunological
evidence, particularly immunocytochemical,
suggesting that insects and other invertebrates
contain "vertebrate peptides". Immunological
methods will be of help not only in localizing
potentially interesting compounds in insects but
also in developing purification procedures. The
major task now is to pursue these leads to
complete structural determination of the
vertebrate-like peptides and then to assign a
physiological role for the newly characterized
insect peptides.

APPLICATIONS

How can our understanding of insect
peptidergic systems be applied to the development
of new control methods ? The first question
which is normally asked in this regard is whether
the peptide itself might have some useful
application. Although the most usual answer is
no, we should not be too hasty in ruling out the
possible use of peptides as insecticides. It is
easy to think of reasons why peptides could not
be insecticides. For example, they cannot
penetrate insect cuticle, if injested they are
rapidly digested and do not enter the
circulation, they are unstable and expensive to
produce, etc. But having said this there is an
example of a naturally occurring peptide which
has insecticidal action when injested orally.

This example is provided by the Colorado beetle. The beetle is protected against predators by a toxic dipeptide secreted from the defense glands located on the pronotum and elytra (DALOZE et al., 1986). The peptide, γ-L-glutamyl-L-2-amino-3 (Z), 5-hexadienoic acid, contains a nonprotein amino acid and so it represents a special case. But it's mode of action is unknown and we cannot rule out the possibility therefore that it interacts with a neuropeptide receptor. This possibility suggests that whereas a neuropeptide itself may not be a suitable insecticide, modifications of it or non-peptide analogs might indeed have insecticidal action when injested.

In order to help organize my thoughts on applications of basic knowledge I will briefly describe five aspects of peptidergic systems. Within each category (Biosynthesis, Release, Receptor, Response, Inactivation) we may be able to identify possible targets and vulnerable points for the development of insecticides that interfer with the system.

Almost nothing is known about peptide biosynthesis in insects. By comparison to other systems, however, we can be sure that neuropeptides represent only part of a gene product. Their biosynthesis depends on the enzymatic processing of a precursor protein. Processing occurs presumably in the lumen of the endoplasmic reticulum and in secretory granules. What controls peptide gene expression and what developmental or functional controls operate on the mechanism of transcription, translation, or precursor processing ? These questions have not yet been addressed in insects! All that can be said now is that important possibilities for interference would be opened if such knowledge were obtained. It could lead for example to the development of enzyme inhibitors or compounds that interact with control systems of peptides expression.

Concerning neuropeptide release we know that this is controlled ultimately by the level of

intracellular free calcium in the peptide-
containing neuron. This level may itself be
regulated by "releasing factors". Such releasing
factors may be peptides, amines or other
biologically active compounds. Releasing factors
may act presynaptically on the nerve terminals of
peptidergic neurons. It is known that the
release of peptide neurohormones from the locust
corpus cardiacum is controlled by other
transmitter substances. For example, ORCHARD and
LOUGHTON (1981) have shown that octopamine can
trigger the release into the hemolymph of
adipokinetic hormone activity. In the moth
Manduca sexta the release of eclosion hormone, an
as yet unsequenced peptide of about 4.3 Kd, is
developmentally regulated by circulating levels
of ecdysteroids (TRUMAN, 1981). Clearly
synthetic compounds with potent effects on the
release of peptide neurohormones or transmitters
could function to control insect behavior,
metabolism and development. Unfortunately very
little is currently known about naturally
occurring releasing factors for neuropeptides. A
systematic and intensive search for them could be
fruitful and result in practical applications.

    We currently know almost nothing about
receptors for insect peptides. This is
unfortunate because in vertebrate pharmacology
the analysis of receptor binding has had a
central importance in the development of
important synthetic ligands and pharmaceuticals.
The development of receptor binding assays for
insect neuropeptides will depend on the
availability of synthetic peptides radio labelled
to a high specific activity. Currently among the
sequenced insect peptides only proctolin is
supplied as an isotope (^{3}H-Tyr Proctolin).
Peptide receptor binding assays would of course
have great utility in the quest of active
compounds that might be useful in insect control.
In conjunction with a bioassay this approach
could be used to screen peptide analogs for both
binding and bioactivity and their ability to act
as physiological antagonists. Certainly non-
peptide ligands of peptide receptors exist but

potentially useful examples, active in insects,
will be found only if more natural insect
peptides are characterized and binding assays
developed. This surely offers one of the best
hopes of providing a rational approach to
synthesis of potentially useful compounds.

Concerning the induction of a physiological
response by insect peptides we can consider
insecticidal actions that might interfer with
second messenger transduction systems. An
important second messenger system in mediating
hormone action involves the hydrolysis of
membrane phospholipids. Two products of this
hydrolysis, inositoltrisphosphate and
diacylglycerol can initiate a complex cascade of
events which regulate cell physiology. Currently
we lack detailed knowledge in insects concerning
which second messengers are involved and how
their actions are linked to peptide-induced
physiological responses. If compounds that
interfer with peptide-induction of second
messengers or with the multiple events that are
subsequently initiated are to be developed, a far
more intensive effort in this field must be
supported.

Peptides do not retain the same biological
activity indefinitely. Yet again, however, we
must say that little is known about the process
of neuropeptide inactivation in insects.
Borrowing ideas developed by experiments
primarily on vertebrates, we can identify
enzymatic hydrolysis as an important means of
neuropeptide inactivation. In fact some
experiments investigating insect peptide
metabolism have been performed (QUISTAD et al.,
1984) and they support the role of an
endopeptidase in the biological inactivation of
proctolin. The enzymes that alter the
physiological potency of released peptides are
clearly possible targets for control compounds.
But studies on the nature of such enzymes in
insects are in their infancy and much more basic
research is required if this potentially fruitful
approach is to be exploited in developing new
insect control compounds.

CONCLUSION

There is no doubt that the near future will bring many exciting developments in insect neurobiology in general and in our knowledge of insect neuropeptides in particular. Insects have and will continue to provide model species and preparations for investigating basic biological problems. For this reason, quite independently of their economic significance, insects will always have an important role to play neuroscience. Many of us using insect preparations as convenient model systems rarely consider the possible practical application of our work. This is unfortunate and in contrast to the attitude taken by our colleagues working on mammalian species where the medical implications of basic research are more often exploited. If we are to find safer and more effective ways of dealing with our main competitors on this planet, we should I believe devote more thought to the application of basic neurobiological knowledge. In this short essay I have attempted to describe both pure and applied aspects of insect neuropeptides research. Although the latter has been necessarily hypothetical, I am confident that in the future pure research in this field will find utility.

REFERENCES

ADAMS M. E. and O'SHEA M. (1983) Peptide co-transmitter at a neuromuscular junction. Science 221, 286-289.

BISHOP C. A., O'SHEA M. and MILLER R. J. (1981) Neuropeptide proctolin (H-Arg-Tyr-Leu-Pro-Thr-OH): Immunological detection and neuronal localization in the insect central nervous system. Proc. Natl. Acad. Sci. USA 78, 5899-6002.

BISHOP C. A. and O'SHEA M. (1982) Neuropeptide proctolin (H-Arg-Tyr-Leu-Pro-Thr-OH): Immuno-cytochemical mapping of neurons in the central nervous system of the cockroach. J. Comp. Neurol. 207, 223-238.

BISHOP C. A., WINE J. J. and O'SHEA M. (1984)
    Neuropeptide proctolin in postural motoneurons
    of the crayfish. J. Neurosci 4, 2001-2009.
BROWN B. E. (1967) Neuromuscular transmitter
    substance in insect visceral muscle.
    Science 155, 595-597.
BROWN B. E. and STARRATT A. N. (1975) Isolation
    of proctolin, a myotropic peptide from
    Periplaneta americana. J. Insect Physiol.
    21, 1879-1881.
BROWN B. E. (1967) Occurrence of proctolin in six
    orders of insects. J. Insect Physiol.
    23, 861-864.
CARLSEN J., CHRISTENSEN M. and JOSEFSSON L. (1976)
    Purification and chemical structure of the red
    pigment-concentrating hormone of the prawn
    Leander adspersus. Gen. Comp. Endocrinol.
    30, 327-331.
CARLSEN J., HERMAN W. S., CHRISTENSEN M. and
    JOSEFSSON L. (1979) Characterization of a
    second peptide with adipokinetic and red
    pigment-concentrating activity from the locust
    corpora cardiaca. Insect Biochem. 9,
    497-501.
DALOZE D., BRAEKMAN J. C. and PASTEELS J. M.
    (1986) A toxic dipeptide from the defense
    glandsof the Colorado beetle. Science
    221, 221-223.
DUVE H., THORPE A., LAZARUS N. R. and LOWRY P. J.
    (1982) A neuropeptide of the blowfly
    Calliphora vomitoria with an amino acid
    composition homologous with vertebrate
    pancreatic polypeptide. Biochem. J. 201,
    429-432.
ECKERT M. H., AGRICOLA H. and PENZLIN H. (1981)
    Immunocytochemical identification of proctolin-
    like immunoreactivity in the terminal ganglion
    and hindgut of the cockroach Periplaneta
    americana (L). Cell Tissue Res. 217, 633-645.
ERSPAMER V. and ANASTASI A. (1962) Structure of
    pharmacological actions of eledoisin, the
    active endecapeptide of the posterior salivary
    gland of Eledone. Experientia 18, 58-61.
EVANS P. D. and O'SHEA M. (1977) An octopaminergic
    neurone modulates neuromuscular transmission in
    the locust. Nature 270, 257-259.

GÄDE G., GOLDSWORTHY G.J., KEGEL G. and KELLER R.
    (1984) Single step purification of locust
    adipokinetic hormones I and II by reversed-
    phase high-performance liquid chromatography,
    and amino-acid composition of the hormone II.
    Hoppe Seylers Z. Physiol. Chem. 365,
    393-398.
HEKIMI S. and O'SHEA M. (1985) Adipokinetic
    hormones in locusts: synthesis and develop-
    mental regulation. Soc. Neurosci. Abstr.
    11, 959.
HEKIMI S. and O'SHEA M. (1986) in preparation.
HOLETS V. R., HOKFELT T., UDE J., ECKERT M. and
    HANSEN S. (1984) Coexistence of proctolin
    with TRH and 5-HT in the rat CNS.
    Soc. Neurosci. Abstr. 10, 692.
KESHISHIAN H. and O'SHEA M. (1985) The distri-
    bution of a peptide neuropeptide in the post
    embryonic grasshopper CNS. J. Neurosci 5,
    992-1004.
KREINER T., ROTHBARD J. B., SCHOOLNIK G. K.,
    SCHELLER R. H. (1984) Antibodies to synthetic
    peptides defined by cDNA cloning reveal a
    network of peptidergic neurons in Aplysia.
    J. Neurosci. 4(10), 2581-2589.
LEUNG M. K. and STEFANO G. B. (1984) Isolation
    and identification of enkephalins in pedal
    ganglia of Mytilus edulis (mollusca).
    Proc. Natl. Acad. Sci. USA 81, 955-958.
LI C. and CALABRESE R. (1983) Evidence for
    proctolin-like substances in the central
    nervous system of the leech. Soc. Neurosci.
    Abstr. 9, 76.
MAY T.E., BROWN B. E. and CLEMENTS A. N. (1979)
    Experimental studies upon a bundle of tonic
    fibers in the locust extensor tibialis muscle.
    J. Insect Physiol. 25, 169-181.
MAYER R. J. and CANDY D. J. (1969) Control of
    hemolymph lipid concentration during locust
    flight: An adipokinetic hormone from the
    corpora cardiaca. J. Insect Physiol.
    25, 169-181.
MILLER T. (1979) Nervous vs. neurohormonal
    control of insect heart beat. Am. Zool.
    19, 77-86.

MORDUE W. and MORGAN P. J. (1984) Diuretic
    hormones in locusts. Abstr 17th Int. Congr.
    Entomol., Hamburg, p. 263.
NAMBU J. R., TAUSSIG R., MAHON A. C. and SCHELLER
    R. H. (1983) Gene isolation with cDNA probes
    from identified Aplysia neurons:
    Neuropeptide modulators of cardiovascular
    physiology. Cell 35, 47-56.
ORCHARD I. and LOUGHTON B. G. (1981) Is octo-
    pamine a transmitter mediating hormone release
    in insects ? J. Neurobiol. 12, 143-153.
O'SHEA M. and ADAMS M. E. (1981) Pentapeptide
    (Proctolin) associated with an identified
    neuron. Science 213, 567-569.
O'SHEA M. and BISHOP C. A. (1982) Neuropeptide
    proctolin associated with an identified
    skeletal motoneuron. J. Neurosci. 2,
    1242-1251.
O'SHEA M., WITTEN J. and SCHAFFER M. H. (1984)
    Isolation and characterization of two
    myoactive neuropeptides: further evidence of
    an invertebrate peptide family. J. Neurosci.
    4, 521-529.
PEARSON K. G. and ILES J. F. (1971) Innervation
    of coxal depressor muscle in cockroach,
    Periplaneta americana. J. Exp. Biol. 54,
    215-232.
PIEK T. and MANTEL P. (1977) Myogenic contrac-
    tions in locust muscle induced by proctolin
    and by wasp, Philanthus triangulum venom.
    J. Insect Physiol. 23, 321-326.
QUISTAD G. B., ADAMS M. E., SCARBOROUGH R. M.,
    CARNEY R. L. and SCHOOLEY D. A. (1984)
    Metabolism of proctolin. Life Sci. 34,
    569-576.
ROBINSON N. L. and GOLDSWORTHY G. J. (1977)
    Adipokinetic hormone and the regulation of
    carbohydrate and lipid metabolism in a working
    flight muscle preparation. J. Insect Physiol.
    23, 9-16.
ROTHMAN B.S., MAYERI E., BROWN R.O., YUAN P.M.,
    and SHIVELY J. E. (1983) Primary structure
    and neuronal effects of ∝-bag cell peptide,
    a second candidate neurotransmitter encoded by
    a single gene in bag cell neurons of Aplysia.
    Proc. Natl. Acad. Sci. USA 80, 5753-5757.

SCHELLER R. H., JACKSON J. F., McALLISTER L. B., ROTHMAN B. S., MAYERI E. and AXEL R. (1983) A single gene encodes multiple neuropeptides mediating a stereotyped behavior. Cell 32, 7-22.

SCHOONEVELD H., TESSER G. I., VEENSTRA J. A. and ROMBERG-PRIVEE H. (1983) Adipokinetic hormone and AKH-like peptide demonstrated in the corpora cardiaca and nervous system of Locusta migratoria by immunocytochemistry. Cell Tissue Res. 230, 67-76.

SCHWARZ T. L., HARRIS-WARRICK R. M., GLUSMAN S. and KRAVITZ E. A. (1980) A peptide action in a lobster neuromuscular preparation. J. Neurobiol. 11, 623-628.

SCHWARZ T. L., LEE G. M. H., SIWICKI K. K., STANDAERT D. G. and KRAVITZ E. A. (1984) Proctolin in the lobster: The distribution release and characterization of a likely neurohormone. J. Neurosci. 4, 1300-1311.

SIGVARDT K., ROTHMAN B. S., MAYERI E. (1983) Analysis of inhibition produced by the candidate neurotransmitter, α-bag cell peptide, in identified neurons of Aplysia. Soc. Neurosci. Abstr. 9, 311.

SIWICKI K. K. and BISHOP C. A. (1985) Mapping of proctolin-like immunoreactivity in the nervous systems of lobster and crayfish. J. Neurosci, submitted.

STARRATT A. N. and BROWN B. E. (1975) Structure of the pentapeptide proctolin, a proposed neurotransmitter in insects. Life Sci. 17, 1253-1256.

STONE J. V., MORDUE W., BATLEY K. E. and MORRIS H. R. (1976) Structure of locust adipokinetic hormone that regulates lipid utilisation during flight. Nature 263, 207-211.

STONE J. V. and MORDUE W. (1980) Adipokinetic hormone. In Neurohormonal Techniques in Insects (Ed by MILLER T. A.) Springer Verlag, New York, pp. 31-80.

TRUMAN J. W. (1981) Interaction between ecdysteroid, eclosion hormone, and bursicon titers in Manduca sexta. Am. Zool. 21, 655-661.

WALKER R. J., JAMES V. A., ROBERTS C. J. and
    KERKUT G. A. (1980) Neurotransmitter
    receptors in invertebrates. In Receptors for
    Neurotransmitters, Hormones and Pheromones in
    Insects (Ed. by SATTELLE D. B., HALL L. M.
    and HILDEBRAND J. G.) Elsevier North Holland,
    Amsterdam, pp. 41-57.
WITTEN J. (1984) The identification of pepti-
    dergic neuromuscular systems in iinsects.
    Ph.D. Thesis, University of Chicago.
WITTEN J. and O'SHEA M. (1985) Peptidergic
    Innervation of Insect Skeletal Muscle:
    Immunochemical Observations.
    J. Comp. Neurol. 242, 93-101.
WITTEN J., SCHAFFER M. H. and O'SHEA M. (1983)
    Structure and biology of two new related
    neuropeptides from insect: Further evidence
    for a peptide family. Soc. Neurosci. Abstr.
    9, 313.

PROTHORACICOTROPIC HORMONES AND NEUROHORMONES

IN BOMBYX MORI

Akinori SUZUKI

Department of Agricultural Chemistry
The University of Tokyo

Bunkyo-ku, Tokyo 113, JAPAN

INTRODUCTION

Since Kopec's discovery of the endocrinological function of insect brain in the early-1920's, numbers of events
in the insect's life cycle have been proved under the control of the neurohormones, whose chemical structures, however, have been unknown with some exceptions such as adipokinetic hormones (Stone et al., 1976) and periplanetins
(Scarborough et al., 1984). The lag in chemistry of
insect neurohormones seems mainly due to the fact that only
a minute quantity of hormones is present in an individual
insect. A clue to eliminate this problem might be the
collection of a large amount of insects as starting materials and/or the application of updated 'micro-chemistry' as
well as technologies of modern bioscience such as monoclonal antibody and gene manipulation.

Japan has a long history of sericulture and it, therefore, has been able to supply a large amount of silkworm
for extracting insect hormones as well-known in the history
of bombykol and ecdysone. Up to nowadays, four kinds of
neurohormones, prothoracicotropic hormone (PTTH), eclosion
hormone (EH), diapause hormone (DH), and melanization and
reddish coloration hormone (MRCH) have been extracted from
B. mori.

PTTH is possibly synthesised in the brain neurosecretory cells, released from corpora allata and stimulates the
prothoracic glands to produce ecdysone.

In 1958, Kobayashi and Kirimura succeeded in preparing
the extract possessing PTTH activity to brainless Bombyx
pupae from pupal brains of B. mori (Kobayashi and Kirimura,
1958).   Susequently Ichikawa and Ishizaki also reported
that the aqueous extract of Bombyx pupal brains stimulated
the adult development of debrained dormant pupae of Samia
cynthia ricini (Ichikawa and Ishizaki, 1961) and concluded
that the active principle must be a protein (Ichikawa and
Ishizaki, 1963).   These were literally the dawn of chemistry of Bombyx PTTH.   Since then, both groups had continued the efforts to purify PTTH and established the proteinaceous or peptidal nature of PTTH until the late-1960's
(Ishizaki and Ichikawa, 1967; Yamazaki and Kobayashi, 1969).
     The observation that the timing of adult eclosion in
B. mori is under the control of enviromental photoperiod
was described by a Japanese sericultural scientist as early
as 1910 and this phenomenon has been utilized effectively
for silkworm egg production in sericulture.   Several years
after Truman and Riddiford demonstrated that a series
of the adult eclosion behavior can be induced by a neurohormone, EH, in the brain homogenates (Truman and Riddiford,
1970), the presence of Bombyx EH in both pupal and adult
Bombyx heads was reported (Morohoshi and Fugo, 1977).
     B. mori hibernates at the embryonic stage and the en-
trance into diapause of the embryo is determined beforehand
by the secretion of DH from the suboesophageal ganglion at
the pupal stage of the mother moth.   Bombyx DH was first
extracted from the brain-suboesophageal ganglion complex in
1957 (Hasegawa, 1957) and the purification of DH has been
done by two independent groups (Isobe and Goto, 1980; Sonobe
and Ohnishi, 1971).   Nowadays DH is believed to consist
of at least 2 molecular species, DH-A (Mr.3300) and DH-B
(Mr. 2000) (Kubota et al., 1976).
     Interestingly, it was found that 80% ethanol extract
of Bombyx brain-suboesophageal ganglion complex and also
adult heads possessed MRCH activity inducing cuticular melanization of armyworm, Leucania separata (Suzuki et al.,
1976).   Some armyworm larvae are known to show polymorphism depending on their population density, so-called phase
variation and the larval color changes are the most conspic-
uous.   The larvae of L. separata shows more or less
black coloration under the crowded conditions and the black
coloration is provoked by MRCH (Ogura, 1975).   On the
role of MRCH in B. mori, Ogura proposed the hypothesis that
DH and MRCH might be the same hormone, based on transplanta-

tion experiments of cephalic organ of B. mori and L. separata (Ogura and Saito, 1973; Ogura 1975b).   This hypothesis, however, has been denied through the purification of MRCH (Matsumoto et al., 1981).

Succeeding these pioneering works, especially those of Ichikawa and Ishizaki, Morohoshi and Fugo, and Ogura, we started investigating the chemistry of Bombyx neurohormones in the early-1970's.   Soon after we started, we exploited a route by which more than millions of Bombyx male adults can be supplied free of charge after having been used for breeding by a sericultural industry (Gunma Kenze Sanshu Co., Maebashi, Gunma, Japan).   Recently, PTTHs (Suzuki et al., 1982; Nagasawa et al., 1984), EH (Nagasawa et al., 1985) and MRCHs (Matsumoto et al., 1984, 1985) have been isolated after spending more than twenty millions of Bombyx heads, and their amino-terminal amino acid sequences have been partially clarified.   In this paper, I would like to describe the recent progress of chemistry of Bombyx neurohormones done by my collegues, focusing on PTTH chemistry.

NEUROHORMONES EXTRACTED FROM BOMBYX HEADS

The Bombyx heads were collected by cutting with a razor blade and stored below -20°C until used.   The acetone powder prepared from the heads was extracted with 80% ethanol.   After removal of ethanol followed by heat-treatment on a boiling water bath for 10 min., with 50-90% acetone were yielded the precipitates possessing MRCH activity and termed 'crude MRCH'   The residual powder in 80% ethanol extraction was extracted with 2% saline.   After heat-treatment in a boiling water bath and then removal of the yielded precipitates, the saline solution was mixed with ammonium sulfate.   The precipitates with 80% saturated ammonium sulfate were dissolved in water and the resulted aqueous solution was subjected to fractional precipitation with acetone.   As described in later section, the PTTH activity to brainless Bombyx pupae was precipitated with 35-55% acetone and that to Samia with 55-75% acetone. The former was termed 'crude 22K-PTTH' (vide infra).   The precipitates with 55-75% acetone possessing EH activity and PTTH activity to Samia were dissolved in water and the aqueous solution was mixed with picric acid to give 90% saturation.   After the yielded precipitates were dissolved in Tris-HCl buffer (pH 7.8, 0.1M), cold acetone was added.   The precipitates with 80% acetone were used as

starting materials for purification of both 4K-PTTH (vide
infra) and EH, and termed 'crude PTTH'.
  The yield of 'crude MRCH' from 1000 <u>Bombyx</u> adult heads
was ca. 120 mg and the cuticular melanization of an isolated
abdomen of a larva of <u>L</u>. <u>separata</u> was induced by injecting 120 ug of 'crude MRCH' derived from 10 <u>Bombyx</u> heads.
Approximately 80 mg of 'crude 22K-PTTH' was obtained from
1000 <u>Bombyx</u> adult heads, 8 µg of which could cause the adult
development of a brainless Bombyx pupa.   In approximate
yield of 30 mg/1000 adult heads, 'crude PTTH' was obtained,
which showed PTTH activity to a brainless <u>Samia</u> pupa at a

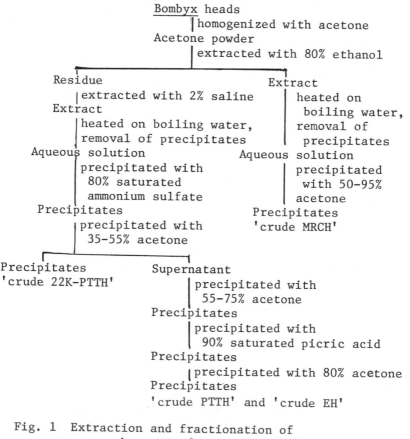

Fig. 1 Extraction and fractionation of
        neurohormones from <u>Bombyx</u> heads

dose of 3 μg (equivalent to 0.1 <u>Bombyx</u> heads) and also stimulated precocious eclosion behavior in a <u>Bombyx</u> pharate
adult at a dose of 150 μg (equivalent to 5 heads).

For isolation of EH, the heads of <u>Bombyx</u> pupae were
extracted and processed through the same procedure as that
for 'crude PTTH' to yield 'crude EH' in the yield of 54 mg/
1000 pupal heads. The 'crude EH' thus obtained showed
distinct EH activity at a dose of 39 μg/insect.

The procedures for preparing these 'crude' preparations
are very simple and highly reproducible, which overcome to
some extent the disadvantage of 'mechanical purification'
by using the heads instead of brains. The procedures are
summarized in Fig. 1.

Using these 'crude' preparations as starting materials,
PTTHs (4K-PTTHs and 22K-PTTH), EH and MRCHs have been isolated. Only 10 μg of EH was isolated from ca. 180,000
<u>Bombyx</u> pharate adult heads with an overall yield of ca.
3.3% and the isolated EH is active at a dose of 1 ng/insect
(Nagasawa et al., 1985). Three molecular species of
MRCH-I, -II and -III have been isolated in the yield of 1.8,
0.2 and 0.3 μg, respectively, from ca. 210,000 adult heads.
They are active at a dose of 6 ng to the cuticular melanization of an isolated abdomen of. <u>L.separata</u> larva, and overall recovery of the activity was approximately 4% (Matsumoto
et al., 1984, 1985). The amino-terminal amino acid sequences of EH and MRCHs partially determined are shown in
Fig. 2.

```
Eclosion hormone (EH)
   1                5                         10
   H-Ser-Pro-Ala-Ile-Ala-Ser-Ser-Tyr-Asp-Ala-Met-Glu-Ile-

Melanization and reddish coloration hormone (MRCH)
   5                       10                        15
   R-Asp-Met-Pro-Ala-Thr-Pro-Ala-Asp-Gln-Glu-Met-Tyr-

        MRCH-I     R=        H-Leu-Ser-Glu-
        MRCH-II    R=        H-???-???-Glu-
        MRCH-III   R=    H-Pro-Leu-Ser-Glu-
```

Fig. 2   Amino-terminal amino acid sequences of
          eclosion hormone (EH),and melanization
          and reddish coloration hormone (MRCH)

PRESENCE OF BOMBYX- AND SAMIA-SPECIFIC
PROTHORACICOTROPIC HORMONES IN B. MORI

Since crude extracts of Bombyx brains and also heads could provoke the adult development when injected into brainless pupae of B. mori and also S. cynthia ricini, PTTH extracted from Bombyx brains is considered to be species nonspecifically active to Samia.   During the purification of PTTH, this idea had come in conflict with that fact that the ratio of PTTH titer measured by brainless Samia pupae and to that measured by brainless Bombyx pupae varied in each preparation, indicating the hypothesis that two species specific PTTHs, one which is specifically active to Bombyx

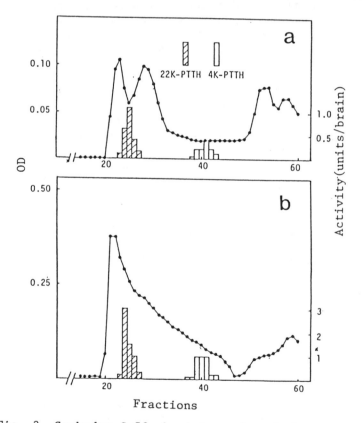

Fig. 3   Sephadex G-50 chromatography of the extracts
         from pupal (Day-0) (a) and larval
         (5th instar, Day-0) (b) brains

and the other active to Samia.

To confirm the above hypothesis, the extracts of Bombyx pupal and larval brains were subjected to Sephadex G-50 chromatography and, as shown in Fig. 3, the fractions active to Bombyx were eluted in positions corresponding Mr. 22,000 and those to Samia in positions corresponding to Mr. 4,400. The extract of Bombyx adult heads, after partial purification, was also chromatographed in the same conditions and 2 kinds of PTTH activity were recovered in the same positions as the cases of pupal and larval extracts. From these results, PTTH active to Bombyx was temporarily termed 22K-PTTH and the other active to Samia termed 4K-PTTH (Ishizaki et al., 1983; Nagasawa et al, 1984). Further, 22K- and 4K-PTTHs were shown to be separable by fractional precipitation with acetone and DEAE Sepharose chromatography.

The facts mentioned above remind me of the early history of Bombyx PTTH. Kobayashi et al. purified PTTH with bioassay using brainless Bombyx pupae, whereas Ichikawa and Ishizaki with that using brainless Samia pupae. Each group achieved considerable purification and examined some properties of PTTH. Although both agreed that PTTH was proteinaceous, the properties of PTTH reported by each group were distinctly different from each other. The PTTH active to Bombyx seemed rather basic, whereas that active to Samia seemed acidic. These are now well-explained by the fact that 2 kinds of molecular species of PTTH exist in B. mori.

Kobayashi et al. have recently obtained the PTTH preparation specifically active to Bombyx; the finally purified sample could cause adult development of brainless Bombyx pupa at a dose of 10 ng (Matsuo et al., 1985). Judging from the specific activity of the finally purified sample, the preparation seems still impure and further 100-fold purification might be necessary for the isolation. The active principle in the preparation might be similar molecule to 22K-PTTH in view of its basic nature and molecular weight (Mr. 20,000) estimated from Sephadex G-75 chromatography.

PROTHORACICOTROPIC HORMONE ACTIVE TO BOMBYX
(22K-PTTH)

For the bioassay of 22K-PTTH, a racial hybrid J-122 x C-115 of B. mori was used, which is known, in contrast to

most other races of B. mori, to give dormant pupae in a high
percentage, when debrained soon after pupation and the test
solution was injected into the debrained male pupae at 10
days after removal of the brain.   To quantify the PTTH
activity was defined the apolysis index as follows.   The
insect that took 3, 4, 5 and 6 days after injection to undergo wing apolysis were assinged scores 4, 3, 2, 1 and zero.
The mean score was designated as the apolysis index and the
dose that gives the index 2 is defined as 1 Bombyx unit
(Ishizaki et al., 1983b).

     The 22K-PTTH was extracted from Bombyx adult heads and
purified through the procedure illustrated in Fig. 4.
At step 5th, 22K-PTTH was separated from 4K-PTTH and EH.
At step 15th, the activity was recovered in 4 fractions corresponding to the peaks in HPLC using TSK-5PW column, sugges-

Step   Bombyx heads
             |
  1st   Acetone powder
             |
  2nd   Washing with 80% ethanol
             |
  3rd   Extraction with 2% saline
             |
  4th   Heat treatment
             |
  5th   precipitation with 80% saturated ammoniun sulfate
             |
  6th   precipitation with 35-55% acetone,'crude 22K-PTTH'
             |
  7th   Sephadex G-50
             |
  8th   DEAE-Sepharose CL-6B (stepwise)
             |
  9th   CM-Sepharose CL-6B (gradient)
             |
 10th   Octyl-Sepharose CL-4B (gradient)
             |
 11th   Sephadex G-75,'highly purified 22K-PTTH'
             |
 12th   Develosil C_8 HPLC (20-50% acetonitrile/0.08% TFA)
             |
 13th   Hi-Pore RP-304 HPLC (20-40% acetonitrile/0.08% TFA)
             |
 14th   Hi-Pore RP-304 HPLC (20-40% acetonitrile/0.1%HFBA)
             |
 15th   TSKgel SP-5PW HPLC
             |
 16th   Hi-Pore RP-304 HPLC (20-32% acetonitrile/0.1%HFBA)
             |
          22K-PTTH

Fig. 4. Purification procedure for 22K-PTTH

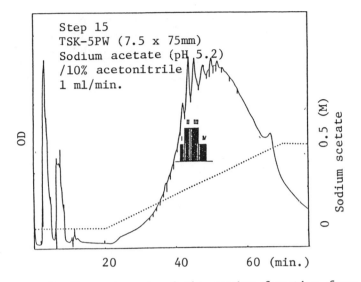

Fig. 5 Chromatogram of the active fraction from step 14th

ting the existence of at least 4 molecular species of 22K-PTTH (Fig. 5).   The active fraction recovered from 2nd peak was subjected to Hi-Pore RP304 HPLC (step 16th) and the activity was eluted in a rather broad area.   The central fraction showing the most marked activity (0.2 ng/unit) gave a single peak in rechromatography on the same HPLC, indicating that 1 molecular species of 22K-PTTH was isolated.   The brief for the purity was further substantiated by amino-terminal analysis as described below.   The overall purification is shown in Table 1.   Only 5.2 μg of

Table 1   Overall purification of 22K-PTTH

Purification	Weight (μg)	Total activity (Bombyx units)	Specific activity (ng/unit)
3rd saline ex.	8.5×10^8	1.5×10^6	5.7×10^5
6th Crude 22K-PTTH	4×10^7	1.5×10^6	2.6×10^4
11th Sephadex G-75	1×10^5	1×10^6	1×10^2
15th TSK-5pW	43	1.5×10^5	0.28
16th Hi-Pore RP-304	5.4	5×10^4	0.11

pure 22K-PTTH was obtained from ca. 500,000 Bombyx adult
heads in the yield of 3.6% after 5.7 x 10^6-fold purification
from 2% saline extract.   In addition to the positive response to Bombyx pupal assay, the isolated 22K-PTTH could
also stimulate the larval ecdysis in Bombyx larval assay
(Sakurai, 1983) at a dose of 0.25-0.50 ng, whereas 22K-PTTH
failed the adult development of debrained pupae at a dose
of more than 1 ug.

When the partially purified sample obtained at step 14
was added at a concentration of 0.06 heads equivalent (0.17
ng)/50 μl to the in vitro culture medium for the prothoracic glands taken out from Bombxy Day-0 pupae, the concentration of ecdysone in the culture medium was markedly increased (2-10 times more than control).   This data suggests that the isolated 22K-PTTH meets with a criteria expected of PTTH of B. mori.

The molecular weight of the isolated 22K-PTTH was estimated to be Mr. 22,000 ± 4,000 by Sephadex G-75 chromatography
and Mr. 29,000 ± 2,000 by SDS-polyacrylamide gel electrophoresis.   The isoelectric electrophoresis indicated the basic
nature of 22K-PTTH (pI 7.5-8.5).   Treatment with proteolytic enzyme such as trypsin and α-chymotrypsin, and the
reductive alkylation with dithiothreitol-iodoacetamide inactivated 22K-PTTH, indicating that 22K-PTTH is peptidal and
contains disulfide bond(s).

Amino acid analysis on the acid hydrolysate of isolated
22K-PTTH gave the amino acid composition as shown in Table 2,
characterized by the more content of Asx, Glx and Pro and
the fewer of Met, His and Phe.   The fractions possessing

Table   2   Amino acid composition of 22K-PTTH

Amino acid	Percent	Amino acid	Percent
Asx	11.8	Tyr	3.9
Glx	17.1	Val	5.8
Ser	7.0	Met	0.2
Gly	4.6	Cys_2	0.5
His	2.0	Ile	6.5
Thr	7.2	Leu	7.5
Ala	4.4	Phe	0.8
Arg	6.8	Lys	5.8
Pro	9.0		

PTTH activity obtained step 15 were shown to have almost similar amino acid composition as isolated 22K-PTTH.   From
this fact it is possible that 22K-PTTH in <u>Bombyx</u> brains
might be composed of several molecular species possessing
similar amino acid composition.
     The Edman degradation of isolated 22K-PTTH has revealed
that Gly is a sole amino terminal amino acid and that the
amino terminal amino acid sequence is H-Gly-Asn-xxx-Gln-Val-,
indicating the purity of the isolated sample.   Further
sequence analysis is now in progress.

PROTHORACICOTROPIC HORMONE ACTIVE TO <u>SAMIA</u>
(4K-PTTH)

Isolation and Characterization of 4K-PTTH

     The bioassay was performed using the debrained dormant
pupae of Eri-silkworm, <u>Samia</u> <u>cynthia</u> <u>ricini</u>.   Titers of
PTTH activity were measured by Samia unit; 1 unit was de-
fined as the minimum dose of PTTH necessary to cause adult
development in more than 50% of assay pupae (Ishizaki and
Suzuki, 1980).
     The isolation procedure for 4K-PTTHs is illustrated in
Fig. 6.   The 'crude PTTH' (ca. 19 g) (Suzuki et al., 1975)
was prepared from 648,000 <u>Bombyx</u> heads in 3 batches, which
was further purified to yield ca. 38 mg of 'highly purified

Step   <u>Bombyx</u> heads

8th   'Crude PTTH'

9th   Sephadex G-50

10th   DEAE-Sepharose CL-6B (stepwise)

11th   SP-Sephadex C-25 (stepwise)

12th   Sephadex G-50, 'highly purified 4K-PTTH'

13th   DEAE-Sepharose CL-6B (gradient)

14th   Sephadex G-50

15th   Reversed phase HPLC

           4K-PTTH-I, -II and -III

Fig. 6   Purification procedure for 4K-PTTHs

4K-PTTH' through conventional chromatographies (Nagasawa et al., 1979).   The methods for preparing 'highly purified 4K-PTTH' is characteristic of their simplicity and reproduciblity so that they can be run without bioassay after repeated several times.

The 'highly purified 4K-PTTH' derived from 648,000 heads was subjected to DEAE-Sepharose Cl-6B column. Elution was made by linear gradient of 0.1 to 0.5 M sodium chloride in 0.05 M Tris-HCl buffer (pH 7.8) and 3 molecular species of 4K-PTTH, 4K-PTTH-I, -II and -III, were eluted by 0.33, 0.35 and 0.38 M chloride ion, respectively. Finally 4K-PTTH-I (50 ug) was isolated by the reversed phase HPLC using Partisil ODS-3 column with linear gradient of 10 to 40% acetonitrile in 10 mM ammonium acetate (Suzuki et al., 1982; Nagasawa et al., 1984b).   4K-PTTH-II and -III (36 ug and 63 ug) were also isolated by the similar HPLC using Develosil ODS-5 column (Nagasawa et al., 1984). 4K-PTTH-I, -II and -III showed the activity to a debrained Samia pupa at a dose of 0.1, 0.4 and 0.1 ng, respectively; the effective doses are as low as 3-11x10^{-11} M, taking account that the brainless pupa contained 0.73 ml of haemolymph and the molecular weight of 4K-PTTH is Mr. ca. 5,000.

The prothoracic glands taken out from freshly ecdysed Samia pupae continued to release ecdysone in Grace medium at low release rate.   When 4K-PTTH was added to the medium at any time during initial several hours of incubation, prothoracic glands readily activated to release ecdysone at recognizable rate; a marked about 3-fold increase was observed as compared to control, when 5 to 500 pg of 4K-PTTH-I was added to 100 ul of the medium after first 2 hr incubation (Nagasawa et al., 1984b). 4K-PTTH appeared acting directly on prothoracic glands to stimulate ecdysone release at a very low dose of 5 pg/gland (1x10^{-11} M).   It is worthy of note that only 1 ng ecdysone was extractable from the prothoracic gland prior to

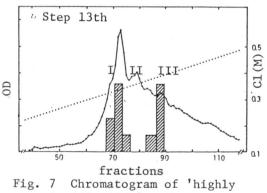

Fig. 7 Chromatogram of 'highly purified 4K-PTTH' on DEAE-Sepharose column

Table 3 Overall purification of 4K-PTTH-I,-II and -III

Purification	Weight (μg)	Total activity (Samia units)	Specific activity (ng/unit)
3rd Saline ex.	1.1×10^6	6.5×10^6	1.7×10^5
8th Crude PTTH	2×10^4	6.5×10^6	3×10^3
12th Sephadex G-50	37.6	6.5×10^6	5.8
15th HPLC			
4K-PTTH-I	0.050	520	0.1
4K-PTTH-II	0.036	90	0.4
4K-PTTH-III	0.063	630	0.1

PTTH addition, suggesting that most of ecdysone released in the presence of 4K-PTTH was newly produced during incubation.

Amino acid analyses on the hydrolysates of 4K-PTTH-I gave amino acid composition (molar ratios); Asx(4), Thr(3), Ser(1), Glx(4), Pro(1), Gly(4), Ala(3), Cys(4), Val(4-5), Leu(5-6), Phe(2), Tyr(1-2), His(1) and Arg(3), which are consistent with the previous estimation made by chemical and enzymatical inactivation (Ishizaki et al., 1978; Nagasawa et al., 1984b).   Inactivation of 4K-PTTH with trypsin and α-chymotrypsin is well explained by the presence of Arg., Phe., and Tyr.   Resistance against cyanogen bromide treatment is attributed to the absence of Met.   Detection of Cys by amino acid anlaysis and the inactivation of 4K-PTTH by reductive alkylation with dithiothreitol-iodoacetamide strongly suggested the presence of disulfide bond(s) in the molecule.

The amino terminal amino acid sequence of 4K-PTTH-I, -II and -III were determined by automated Edman degradation with use of an Applied Biosystem model 470A gas phase protein sequencer (Nagasawa et al., 1984).   The phenylthiohydanthoin derivative of amino acid formed at each cycle of the Edman degradation was identified by HPLC using Ultrasphere ODS column with a gradient elution of 10 to 50% acetonitrile in 10 mM sodium acetate buffer (pH 4.5). The analysis was performed with 2.0 nmol each of 4K-PTTH-I, -II and -III, revealing the 19 amino acid residues from amino terminal, although any PTH amino acid was not detected at cycles more than 20.   The amino acid residues at positions 6, 7 and 11 in 4K-PTTH-I and -II, and those at positions 6, 7, 11 and 13 in 4K-PTTH-III were not identified and suggested the presence of Cys residues.

To establish the position of Cys residues, 4K-PTTH-II was treated with dithiothreitol-iodoacetamide to afford S-carboxamide methyl (CAM) derivative, which was subjected to Edman degradation.   The PTH-CAM-Cys was detected at cycles 6, 7 and 11 as expected.   In addition, PTH-CAM-Cys was identified at cycle 20 and any PTH-amino acids were not detected at further cycles than cycle 20.   Thus Cys residues were confirmed at positions 6, 7, 11 and 20 in amino terminal of 4K-PTTH-II, indicating the presence of Cys residues at same positions of 4K-PTTH-I and -III as those of 4K-PTTH-II.

Almost half of amino acid residues in the amino terminal are common in these three molecules.   Amino acid replacement apparently occurs in such a way that the hydrophobic or hydrophylic nature of the amino acid residues is retained at their respective positions.   Val is substituted by Ile, Phe by Leu, Arg by Gln, Thr by Ser, Leu by Val and Leu by Ala.   All these substitutions, except for Leu by Ala at position 17, would be possible by single base change in the genetic code.   Because these molecules have approximately same level of biological activity, 4K-PTTHs could tolerate the substitutions at these positions to maintain the biological functions.

Most striking feature of amino acid sequences of 4K-PTTHs is the homology with the insulin family peptides; an approximate half of amino acid residues in amino terminal of 4K-PTTH-II is identical to human insulin A-chain (from positions 1 to 20) and insulin-like growth factor I (from position 42 to 62) .

(A)

```
                  1              5             10            15            20
4K-PTTH-I    H-Gly-Val-Val-Asp-Glu{Cys{Cys}Phe-Arg-Pro{Cys}Thr-Leu-Asp-Val-Leu-Leu-Ser-Tyr{Cys}
4K-PTTH-II   H-Gly-Ile-Val-Asp-Glu-Cys-Cys-Leu-Arg-Pro-Cys-Ser-Val-Asp-Val-Leu-Leu-Ser-Tyr-Cys-
4K-PTTH-III  H-Gly-Val-Val-Asp-Glu{Cys{Cys}Leu-Gln-Pro{Cys}Thr-???-Asp-Val-Val-Ala-Thr-Tyr{Cys}
```

(B) The boxed sections indicate identical amino acids.

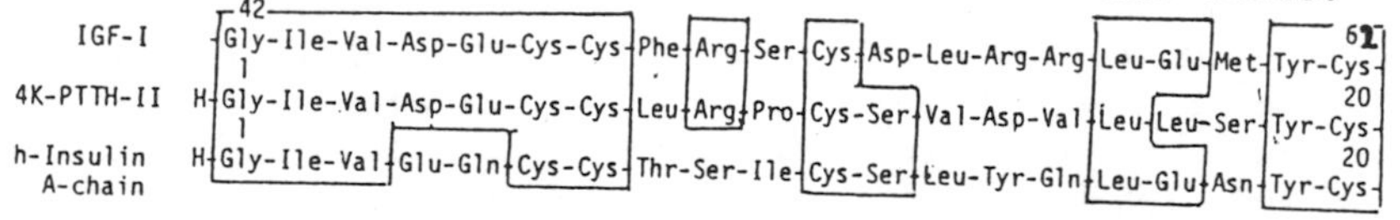

Fig.   8   (A) Amino terminal amino acid sequence of 4K-PTTH-I, -II and -III   (B) Homology between 4K-PTTH and insulin-like growth factor I (IGF-I) and human insulin A-chain

The monoclonal antibodies were produced by immunizing mice with a synthetic decapeptide, H-Gly-Val-Val-Asp-Glu-Cys-Cys-Phe-Arg-Pro-OH, corresponding amino terminal of 4K-PTTH-I, complexed with BSA (5:1 molar ratio). A hybridoma clone secreted an antibody that recognized specifically the 4K-PTTH with reduced disulfide bond(s) but did not react to an intact molecule. Immunohistochemistry using this antibody showed that the dorsomedial neurosecretory cells of pars intercerebralis and nerve fibers in periphery of corpora allata in 5th instar Day-0 larva of B. mori were strongly immunoreactive, indicating that 4K-PTTH is possibly synthesized in neurosecretory cells of brain and transfered to and released from corpora allata. The neurosecretory cells and corpora allata of Samia 5th instar larva were also immmunoreactive to this antibody. Therefore, PTTH molecule of Samia might contain the amino acid sequence same as, or at least similar to that of 4K-PTTH isolated from B. mori (Ishizaki et al., 1986).

Amino Acid Sequence of 4K-PTTH-II

The automated Edman degradation of intact 4K-PTTH-II and its CAM-derivative afforded a single amino acid sequence through position 20, which could not explain the estimated molecular weight and the results of amino acid analyses on the acid hydrolysate. Further, incubation of intact 4K-PTTH-II with carboxypeptidase A (CPA) resulted in a very slow release of approximately equimolar Asp and Val, indicating the carboxyl terminal sequence was -Val-Asp-OH rather than -Asp-Val-OH, if the substrate-specificity was taken into consideration (Hartsuck et al., 1971). These results strongly supported the idea that 4K-PTTH-II might be a double chain peptide connected with disulfide bond(s), one which constituted of 20 amino acid residues possessing Gly in its amino terminal and the other whose amino terminal blocked and carboxyl terminal -Val-Asp-OH.

Then CAM derivative of 4K-PTTH-II was thoroughly investigated by HPLC and finally separated into 2 peptides (CAM-A and CAM-B chains). Amino acid compositions of CAM-A and CAM-B chains were different from each other and the sum of them was almost equal to the amino acid composition of intact 4K-PTTH-II. Therefore, 4K-PTTH-II **consists** of 2 peptide chains, A and B chains, connected with disulfide bond(s) to each other as expected above.

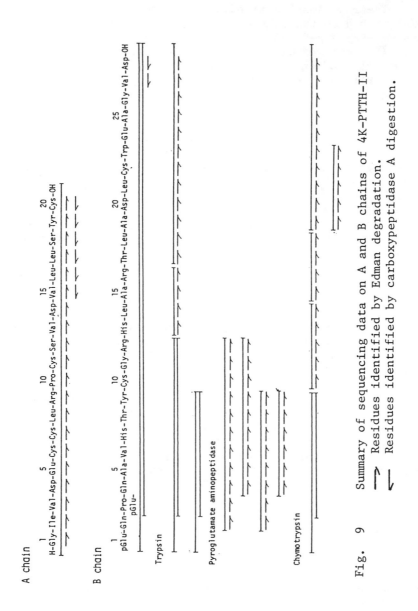

Fig. 9   Summary of sequencing data on A and B chains of 4K-PTTH-II.
          ⌐ Residues identified by Edman degradation.
          └ Residues identified by carboxypeptidase A digestion.

The separated CAM-A chain was subjected to automated
Edman degradation and the sequence composed of 20 amino
acid residues was confirmed.   The carboxy terminal amino
acid sequence from positions 15 to 20 was reconfirmed by
the results obtained from CPA digestion of CAM-A chain.
Any amino terminal amino acid of the CAM-B chain was
not detected by Edman degradation.   In consideration of
the fact that the blocked amino terminal of vertebrate bioactive peptides is often pyroglutamic acid residue (pGlu)
and that CAM-B chain afforded 4 molar Glu in the amino
acid analysis on its acid hydrolyzate, the presence of pGlu
in amino terminal of CAM-B chain was suggested.   Then,
CAM-B chain was treated with pyroglutamic acid aminopeptidase to afford a free amino terminal(Gln), though pGlu removed by pyroglutamic acid amino peptidase could not be detected due to the interruption by a large amount of reagents
originally contained in the enzyme complex available.
Whole amino acid sequence of CAM-B chain was determined
by Edman degradation on tryptic and α-chymotryptic peptides,
which were obtained by trypsin and α-chymotrypsin digestion,
respectively, followed by HPLC separation.   The carboxyl
terminal sequence, -Val-Asp-OH, was confirmed by CPA digestion.   In conclusion, CAM-B chain is a mixture of 4
microheterogeneous peptides, with 28 and 26 amino acid
residues, whose position 5 is Ala or Gly.
The amino acid sequence analyses are summarized in Fig.
9 and 4K-PTTH-II is considered to be a double chain peptide
connected with 2 disulfide bonds in such a way as insulin
molecule.   The confirmation of connection with disulfide
bonds and the amino acid sequence analyses of other 4K-PTTH
are now in progress.   It seems interesting to know the
reasons and biological significance for the presence of
such a lot of molecular species of 4K-PTTH in B. mori.

Homology with Insulin Family Peptides

The complete amino acid sequence of 4K-PTTH-II is compared with those of human insulin and porcine relaxin.
As shown previously, considerable homology was found in the
amino acid sequence of A chain between 4K-PTTH-II and human
insulin.   The sequence of B chain is less homologous (38
%) between 4K-PTTH-II and human insulin.   4K-PTTH-II is
slightly homologous in amino acid sequence with relaxin, a
hormone belonging to the insulin family, but it still has

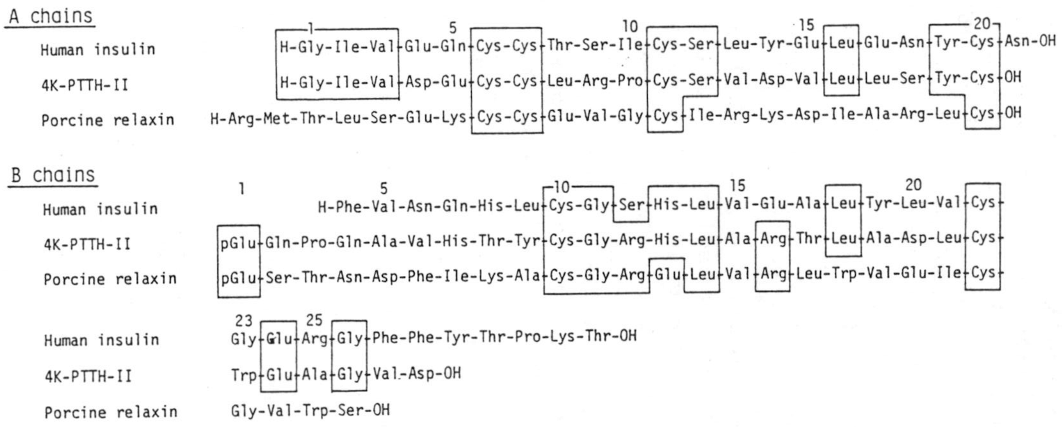

A chains

```
                    1                   5                      10                        15                      20
Human insulin    H-Gly-Ile-Val-Glu-Gln-Cys-Cys-Thr-Ser-Ile-Cys-Ser-Leu-Tyr-Glu-Leu-Glu-Asn-Tyr-Cys-Asn-OH
4K-PTTH-II       H-Gly-Ile-Val-Asp-Glu-Cys-Cys-Leu-Arg-Pro-Cys-Ser-Val-Asp-Val-Leu-Leu-Ser-Tyr-Cys-OH
Porcine relaxin  H-Arg-Met-Thr-Leu-Ser-Glu-Lys-Cys-Cys-Glu-Val-Gly-Cys-Ile-Arg-Lys-Asp-Ile-Ala-Arg-Leu-Cys-OH
```

B chains

```
                    1                   5                      10                        15                      20
Human insulin    H-Phe-Val-Asn-Gln-His-Leu-Cys-Gly-Ser-His-Leu-Val-Glu-Ala-Leu-Tyr-Leu-Val-Cys
4K-PTTH-II       pGlu-Gln-Pro-Gln-Ala-Val-His-Thr-Tyr-Cys-Gly-Arg-His-Leu-Ala-Arg-Thr-Leu-Ala-Asp-Leu-Cys
Porcine relaxin  pGlu-Ser-Thr-Asn-Asp-Phe-Ile-Lys-Ala-Cys-Gly-Arg-Glu-Leu-Val-Arg-Leu-Trp-Val-Glu-Ile-Cys
```

```
                  23    25
Human insulin    Gly-Glu-Arg-Gly-Phe-Phe-Tyr-Thr-Pro-Lys-Thr-OH
4K-PTTH-II       Trp-Glu-Ala-Gly-Val-Asp-OH
Porcine relaxin  Gly-Val-Trp-Ser-OH
```

Fig. 10   Comparison of the amino acid sequence of 4K-PTTH-II with
those of human insulin and of porcine relaxin
Boxes show the identical amino acid residues.

homology with the carboxyl terminal Cys residue in A chain
and the amino terminal pGlu residue in B chain, both of
which are characteristic features of the amino acid sequence
of relaxin.   The locations of 6 Cys residues are completely identical with those of insulin and relaxin.
The residues at positions 2 and 16 in A chain and positions
14 and 18 of B chain, probably involved in hydrophobic core,
along with the residue at position 11 in B chain(Gly) are
homologous with those of insulin, suggesting that 4K-PTTH-II
might be able to have a higher structure principally similar to that of insulin (Blundel and Humbel, 1980).   All
these data suggest that 4K-PTTH-II belongs to the insulin
family (Nagasawa et al., 1986).

Up to nowadays, the presence of insulin-like peptides
has been indicated by immunological, immunohistochemical and
biological assays in more than 10 species of insects.
Recently, the insulin-like peptide of Manduca was isolated
showing the similarities to mammalian insulin in its solu-
bilities, estimated molecular size, immunoreactivity to
guinea pig anti-insulin and amino acid composition (Kramer
et al., 1982).   However, their amino acid sequences have
not been known yet.   Therefore, 4K-PTTH-II seems to be
the first example of insect peptide belonging to the insulin
family, whose whole amino acid sequence has been clarified.

The homology in the amino acid sequences of 4K-PTTH
and insulin suggests that the genes for insulin family peptides have arisen from a common ancestral gene and
that these genes must be conserved during evolution to
insects and mammals.

REFERENCES

Blundell T. L. and Humbel R. E. (1980) Hormonal families:
    pancreatic hormones and homologous growth factors.
    Nature 278, 781-787.

Hartsuck J. A. and Lipcomb W. N. (1971) Carboxypeptidase A.
    In The Enzymes. vol. III. ed. Boyer P. D. pp1-56. Acade-
    mic Press. New York.

Hasegawa K. (1957) The diapause hormone of the silkworm,
    Bombyx mori.   Nature 179, 1300-1301.

Ichikawa M. and Ishizaki H. (1961) Brain hormone of the
   silkworm, Bombyx mori. Nature 191, 933-934.

Ichikawa M. and Ishizaki H. (1963) Protein nature of the
   brain hormone of insects. Nature 191, 308-309.

Ishizaki H. and Ichikawa M. (1967) Purification of the
   brain hormone of the silkworm, Bombyx mori. Biol. Bull.,
   133, 355-368.

Ishizaki H., Suzuki A., Isogai A., Nagasawa H. and Tamura S.
   (1978) Enzymatic and chemical inactivation of partially
   purified prothoracicotropic hormone of the silkworm,
   Bombyx mori. J. Insect Physiol.18, 1621-1627.

Ishizaki H. and Suzuki A. (1980) Prothoracicotropic hormone.
   In Neurohormonal Techniques in Insects ed. Miller T. A.
   pp244-276. Springer-Verlag, New York.

Ishizaki H., Mizoguchi A., Fujishita M., Suzuki A., Moriya
   I., Ooka H., Kataoka H., Isogai., Nagasawa H., Tamura S.
   and Suzuki A. (1983) Species specificity of the insect
   prothoracicotropic hormone (PTTH): the presence of Bombyx
   and Samia specific PTTHs, in the brain of Bombyx mori.
   Devel. Growth and Differ. 25, 593-600.

Ishizaki H., Suzuki A., Moriya I., Mizoguchi A., Fujishita
   M., Ooka H., Kataoka H., Isogai A., Nagasawa H. and
   Suzuki A. (1983b) Prothoracicotropic hormone bioassay:
   pupal adult Bombyx assay. Devel. Growth and Differ. 25,
   585-592.

Ishizaki H., Mizoguchi A., Hatta M., Suzuki A., Nagasawa H.,
   Kataoka H., Isogai A., Tamura S., Fujino M. and Kitada
   C. (1986) Prothoracicotropic hormone (PTTH) of the silk-
   worm, Bombyx mori: 4K-PTTH. In Proceeding of the UCLA
   Symposium, Molecular and Cellular Biology, in press.

Isobe M. and Goto T. (1980) Diapause hromone. In Neurohormonal Techniques in Insects. ed. Miller . A. pp216-243.
   Springer-Verlag. New York.

Kramer k. J., Childs C. N., Speirs R. D.and Jacobs R. M.
   (1982) Purification of insulin-like peptides from
   insect hemolymph and royal jelly. Insect. Biochem. 12,
   91-98.

Kobayashi M. and Kirimura J. (1958) The brain hormone in the silkworm, Bombyx mori. Nature 181, 1217.

Kubota I., Isobe M., Goto T. and Hasegawa K. (1975) Molecular size of the diapause hormone of the silkworm, Bombyx mori. Z. Naturf. 31c, 132-134.

Matsumoto S., Isogai A., Suzuki A., Ogura N., Sonobe H. (1981) Prification and properties of the melanization and reddish colouration hormone (MRCH) in the armyworm, Leucania separata (Lepidoptera). Insect Biochem. 11, 725-733.

Matsumoto S., Isogai A. and Suzuki A. (1984) Isolation of the melanization and reddish coloration hormone (MRCH) of the armyworm, Leucania separata, from the silkworm, Bombyx mori. Agric. Biol. Chem. 48, 2401-2403.

Matsumoto S., Isogai A. and Suzuki A. (1985) N-Terminal amino acid sequence of an insect neurohormone, melanization and reddish coloration hormone (MRCH): heterogeneity and sequence homology with human insulin-like growth factor II. FEBS Letter 189, 115-118.

Matsuo N., Aizono Y., Funatsu G., Funatsu M. and Kobayashi M. (1985) Purification and some properties of prothoracicotropic hormone in the silkworm, Bombyx mori. Insect Biochem. 15, 189-195.

Morohoshi S. and Fugo H. (1977) Some aspect on a hormone controlling adult eclosion of Bombyx mori. Proc. Japan Acad. 53, 75-78.

Nagasawa H., Isogai A., Suzuki A., Tamura S. and Ishizaki H. (1979) Purification and properties of the prothoracicotropic hormone of the silkworm, Bombyx mori. Devel. Growth and Differ. 21, 29-38.

Nagasawa H., Kataoka H., Isogai A., Tamura S., Suzuki A., Ishizaki H., Mizoguchi A., Fujishita Y. and Suzuki A. (1984) Amino-terminal amino acid seouence of the silkworm prothoracicotrpic hormone: homolgy with insulin. Science 226, 1344-1345.

Nagasaw H., Kataoka H., Hori Y., Isogai A., Tamura S.,
  Suzuki A., Guo F., Zhong X., Mizoguchi A., Fujishita M.,
  Takahashi S. Y., Ohnishi E. and Ishizaki H. (1984b)
  Isolation and characterization of the prothoracicotropic
  hormone from Bombyx mori. Gen. Comp. Endcrinol. 53, 143-
  152.

Nagasawa H., Kamito T., Takahashi S., Isogai A., Fugo H.
  and Suzuki A. (1985) Eclosion hormone of the silkworm,
  Bombyx mori. Purification and determination of the N-
  terminal amino acid sequence. Insect Biochem. 15, 573-
  578.

Nagasawa H., Kataoka H., Isogai A., Tamura S., Suzuki A.,
  Mizoguchi A., Fujishita Y., Suzuki A., Takahashi S. Y.,
  and Ishizaki H. (1986) Amino acid sequence of a prothoracicotropic hormone of the silkworm Bombyx mori. Proc.
  Natl. Acad. Sci. USA. in press.

Ogura N. and Saito T. (1973) Induction of embryonic diapause
  in the silkworm, Bombyx mori, by implantation of ganglia
  of the common armyworm larvae, Leucania separata Walker.
  Appl. Ent. Zool. 8, 46-48.

Ogura N. (1975) Hormonal control of larval coloration in
  the armyworm, Leucania separata. J. Insect Physiol. 21,
  559-576.

Ogura N. (1975b) Induction of cuticular melanization in
  larvae of armyworm, Leucania separata Walker(Lepidoptera:
  Noctinidae), by implantation of ganglia of the silkworm,
  Bombyx mori (Lepidoptera: Bombycidae). Appl. Ent. Zool.
  10, 216-219.

Scarborrough R. M., Jamieson G. C. Kalish F., Kramer S. J.,
  McEnroe G. A., Miller C. A. and Schooley D. A. (1984)
  Isolation and primary structure of two peptides with
  cardioacceleratory and hyperglycemic activity from the
  corpora cardiaca of Periplaneta americana. Proc. Natl.
  Acad. Sci. USA. 81, 5575-5579.

Stone J. V., Mordue W., Batley K. E. and Morris H. R. (1976)
  Structure of larval adipokinetic hormone, a neurohormone
  that regulates lipid utilization during flight. Nature
  263, 207-211.

Sonobe H. and Ohnishi E. (1971) Silkworm Bombyx mori L.,
    nature of diapause hormone. Science 174, 835-838.

Suzuki A., Isogai A., Horii T.,Ishizaki H. and Tamura S.
    (1975) A simple procedure for partial purification of
    silkworm brain hormone. Agric. Biol. Chem. 39, 2157-2162.

Suzuki A., Matsumoto S., Ogura N., Isogai A. and Tamura S.
    (1976) Extraction and partial purification of the hor-
    mone inducing cuticular melanization in armyworm larvae.
    Agric. Biol. Chem. 40, 2307-2309.

Suzuki A., Nagasawa H., Kataoka H., Hori Y., Isogai A.,
    Tamura S., Guo F., Zhong X., Ishizaki H., Fujishita M.
    and Mizoguchi A. (1982) Isolation and chracterization
    of prothoracicotropic hormone from silkworm, Bombyx mori.
    Agric. Biol. Chem. 46, 1107-1109.

Sakurai S. (1983) Temporal organization of endocrine events
    underlying larval-larval ecdysis in the silkworm, Bombyx
    mori. J. Insect Physiol. 29, 919-932.

Truman J. W. and Riddiford L. M. (1970) Neuroendocrine
    control of ecdysis in silkmoths. Science 167, 1624-1626.

Yamazaki M. and Kobayashi M. (1969) Purification of the
    proteinic brain hormone of the silkworm, Bombyx mori.
    J. Insect Physiol. 15, 1981-1990.

ENDOCRINE TIMING SIGNALS THAT DIRECT ECDYSIAL PHYSIOLOGY AND BEHAVIOR

Stuart E. Reynolds

School of Biological Sciences,
University of Bath, Claverton Down,
Bath   BA2 7AY, U.K.

INTRODUCTION

Although the event of ecdysis, when the old cuticle is shed, is brief compared to the morphogenetic and synthetic developmental program that must precede it, nevertheless ecdysis is a particularly vulnerable time in the insect's life, a periodically repeated crisis when the risks of injury and predation are high, (see Reynolds, 1980).

The insect is vulnerable at ecdysis because mechanical considerations demand that the new cuticle is soft and flexible as it is extricated from the remains of the old one. Often it must also be extensible, so that the insect may attain a larger body size as it molts. This means that the new cuticle is ill-suited for its job as an exoskeleton, and the insect's powers of locomotion at the time of ecdysis are usually very limited. The role of the cuticle as a defence system is also compromised, and predators, parasites and pathogens all find that the body armor of a molting host is easier to penetrate. If the ecdysing insect falls from its perch, it is easily damaged.

Newly ecdysed insects often lack their normal protective coloration, whether this is cryptic or aposematic, because darkening of the cuticle frequently accompanies its mechanical stabilization. For reasons that are less well understood, the new cuticle is briefly less

53

waterproof and impermeable immediately after ecdysis than
subsequently, and so this time is associated with an
unusually high rate of water loss.

For many insects that are r-selected to exploit
transient resources, speed of growth and rapidity of
development are of paramount importance in the race to
reproduce. For these insects, the unfavourable mechanical
properties of the cuticle immediately before and after
ecdysis mean a period of enforced fasting.

For all these reasons, it is important to insects
that molting, and particularly the event of ecdysis, should
be brief. In particular, the new cuticle must be stiffened
and hardened as soon as possible.

While it has been known for many years that the
developmental program of molting is initiated by the
secretion from the endocrine prothoracic glands of the
steroid molting hormone, ecdysone (see review by Richards,
1981), it is only more recently that it has been realised
that the events of ecdysis itself are separately regulated
by two peptide neurohormones, which apparently have no
function at any other time in the insect's life. These two
hormones, eclosion hormone (EH) and bursicon, act
sequentially to ensure that behaviour is properly
integrated with physiological change during ecdysis. In
doing so, they seem to act principally as timing signals
that trigger the initiation of the processes they direct.

Even more recently it has been realised that
ecdysteroid molting hormones themselves may play an
important but negative role in timing events late in the
molting cycle, such that these events can occur only when
the ecdysteroid titer has fallen below a threshold level.

ECDYSTEROIDS: HORMONES THAT REGULATE PRE-ECDYSIAL
EVENTS.

As has just been pointed out, there are many reasons
for supposing that it will be of selective advantage to a
molting insect to stabilize the new cuticle as quickly as
possible.

In fact, in many insects at least, part of the new cuticle is tanned before ecdysis. The extent of this pre-ecdysial tanning varies a good deal. In cockroaches, for example, hardly any of the new cuticle is stabilized before ecdysis, whereas in adult beetles and moths most is at least partially tanned well before ecdysis. In general, where pre-ecdysial tanning occurs, the structures which are stabilized before ecdysis are those that can easily be extracted from their ecdysial sheaths, and which will not change greatly in shape or size during or immediately after ecdysis.

A good deal of evidence has now accumulated to show that pre-ecdysial tanning is probably controlled by the declining titre of ecdysteroid. For example in Tenebrio, the progress of pharate adult development, as monitored by the state of pre-ecdysial tanning is slowed, or even halted by injections of 20-hydroxyecdysone (20-HE), (Reynolds, S.E. and Williams-Wynn, C.A., unpublished results). When fragments of integument are taken from the dorsal thorax of pharate adult beetles at a stage just before pre-ecdysial tanning would normally begin, the cuticle fragements will undergo normal tanning when cultured in vitro. This tanning is inhibited when modest concentrations of 20-HE are added to the medium (Fig 1).

The work of Riddiford and her associates on the pre-ecdysial melanization that occurs in allatectomised or black mutant larvae of Manduca provides a similar insight. Curtis et al (1984) have shown that this melanization occurs at a time when the ecdysteroid titre is falling, and that injections of 20-HE delay the onset of melanin formation in the already formed premelanin granules. Hiruma and Riddiford (1984) have shown that 20-HE inhibits the activation of the prophenoloxidase in these melanin granules.

Pre-ecdysial tanning clearly solves a part of the problem of ensuring that the interlude of ecdysis is a brief one, but there are always parts of the cuticle that cannot be stabilized before ecdysis (the wings of adult insects, for example) and it is necessary to delay sclerotization of these parts until they have been freed of exuvial constraints.

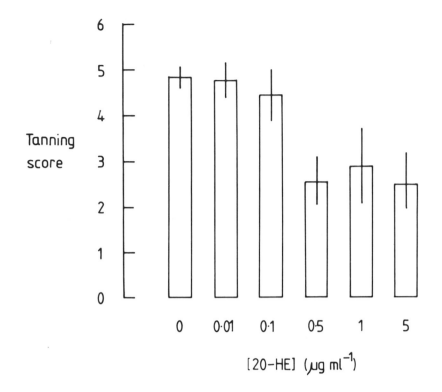

Fig. 1   Inhibitory effect in Tenebrio on pre-ecdysial
         tanning of 20-HE

Developing pharate adults were selected on day 5,
just before the onset of pre-ecdysial tanning,   and
surface sterilised in 70% ethanol prior to dissection
in sterile saline solution.   After removal of the
pupal cuticle,   fragments of pharate adult integument
from the dorsal thorax (pronotum) were washed in
medium (TC199 with Hanks salts and 20 mM HEPES) and
transferred to 1 ml of fresh medium in a Bijou bottle
equipped with a cotton plug. The final culture medium
contained varying concentrations of 20-HE as
indicated below the bars. Tanning was scored visually
after 5d on a scale from 0-6 according to the
intensity of color (0 = no tanning, 6 = uniform, deep
golden brown).   Means $\pm$ S.E.   ($\underline{n}$ = 10-15 for each
concentration).

It is probable that the declining titer of ecdysteroid molting hormones also regulates other events that occur in this immediately pre-ecdysial phase of the molting cycle, an idea that was first advanced by Slama (1980), who found that injections of 20-HE delayed or prevented the ecdysis of adult Tenebrio from the pupal case (see below). It seems very likely for instance that activation of the molting fluid, and thus breakdown of the old endocuticle, is an event regulated by the decline in ecdysteroid titer. Schwartz and Truman (1983) noted that 20-HE injections and infusions delayed or prevented both moulting fluid activation and resorption in pharate adult Manduca. A similar phenomenon occurs in both larvae and pupae of the same insect (our unpublished data).

As we shall see below, the falling ecdysteroid titer also regulates the timing of ecdysis itself, either directly (as may be the case in Tenebrio) or indirectly, via the release of another hormone (as is certainly the case in Manduca - Truman et al, 1983).

Thus the first timing signal that directs ecdysial physiology and behaviour is a fall in the ecdysteroid titre. It is not yet clear how this signal is used to control the sequence of changes occurring immediately before ecdysis itself begins. One possibility is that successive events have progressively lower ecdysteroid thresholds. Once a threshold is passed, that event is triggered. Another possibility is that the rates of individual developmental processes are quantitatively and continuously regulated by the ecdysteroid titer. Succession of developmental change here would be the result of differing sensitivities of various processes to the inhibitory influence of ecdysteroid. Schwartz and Truman (1983) considered this possibility most probable because the effects of 20-HE on the rate of development are dose-dependent. It is possible that both mechanisms operate.

ECLOSION HORMONE: A HORMONE THAT INITIATES ECDYSIS

It is important that ecdysis occurs at the right time. The insect must be able to assess the progress of its own development so that ecdysis is initiated neither too

soon nor too late. This is the recognition of what I have
previously called "developmental readiness" (Reynolds,
1980). Additionally, environmental conditions must be
favourable. In many species this means that ecdysis
typically occurs at a certain time of day, in a circadian
gated rhythm.

There is now good evidence that the initiation of
ecdysis at least in Lepidoptera is a consequence of the
release of a peptide hormone, the eclosion hormone (EH).
This was first shown in adult silkmoths, Hyalophora
cecropia and Antheraea pernyi (Truman and Riddiford, 1970),
and has since been shown in other adult Lepidoptera
(Manduca sexta - Reynolds, 1977; Bombyx mori - Fugo et al
1984; Lymantria dispar - Schnee et al, 1984). At first
thought to be used in the control only of adult eclosion,
EH is now known to be released prior to ecdysis in Manduca
pupae (Truman et al, 1980), and larvae (Copenhaver and
Truman, 1982). EH may even be involved in embryonic ecdyses
(Truman et al, 1981) and hatching (Fugo et al, 1985).

In all adult moths studied, EH is synthesised in
neurosecretory cells of the brain (Truman, 1973a; Fugo and
Iwata, 1983; Schnee et al, 1984). The site of storage and
release apparently varies; in Antheraea EH is present in
the corpora cardiaca (CC) (Truman, 1973a), whereas in
Bombyx it is present in the CA. Copenhaver and Truman
(1984) have shown by microdissection and bioassay that in
adult Manduca, EH originates from a paired group of
neurosecretory neurons in the dorso-lateral protocerebrum.
A polyclonal antiserum raised against HPLC-purified EH
labels 5-6 cells in each group (Copenhaver and Truman,
1985).

EH has been purified from Manduca (Reynolds and
Truman, 1980) and from Bombyx (Nagasawa et al, 1983). The
properties of EH from either source are very similar, the
hormone being an approximately 8.5 kDa peptide. Enzyme
experiments indicate that the hormone is not glycosylated
(or at least this is not essential to its function) and
that it probably possesses at least one disulfide bridge
within its structure that is required for bioactivity. So
far, the amounts of peptide obtained by purification have
not permitted the determination of a complete sequence, but
partial sequence data have been obtained by the Japanese
group (pers. comm.) and this may be expected to lead to the

prediction of the full amino acid sequence from a recombinant cDNA clone.

EH seems to play a complex role in co-ordinating ecdysial physiology and behavior (Reynolds, 1980; Truman 1980). Its most obvious function is to initiate ecdysis behavior, a more-or-less stereotyped fixed action pattern that serves to extricate the insect from the exuvia (Truman and Sokolove, 1972; Truman, 1978). If this were all that EH did, one might wonder why such a neuroendocrine control system were needed. Since the original decision to release EH must be taken within the CNS, it would seem unnecessary to communicate this decision to the rest of the nervous system via the circulation. The answer seems to be that EH release serves as a timing signal that synchronises the initiation of the ecdysis motor program with other behavioral and physiological changes that occur at the time of ecdysis (Fig 2). Among these are the flipping of "behavioral switches" (Reynolds, 1980) that "turn on" (ie make available) new motor patterns (Truman, 1973b; 1976) and reflexes (Truman and Levine, 1982), probably due to the removal of tonic inhibition (Levine and Truman 1982). Interestingly, one of the "behavioral switches" so flipped is one that permits the release of bursicon, the last of the three hormones considered in this paper. Bursicon cannot be released unless the CNS has previously been exposed to EH (Truman, 1973c), a situation that may be regarded as a "failsafe" device that prevents the premature onset of postecdysial cuticle tanning.

Other ecdysial events that are directed by EH release include the plasticization of certain regions of the new cuticle (Reynolds, 1977), and the initiation of cell death in certain muscles (Schwartz and Truman, 1982) and neurons (Truman and Schwartz, 1982).

This concept of EH's role as being primarily that of a timing signal requires that the hormone be released into the hemolymph in a single massive surge. Our measurement of EH titres in individual eclosing adult Manduca (Reynolds et al 1979) support this. The amount of EH in the blood of the pharate adult moth rises abruptly about 2h before the moth emerges from the pupal case, the period of hormone release being very brief (less than the 20 min sample interval). The titer of active EH then falls rapidly (half-life about 45 min), either due to removal or inactivation.

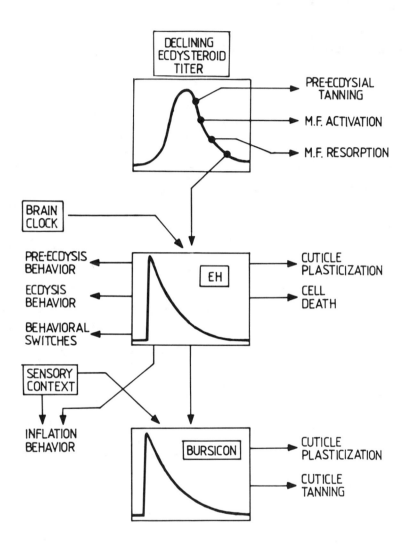

Fig. 2   Integration of behaviour and physiology during
         adult eclosion

         Based principally on Manduca. See text.

The rather long delay between EH release and the escape of the adult moth in Manduca may not be typical. Manduca initiates typical pre-ecdysial movements long before the actual act of escape is completed, the actual timing of emergence often being determined by sensory input (Reynolds, 1980). Other moths show much more abbreviated and stereotyped ecdysis behaviours. In the silkmoths Hyalophora cecropia and Antheraea pernyi, the whole program normally lasts only about an hour. The experimental evidence from in vivo (Truman, 1973a) and in vitro (Truman, 1978) studies of EH action indicates that EH acts to initiate the readout of the ecdysial motor program within a few minutes.

A high proportion (ca. 90%) of the EH present in the CC is released at this time (Reynolds, 1977). The amount of EH that is released in this single surge of hormone is considerably greater than that which is necessary to cause its effects. Truman (1973a) found in Antheraea that the safety factor for a maximal effect was at least 2x; amounts of hormone considerably less than half of the EH present in a single retrocerebral complex were still able to elicit premature eclosion. Again this is consistent with EH's role as a timing signal, in which it may be supposed that the actual concentration of hormone achieved in the blood during the surge does not matter, provided that this exceeds the triggering threshold.

If EH release is a timing signal, then what determines when this signal is given?

The release of EH from the CC in the adult eclosion of silkmoths is well known to be gated by a circadian clock in the brain (Truman, 1972). The insect apparently assesses its own state of "developmental readiness" by monitoring the declining amount of ecdysteroid moulting hormone in the blood (Truman et al, 1983). Only when the ecdysteroid titer falls can EH be released. This "switching on" of the EH release system resembles the "fail-safe" control exerted by EH itself over the release of bursicon (see above).

A further demonstration of the importance of not initiating ecdysis too soon is seen in the fact that the target tissues for EH do not become responsive to the hormone until just a few hours before the normal time of EH release (Reynolds et al 1979; Truman et al 1980). The acquisition of responsiveness to EH is dependent upon the fall in titer of ecdysteroid (Truman et al, 1983).

In larvae and pupae, EH is not released from the retrocerebral complex as it is in adults, but from the neurohemal transverse nerves of the abdominal CNS (Truman et al, 1980; Copenhaver and Truman, 1982). This different anatomical location of the pre-adult EH release system may reflect the different way in which release is controlled (Truman, 1980) - in contrast to the adult moth, EH release in the larva and pupa is not subject to circadian gating, but is directly controlled by the declining ecdysteroid titer (Truman et al 1983). EH cells in the abdominal CNS have not yet been identified. However it is known that the abdominal system is self-contained and autonomous, since pupae that develop from debrained larvae still release EH normally (Taghert et al 1980). EH activity in pupae (Taghert et al. 1980) and embryos (Fugo et al, 1985) has proved indistinguishable in its chemical properties from that in adult insects, so that it may be supposed that the same hormone is synthesised and released by the pre-adult, abdominal EH release system, and the adult cerebral EH release system. Proper proof of this will of course require that the full chemical structure of EH from both sources by determined.

IS EH A UNIVERSAL ECDYSIS HORMONE?

Whether EH is present in insect orders other than Lepidoptera, and there used to control ecdysis, is not yet certain. While Truman et al (1981) found EH activity, as assessed by a Manduca bioassay in insects from five non-Lepidopteran Orders, it is not known whether these materials really represent EH-like substances, or whether they play a functionally similar hormonal role in these insects. Fraenkel and Su (1984) have shown the existence of a factor that has EH-like effects in promoting adult eclosion in blowflies, but the relation of this factor to Lepidopteran EH is not known. Additionally, not all the criteria required to establish this factor as necessary and sufficient to trigger eclosion have been satisfied.

It remains distinctly possible that many non-Lepidopteran insects do not use an EH-like hormone to

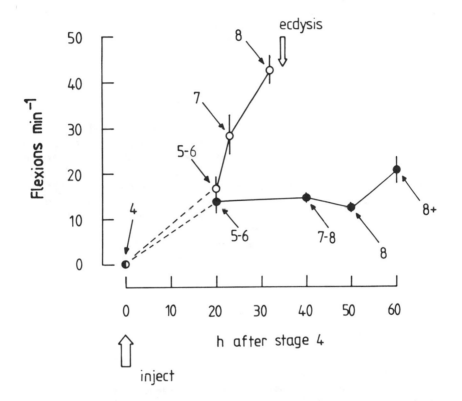

<u>Fig. 3</u> <u>Retardation of the pre-ecdysial program of</u>
<u>abdominal movements in Tenebrio by 20-HE</u>

5d old developing pharate adults (just beginning
pre-ecdysial tanning) were injected either with 5μg
20-HE in 50% ethanol or with 50% ethanol alone
(2 μl injection volume). The insects were kept at
25°C. At various times after treatment the insects
were videotaped for 5 min sample period, from which
the frequency of abdominal flexions was determined.
Means ± S.E. (n = 16). Numbers indicate stages of
development as assessed by pre-ecdysial tanning and
molting fluid resorption.

initiate ecdysis, but instead rely on the declining titre of ecdysteroid to trigger the beginning of ecdysis behavior. For example, despite many attempts (Williams-Wynn, C.A. and Reynolds, S.E., unpublished), we have failed to detect any EH bioactivity in the brain or nerve cord of pharate adult Tenebrio. Furthermore, ecdysis behavior in Tenebrio is initiated on time and executed normally not only in headless individuals, but even in isolated abdomens. Injections of brain or nerve cord extracts consistently fail to affect the time of ecdysis, although 20-HE is capable of delaying, or even completely preventing ecdysis, as was first observed by Slama (1980). We found, as did Farkas (1984), that an articificially high titer of 20-HE leads to inhibition of the progress of the abdominal pumping movements that are part of this insect's program of pre-ecdysis behavior (Fig 3). The most conservative interpretation of all this data seems to be that in Tenebrio, the initiation of ecdysis is not directed by the release of an EH-like neurohormone, but is timed solely by the declining ecdysteroid titer. This does not of course necessarily mean that an EH-like hormone may not play a role other than that of initiating ecdysis behavior.

BURSICON: A HORMONE THAT TRIGGERS POST-ECDYSIAL TANNING

Because post-ecdysial tanning is rapid, it is important that it begins at the right time. If the new cuticle hardens too soon, the molting insect will be trapped inside the exuvia, or at least may find that the new cuticle cannot be properly inflated. If tanning is delayed, then the risk of predation or accidental damage is greatly enhanced. In every insect that has been investigated thoroughly, post-ecdysial tanning has been found to be directed by a neurohormone, bursicon, that is released only at ecdysis. In many senses, bursicon has the best claim to be considered the "universal ecdysis hormone", even though it does not initiate ecdysis.

There are basically two strategies for the control of post-ecdysial tanning, and thus two patterns of bursicon release. For those insects that undergo ecdysis in predictable circumstances which include immediate access to a suitable site for inflation of the new cuticle, there is no point in delaying tanning, and thus bursicon is in these cases released during ecdysis itself, before the insect has completed its escape from the old cuticle. This approach carries the risk that if the progress of ecdysis is delayed or interrupted, the new cuticle may tan before it has been freed and/or expanded. On the other hand, it may well be that when this occurs, the molting insect is doomed anyway. A good example of this kind of control of bursicon release is seen in Manduca pupae (Fig 4A - Reynolds, S.E,, Taghert, P.H. and Truman, J.W., unpublished). Once ecdysis behaviour has begun, post-ecdysial tanning is inevitable, bursicon release beginning even before the old larval cuticle has been split. Other examples of this strategy include Periplaneta (Mills, 1966), Leucophaea (Srivastava and Hopkins, 1975), Locusta (Vincent, 1972) Schistocerca (Padgham, 1976), and Tenebrio (Abboud et al, 1983).

In a second strategy, adopted by insects which ecdyse in situations which do not guarantee ready access to a suitable site for inflation, control over the initiation of tanning is not relinquished until such a site has been attained. Examples of this include adult blowflies (eg the original demonstrations of bursicon by Cottrell, 1962, and Fraenkel and Hsaio, 1962) and adult Manduca (Fig 4B - Truman, 1973; Reynolds et al 1979). In these insects, both of which undergo ecdysis in underground pupation sites, bursicon is not released until the newly emerged adult has successfully dug free of the soil and selected a spot in which to expand the new cuticle.

Like EH, bursicon is released in a single, massive surge and is then quickly metabolised or otherwise inactivated. The two examples of Fig. 4 are typical of bursicon release profiles. As for EH, the amount of bursicon that is released represents a very considerable proportion of what is stored in the neurohemal organs that release it. In adult Manduca, 50% of the bursicon in the abdominal transverse nerves is depleted in the course of a few minutes (Truman, 1973). Like EH, the titer of bursicon at the peak of the surge seems to be considerably greater than is necessary to trigger a response; Reynolds et al (1979) found that active blood could be diluted 8x and still cause tanning in recipient isolated wings. Thus, like EH, the surge of bursicon seems to fulfill all the criteria expected of a timing signal rather than of a continuously regulatory agent.

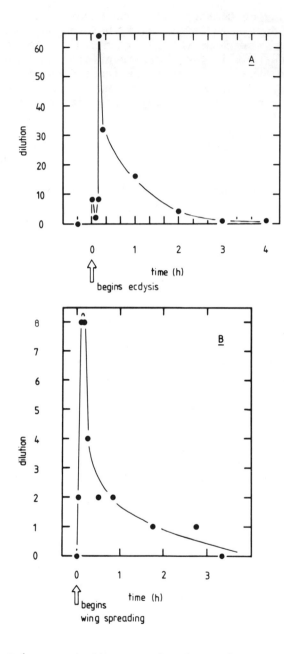

<u>Fig. 4</u> (see opposite page for legend).

Fig. 4 (see opposite page)

Titres of bursicon in Manduca at the time of (A)
pupal ecdysis and (B) adult ecdysis.

Data in (A) are from Reynolds, S.E., Taghert, P.H.
and Truman, J.W. (unpublished). Data points represent
pupal bioassay scores from blood samples pooled from
several individuals. Data in (B) are from Reynolds et
al (1979).   Data points represent adult bioassay
scores from sequential blood samples taken from a
single, cannulated individual. In both (A) and (B)
the points represent the highest dilution that gave a
positive response in the bioassay. The open arrows
mark the onset of ecdysis behaviour (abdominal
peristalsis) in A,and the onset of wing spreading
behaviour in B.

Fig. 5 (see following page)

Pupal and adult bursicon in Manduca are different

(A) Activities of bursicon in different tissues in
pupae and adults. The bars indicate bursicon activity
in the indicated tissue expressed relative to the
structure with the highest bursicon content. Open
bars show bursicon activity before ecdysis, hatched
bars activity after ecdysis. Activity was determined
by serial dilution of an extract of 10 tissues. Each
dilution was tested on at least four isolated wings.
Activity is expressed as the greatest dilution that
gave tanning in 50% of test wings. Pupal tissues were
tested in pharate pupal wings, adult tissues were
tested in pharate adult wings.
(B) Relative activity of pupal and adult bursicons in
pupal and adult bioassays. The bars indicate bursicon
activity (determined as in (A)) of various tissues
measured in both pupal (open bars) and adult (hatched
bars) bioassays.

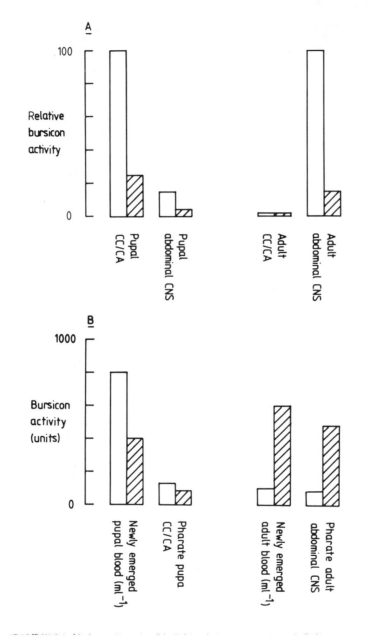

Fig. 5 (see previous page for legend).

If this is the case, then we should expect to find that like EH, bursicon has an integrative role in ecdysis, and controls more than just post-ecdysial tanning. I have reviewed the evidence for this in more detail elsewhere (Reynolds, 1983) so that it will suffice here to say that bursicon has been suggested to control cuticle plasticization (Reynolds, 1976; 1977); post-ecdysial cuticle deposition (Fogal and Fraenkel, 1969); programmed cell death (Seligman and Doy, 1973); post-ecdysial diuresis (Mills and Whitehead, 1970); and tracheal air-filling (Vincent, 1972). It must be admitted that the evidence for some of these claims is a bit shaky. Definitive proof of identity between the hormonal factors controlling all these diverse processes can only come from complete chemical characterisation of the endogenous factor(s).

More specifically, Fraenkel et al (1984) have criticized my suggestion (1976) that bursicon controls cuticle plasticization in flies, on the grounds that whereas injections of cyclic AMP can mimic the effects of bursicon in causing tanning, this treatment does not cause cuticle plasticization. Actually, this is not a valid argument. There are plenty of examples in which different actions of the same regulatory agent are mediated by different second messengers. For example, adrenaline can act at least 3 classes of receptors: α_1 (mediated by increased intracellular Ca); α_2 (mediated by adenylate cyclase inhibition); and β (mediated by adenylate cyclase activation) (Gilman et al, 1985). Serotonin (5-HT) acts at two different types of receptor in the same tissue in blowfly salivary glands, where $5-HT_1$ receptors act via Ca and $5-HT_2$ receptors act via cyclic AMP (Berridge and Heslop, 1981). Nevertheless, it does remain possible that blowflies do not control both plasticization and tanning with the same hormone. To show this would require that the two factors can be separated.

Such difficulties highlight our rather poor knowledge of bursicon's chemical nature. Despite a number of attempts to prepare pure bursicon (reviewed in Reynolds, 1985), all we know about bursicon is that it is a protein of molecular weight of between about 20 and 60 kDa. No sequence data is available. Actually, it is likely that bursicons from different species differ substantially, as cross-reactivity in bioassays is poor. Probably the best prospect for a complete chemical characterization of bursicon will come from recombinant DNA methods. The availability of monoclonal antibodies against bursicon from adult Manduca abdominal nerve cord (Taghert et al, 1983; 1984) may assist in cloning the bursicon gene(s) from a Manduca cDNA library using an expression vector (Taghert, pers. comm).

Actually it is quite likely that even within a single insect, there may be multiple forms of bursicon. Blowflies provide a good example of this. Adult blowflies control post-ecdysial tanning with a typical bursicon-type hormone which is released from the fused thoracic/abdominal ganglion under the control of a center in the brain (Fraenkel and Hsaio, 1965). Fly bursicon has a molecular weight of about 40 kDa (Fraenkel et al, 1966). In blowfly larvae, however, at the time of puparium formation, tanning of the larval cuticle is directed by another hormone, puparium tanning factor (PTF - Sivsubramanian et al 1974) which is clearly distinguishable from adult fly bursicon, since its molecular weight is much greater, and the two hormones do not cross react in their respective bioassays. Although PTF is thus distinguishable from adult fly bursicon, its function and pattern of release is clearly bursicon-like.

Manduca may present a similar case. In pharate adult moths, bursicon is found principally in the abdominal nerve cord, and its associated neurohemal transverse nerves (Truman, 1973). Microdissection and a variety of experimental manipulations allowed Taghert and Truman (1982) to identify four pairs of neurons present in each abdominal ganglion that contain bursicon.

By contrast, in pharate pupae of Manduca, a pupal bioassay reveals most bursicon activity to be present in the CC, as well as a smaller amount in the abdominal CNS (Fig 5A). The main site of release is apparently the CC. Comparisons of the efficacy of the "bursicons" from pupal and adult sources in both pupal and adult bioassays reveals that the two hormones cannot be identical (Fig 5B). Appropriately, pupal bursicon is more effective in the pupal bioassay, and adult bursicon is more effective in the adult bioassay.

This alternation of release sites beween pupa and adult parallels the similar (but opposite) switch between cephalic and abdominal release sites for EH. It is perhaps signficant that the controls over bursicon release in pupal and adult ecdyses are quite different, with the pupal hormone being released during ecdysis as a part of the ecdysis behaviour program, whereas the release of adult bursicon is suppressed by a centre in the head until wing-inflation is underway (Truman and Endo, 1974).

CONCLUSION

In this paper I have focussed on the roles of three hormones in determining the times at which the various physiological and behavioural events associated with ecdysis occur. I would not like anyone to imagine that these three hormones control all of the physiological and behavioural change that occurs at this time - it is unlikely for instance that they regulate the metabolic changes that accompany ecdysis. A recent example of work that shows that ecdysis involves additional hormones is that of Tublitz and Truman (1985) who have shown that the increased rate of beating of the heart in newly-eclosed adult Manduca during wing inflation is due to the release of a pair of small cardioacceleratory peptides from the abdominal CNS.

However, our three hormones may well account for much of what we need to know about the timing of ecdysis-related events. Interestingly the three hormones apparently control physiology and behaviour in two different ways. From the rather scanty evidence so far available, it looks very much as though ecdysteroid molting hormones regulate developmental events late in the molt cycle in a continuous way - ie. the falling ecdysteroid titre does not "trigger" a sequence of events as it passes a series of thresholds, but rather continuously modulates these processes.

On the other hand the successive releases of EH and bursicon appear to be timing signals that trigger all-or-none responses in their target tissues. Given the all-or-none nature of ecdysis itself, this seems appropriate. The principal problem for the molting insect is in ensuring that all the diverse changes required to make ecdysis a success occur together and in a properly co-ordinated sequence. The timing signals of EH and bursicon release appear designed to do just that.

ACKNOWLEDGEMENTS

Unpublished work described here was assisted by grants from the Royal Society and the Science and Engineering Research Council. Work on Manduca was performed under license from the Ministry of Agriculture, Fisheries and Food.

REFERENCES

Abboud, Y.M., Charnley, A.K., Reynolds, S.E. and Williams-Wynn, C.A. (1983) Bursicon in the mealworm, Tenebrio molitor L. and its role in the control of postecdysial tanning. J. Insect Physiol. 29, 947-951.

Berridge, M.J. and Heslop, J.P. (1981) Separate 5-hydroxytryptamine receptors on the salivary gland of the blowfly are linked to the generation of either cyclic adenosine 3',5'-monophosphate or calcium signals. Br. J. Pharmac. 73, 729-738.

Copenhaver, P.F. and Truman, J.W. (1982) The role of eclosion hormone in the larval ecdyses of Manduca sexta. J. Insect Physiol. 28, 695-701.

Copenhaver, P.F. and Truman, J.W. (1984) Spontaneous and induced release of the peptide eclosion hormone from identified neurons in the moth Manduca sexta. Soc. Neurosci. Abstr. 10, 153.

Copenhaver, P.F. and Truman, J.W. (1985) Immunocytochemical identification of the neurosecretory cells that produce the peptide eclosion hormone in the moth Manduca sexta. Soc. Neurosci. Abstr. 11, 326.

Cottrell, C.B. (1962) The imaginal ecdysis of blowflies. The control of cuticular hardening and darkening. J. exp. Biol. 39, 395-411.

Curtis, A.T., Hori, M., Green, J.M., Wolfgang, W.J., Hiruma, K., and Riddiford, L.M. (1984) Ecdysteroid regulation of the onset of cuticular melanization in allatectomised and black mutant Manduca sexta larvae. J. Insect Physiol. 30, 597-606.

Farkas, R. (1984) The effects of 20-hydroxyecdysone on haemolymph pressure pulsations in Tenebrio molitar. J. Insect Physiol. 30, 797-802.

Fogal, W. and Fraenkel, G. (1969) The role of bursicon in melanization and endocuticle formation in the adult fleshfly, Sarcophaga bullata. J. Insect Physiol. 15, 1235-1247.

Fraenkel, G. and Hsaio, C. (1962) Hormonal and nervous control of tanning in the fly. Science, Wash. 151, 91-93.

Fraenkel, G. and Hsaio, C. (1965) Bursicon, a hormone which mediates tanning of the cuticle in the adult fly and other insects. J. Insect Physiol. 11, 513-556.

Fraenkel, G., Hsaio, C. and Seligman, M. (1966) Properties of bursicon: An insect protein hormone that controls cuticular tanning. Science, Wash. 151, 91-93.

Fraenkel, G. and Su, J. (1984) Hormonal control of eclosion of flies from the puparium. Proc. Natn. Acad. Sci. USA. 81, 1457-1459.

Fraenkel, G., Su, J. and Zdarek, J. (1984) Neuromuscular and hormonal control of post-eclosion processes in flies. Archiv. Insect Biochem. Physiol. 1, 345-366.

Fugo, H. and Iwata, Y. (1983) Change of eclosion hormone activity in the brain during the pupal adult development in the silkworm Bombyx mori. J. seric Sci., Tokyo 52, 79-84 (In Japanese with English summary).

Fugo, H., Iwata, Y. and Nakajima, M. (1984) Eclosion hormone activity in haemolymph of eclosing silkmoth, Bombyx mori. J. Insect Physiol. 30, 471-475.

Fugo, H., Saito, H., Nagasawa, H. and Suzuki, A. (1985) Eclosion hormone activity in developing embryos of the silkworm, Bombyx mori. J. Insect Physiol. 31, 293-298.

Gilman, A.G., Goodman, L.S., Rall, T.W., and Murad, F. (1985) The Pharmacological Basis of Therapeutics. 7th edition. MacMillan, New York.

Hiruma, K. and Riddiford, L.H. (1984) Regulation of melanization of tobacco hornworm larval cuticle in vitro. J. exp. Zool. 230, 393-403.

Levine, R.B. and Truman, J.W. (1982) Metamorphosis of the insect nervous system: changes in the morphology and synaptic interactions of identified neurons. Nature, Lond. 299, 250-252.

Mills, R.R. (1966) Hormonal control of tanning in the American cockroach. III. Hormone stability and post-ecdysial changes in hormone titre. J. Insect Physiol. 12, 275-280.

Mills, R.R. and Whitehead, D.L. (1970) Hormonal control of tanning in the American cockroach: changes in blood cell permeability during ecdysis. J. Insect Physiol. 16, 331-340.

Nagasawa, H., Fugo, H., Takahashi, S., Kamito, T., Isogai, A. and Suzuki, A. (1983) Purification and properties of eclosion hormone of the silkworm, Bombyx mori. Agric. Biol. Chem. 47, 1901-1906.

Padgham, D.E. (1976) Control of melanization in first instar larvae of Schistocerca gregaria. J. Insect Physiol. 22, 1409-1419.

Reynolds, S.E. (1976) Hormonal regulation of cuticle extensibility in newly-emerged adult blowflies. J. Insect Physiol. 22, 529-534.

Reynolds, S.E. (1977) Control of cuticle extensibility in the wings of adult Manduca at the time of eclosion: effects of eclosion hormone and bursicon. J. exp. Biol. 70, 27-39.

Reynolds, S.E. (1980) Integration of behaviour and physiology in ecdysis. Adv. Insect Physiol. 15, 475-593.

Reynolds, S.E. (1983) Bursicon. In Endocrinology of Insects
    (eds. Downer, R.G.H and Laufer, H.). Alan R. Liss,
    New York, pp 235-248.

Reynolds, S.E. (1985) Hormonal control of cuticle
    mechanical properties. In Comprehensive Insect
    Physiology, Biochemistry and Pharmacology (eds.
    Kerkut, G.A. and Gilbert, L.I.) Pergamon Press,
    Oxford., vol 8, pp. 335-351.

Reynolds, S.E., Taghert, P.H. and Truman, J.W. (1979)
    Eclosion hormone and bursicon titres and the onset of
    hormonal responsiveness during the last day of adult
    development in Manduca sexta (L). J. exp. Biol. 78,
    77-86.

Reynolds, S.E. and Truman, J.W. (1980) Eclosion hormones.
    In Neurohormonal techniques in insects (ed Miller,
    T.A.) Springer-Verlag, New York, pp 196-215.

Richards, G. (1981) Insect hormones in development. Biol.
    Rev. 56, 501-549.

Schnee, M.E., Ma, M. and Kelly, T.J. (1984) Hormonal basis
    of adult eclosion in the gypsy moth (Lepidoptera:
    Lymantriidae). J. Insect Physiol. 30, 351-356.

Schwartz, L.M. and Truman, J.W. (1982) Peptide and steroid
    regulation of muscle degeneration in an insect.
    Science, Wash. 215, 1420-1421.

Schwartz, L.M. and Truman, J.W. (1983) Hormonal control of
    rates of metamorphic development in the tobacco
    hornworm Manduca sexta. Dev. Biol. 99, 105-114.

Seligman, I.M. and Doy, F.A. (1973) Hormonal regulation of
    disaggregation of cellular fragments in the
    haemolymph of Lucilia cuprina. J. Insect Physiol. 19,
    125-135.

Sivasubramanian, P., Friedman, S., and Fraenkel, G. (1974)
    Nature and role of proteinaceous hormonal factors
    acting during puparium formation in flies. Biol.
    Bull. Mar. Biol. Lab. Woods Hole. 147, 163-185.

Slama, K. (1980) Homeostatic functions of ecdysteroids in ecdysis and oviposition. Acta. ent. bomemosl. 77, 145–168.

Srivastava, B.B.L. and Hopkins, T.L. (1975) Bursicon release and activity in haemolymph during metamorphosis of the cockroach, Leucophaea maderae. J. Insect Physiol. 21, 1985–1993.

Taghert, P.H. and Truman, J.W. (1982) Identification of the bursicon-containing neurones in abdominal ganglia of the tobacco hornworm, Manduca sexta. J. exp. Biol. 98, 385–401.

Taghert, P.H., Truman, J.W. and Reynolds, S.E. (1980) Physiology of pupal ecdysis in the tobacco hornworm, Manduca sexta. II. Chemistry, distribution and release of eclosion hormone at pupal ecdysis. J. exp. Biol. 88, 339–349.

Truman, J.W. (1973a) Physiology of insect ecdysis. II. The assay and occurrence of the eclosion hormone in the Chinese Oak silkmoth, Antheraea pernyi. Biol. Bull. Mar. Biol. Lab. Wood's Hole, 144, 200–211.

Truman, J.W. (1973b) How moths "turn on": a study of the action of hormones on the nervous system. Am. Sci. 61, 700–706.

Truman, J.W. (1973c) Physiology of insect ecdysis. III. Relationship between the hormonal control of eclosion and of tanning in the tobacco hornworm, Manduca sexta. J. exp. Biol. 58, 821–829.

Truman, J.W. (1976) Development and hormonal release of adult behaviour patterns in silkmoths. J. comp. Physiol. 107, 39–48.

Truman, J.W. (1978) Hormonal release of stereotyped motor programmes from the isolated nervous system of the cecropia silkmoth. J. exp. Biol. 74, 151–173.

Truman, J.W. (1980a) Eclosion hormone: its role in coordinating ecdysial events in insects. In Insect Biology in the Future (eds Locke, M. and Smith, D.S.) Academic Press, New York, pp 385–401.

Truman, J.W. and Endo, P.T. (1974) Physiology of insect ecdysis: neural and hormonal factors involved in wing-spreading behaviour of moths. J. exp. Biol. 61, 47-56.

Truman, J.W. and Levine, R.B. (1982) The action of peptides and cyclic nucleotides on the nervous system of an insect. Fedn. Proc. 41, 2929-2932.

Truman, J.W. and Riddiford, L.M. (1970) Neuroendocrine control of ecdysis in silkmoths. Science, Wash. 167, 1624-1626.

Truman, J.W., Rowntree, D.B., Reiss, S.E. and Schwartz, L.M. (1983) Ecdysteroids regulate the release and action of eclosion hormone in the tobacco hornworm, Manduca sexta (L). J. Insect Physiol. 29, 895-900.

Truman, J.W. and Schwartz, L.M. (1982) Insect systems for the study of programmed neuronal death. Neuroscience Commentaries 1, 66-72.

Truman, J.W. and Sokolove, P.G. (1972) Silkmoth eclosion: hormonal triggering of a centrally programmed pattern of behaviour. Science, Wash. 175, 1491-1493.

Truman, J.W., Taghert, P.H., Copenhaver, P.F., Tublitz, N.J. and Schwartz, L.M. (1981) Eclosion hormone may control all ecdyses in insects. Nature, Lond. 291, 70-71.

Truman, J.W., Taghert, P.H. and Reynolds, S.E. (1980) Physiology of pupal ecdysis in the tobacco hornworm, Manduca sexta. I. Evidence for control by eclosion hormone. J. exp. Biol. 88, 327-337.

Tublitz, N.J. and Truman, J.W. (1985) Insect cardioactive peptides. II. Neurohormonal control of heart activity by two cardioacceleratory peptides in the tobacco hawkmoth, Manduca sexta. J. exp. Biol. 114, 381-395.

Vincent, J.F.V. (1972) The dynamics of release and possible identity of bursicon in Locusta migratoria migratorioides. J. Insect Physiol. 18, 757-780.

HORMONAL CONTROL OF DIURESIS IN INSECTS

Simon Maddrell
A.F.R.C. Unit, Department of Zoology
University of Cambridge
Downing Street, Cambridge CB2 3EJ, U.K.

INTRODUCTION

What are the major constraints that affect the water balance of insects? Most insects live in the terrestrial environment and under most conditions (loosely where the relative humidity is less than 98%) this environment is a drying one (Edney, 1977). Insects, because they are very small relative to most other terrestrial organisms, have a higher surface area/volume ratio. This suggests that particularly for them, life on land poses acute problems of maintaining water content; they seem certain to be prone to desiccation. On the other hand, insects living in fresh water or water whose osmotic concentration is less than that of their haemolymph are likely to face exactly the opposite problem – they have to cope with a constant surplus of water as a result of osmotic influx. Understandably, then virtually all accounts of insect osmoregulation have concentrated on adaptations for water conservation in terrestrial insects and for water elimination in insects living in dilute waters. Of the systems that allow insects to cope with these problems we are here mainly concerned with the excretory system.

In essence, the insect excretory system depends on the production of a primary excretory fluid by the Malpighian tubules. This fluid contains ions, whose transport underlies the fluid production, water whose transport is linked to that of the ion transport, possibly by osmosis, together with substances that pass from the haemolymph either by active transport through the tubule cells or by diffusion through the cell-cell junctions (Maddrell, 1981; O'Donnell et al., 1984). This

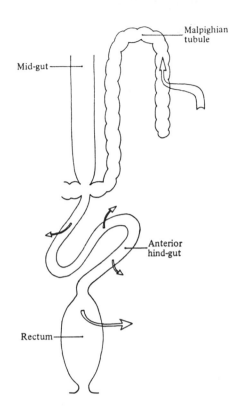

Fig. 1. Fluid movements in the insect excretory system.
Fluid is transported along with solutes into the lumina
of the Malpighian tubules. Much or all of the water is
then recovered in the anterior hind-gut and rectum
together with other useful substances.

fluid is passed from the tubules into the alimentary
canal (Fig. 1), where it is subjected to selective
reabsorption. By this means useful substances are
retained, unwanted materials are eliminated and the
osmotic and ionic content of the excretory fluid is
adjusted appropriately for the ionic and osmoregulatory
needs of the insect. In a few cases, ions and some fluid
may be added to the excretory fluid in the rectum before
it is eliminated (see, for example, Bradley & Phillips,
1977).

DIURESIS IN INSECTS

How might diuresis, an increased rate of elimination of fluid, be achieved by the excretory system? Clearly, this could be produced either by accelerating the rate of fluid secretion into the system and/or by reducing the rate of fluid recovery from the system (Fig. 1). Rather little is known about how diuresis might result from correlated action on both fluid production and fluid recovery sites, though it seems likely that such coordinated activity must often occur. One case where there is evidence for such a process is that of the locust where hormones are known which accelerate Malpighian tubule fluid production and which regulate rectal reabsorption (Phillips et al., 1982).

However, perhaps because it is easier to work on Malpighian tubules, there are now known many insects where there is evidence that the rate of fluid secretion by the tubules can be much increased by the release into the haemolymph of a hormone (Table 1).

THE FUNCTION OF HORMONES ACCELERATING TUBULE ACTION

These hormones have usually been termed diuretic hormones, presumably on the grounds that their release is likely to lead to an increased rate of elimination of excretory fluid by the insect. Only in some cases, however, is it known that such diuresis follows the release of the hormone. This is clear for example in blood sucking insects, such as mosquitoes, the tsetse fly and Rhodnius prolixus and its relatives where the need after a blood meal is for a rapid elimination of surplus fluid (Maddrell, 1980) and in insects such as newly emerged adult lepidopterans where the haemolymph volume is greatly reduced, giving a weight reduction appropriate for insects about to enter a flying phase (see Nicolson, 1976 for example). In others the need for an actual diuresis seems much less clear. A particularly striking case is the recent discovery of a hormone that stimulates Malpighian tubule fluid production in a desert beetle (Nicolson & Hanrahan, 1986), where one would not expect there to be a need for rapid fluid elimination.

For insects where there seems not to be a need for increased fluid elimination and yet they have a hormone to accelerate tubule function, the following argument has

Table 1 Insects for which there is evidence that the rate
of fluid transport by their Malpighian tubules can be
accelerated by a hormone.

Order	Species	Reference
Hymenoptera	Apis mellifera	Altmann (1956)
Diptera	Calliphora vomitoria	Knowles (1976)
	Calliphora erythrocephala	Schwartz & Reynolds (1979)
	Glossina morsitans	Gee (1976)
	Aedes taeniorhynchus	Maddrell & Phillips (1978)
	Anopheles freeborni	Nijhout & Carrow (1978)
	Aedes aegypti	Williams & Beyenbach (1983)
Lepidoptera	Pieris brassicae	Nicolson (1976)
	Calpodes ethlius	Ryerse (1978)
	Danaus plexippus	Dores, Dallman & Herman (1979)
Coleoptera	Anisotarsus cupripennis	Nunez (1956)
	Onymacris plana	Nicolson & Hanrahan (1986)
Hemiptera	Rhodnius prolixus	Maddrell (1962)
	Triatoma infestans	Maddrell (unpub. results)
	Triatoma phyllosoma	" " "
	Dipetalogaster maxima	" " "
	Dysdercus fasciatus	Berridge (1966)
Orthoptera	Periplaneta americana	Mills (1967)
	Locusta migratoria	Cazal & Girardie (1968)
	Schistocerca gregaria	Mordue (1969)
	Carausius morosus	Pilcher (1970)

been advanced to explain the existence of the hormone. It might be that they would need to release a "diuretic" hormone during activity, particularly flight activity, in order to accelerate the rate of operation of the excretory system. Flight in insects involves the most active tissue metabolism known (Weis-Fogh, 1964), and yet, at rest, insects have a relatively much reduced metabolic rate (Alexander, 1971). So the rate of appearance in the haemolymph of metabolic waste products is likely to increase a great deal during flight and this may well require a greatly increased activity of the excretory system. This increased rate of operation is not so that fluid can be eliminated, but so that the haemolymph can be much more rapidly filtered through the Malpighian tubules so as to remove the metabolic wastes which accumulate more rapidly in flight (Maddrell, 1980). A necessary concomitant to this might be that water, ions, and useful substances from the fast flow of primary excretory fluid are rapidly recovered into the haemolymph from the midgut; no extra water need be lost.

Another difficulty has been to explain why the rates of fluid secretion by Malpighian tubules induced by the action of the diuretic hormones should be so high. For insects which actually eliminate fluid at high rates like blood-sucking insects after a meal or newly emerged adult flying insects this might be expected. However, very high rates of tubule fluid production are found in insects which do not evidently need rapidly to rid themselves of fluid. The most impressive case is that of the desert beetle, Onymacris plana, where each of the Malpighian tubules when stimulated can secrete fluid at rates higher than 100 nl min^{-1} (Nicolson & Hanrahan, 1986), a rate which per unit length is not very different from that seen in the tubules of blood sucking insects, thought to be capable of the highest known rates of fluid secretion per unit weight of tissue. A possible explanation is that flight, because it is such an intense activity, also requires intense activity of support systems such as that of the excretory system.

WATER LOSS AND WATER PRODUCTION IN FLIGHT

These relatively acceptable ideas as to the needs of insects for accelerating their excretory systems have recently been greatly shaken and, indeed, the way one

views osmoregulation in terrestrial insects has been
profoundly changed by recent work on water balance in
flying insects. This work by Nicolson & Louw (1982) and
by Bertsch (1984) has measured the rates of respiratory
and evaporative water loss in flying bees and from
measurements of oxygen uptake and carbon dioxide loss,
compared the water losses with water production by
metabolism. Their surprising results show that under
normal environmental conditions, flying carpenter bees
and bumblebees produce metabolic water faster than it is
lost by evaporation from the respiratory system and body
surface. Since these bees use nectar as fuel for flight,
which is about 25-50% sucrose in water, the metabolism of
the sugar from the nectar produces even more water. The
outcome may be a very considerable water load for the
flying bee. Bertsch (1984) calculates that a 220 mg male
bumblebee may produce a water load from its daily flight
activity that equals the whole volume of the body water.

     If other insects during flight also lose water more
slowly than they produce it, then far from having to
conserve water, they may need to eliminate it. This
radically alters what one sees as the major
osmoregulatory problems of such insects.

     These discoveries of course raise several exciting
questions and possibilities but they also provide new
explanations for what had seemed surprising or
unnecessary adaptations of the excretory systems of
terrestrial insects.

RESPIRATORY WATER LOSS

     Perhaps the main question that is raised is how the
respiratory system of insects, or at least of bees, is
organized so that less water is lost than is produced by
the metabolism that the respiration supports. This
ability is quite different from that seen in flying
terrestrial mammals and birds. Both bats and birds lose
water faster than they produce it by metabolism (see
Bertsch, 1984). A detailed analysis of the
water-conserving abilities of the insect respiratory
system has been made by Kestler (1985) but this work was
done for relatively inactive insects. How water is
conserved in flight cannot yet then be answered. It is of
course possible that the insect system has no special
water conserving features during flight, but that birds

and bats, because they are larger with lower surface area to volume ratios, have evolved water evaporative mechanisms to provide extra cooling and this explains the differences in water balance in flight.

METABOLIC WATER PRODUCTION AND WATER BALANCE

Among the possibilities raised by these discoveries of water gain in flying bees is the fascinating one that terrestrial insects might be able to use periods of activity as a method of replacing water lost in a preceding phase of inactivity. This would of course entail using up food reserves but it might give insects an important extra advantage in competing with small vertebrates for niches for small terrestrial animals. It is believed that their success in such niches may well have a lot to do with superior osmoregulatory abilities (Maddrell, 1981, 1982); if they can in addition give themselves a positive water balance by activity (an ability so far unknown for other groups of animals) then this may be crucial to their success.

ELIMINATION OF WATER IN FLIGHT

It has often been observed that insects urinate during flight and it has been difficult to see a reasonable explanation for this. Of course if flying insects generate excess water in flight, the explanation becomes clear. Instances of such fluid elimination are found in small insects, for example aphids (Cockbain, 1961) as well as others, including honeybees (Pasedach-Poeverlein, 1941), Rhodnius (Gringorten & Friend, 1979), carpenter bees (Nicolson & Louw, 1982), and bumblebees (Bertsch, 1984).

One would predict that the rate of water elimination in flight would depend not only on the difference between the rate of evaporative water loss and the rate of metabolic water production, but also on the source of energy for flight. If sugar or diglyceride is mobilised from reserves in the fat body this need not add to the water load. But, for example, if sugar from nectar or honey is used, then the water in which it was dissolved adds very significantly to the excess water production. For example, Bertsch's (1984) calculations show that a

typical bumblebee gains 16 mg of water per day as a
result of metabolic water production in flight exceeding
evaporative water loss but a further 110 mg of water is
produced from the nectar consumed to support that flight.
It follows that one would expect to find the fastest
rates of and the longest continued secretion by
Malpighian tubules in those insects that use as fuel for
flight substances initially in solution in water.

RECTAL ION ABSORPTION

There is a further prediction from the likely need
of flying insects to rid themselves of surplus water.
This is that ion loss in the excreted fluid must be
minimized, because of course flight does not generate
ions. In Phillips' pioneering studies on the physiology
of the insect rectum (Phillips, 1964, 1969), it was clear
that the locust and blowfly could produce urine with very
low levels of ions, as low as 6 mM Cl in the blowfly and
1 mM Na, 22 mM K and 5 mM Cl in the locust. Of course
such an ability would be useful to these insects if they
feed on ion-deficient diets, but this is not likely to be
the case in the wild. A more likely explanation is that
such ion reabsorption would be needed during diuresis in
or after flight. Similarly low levels of Na and K were
found in the rectal fluid of carpenter bees where it is
known that flight produces a water load (Nicolson & Louw,
1982). In some preliminary experiments in this
laboratory, we have found that flying bumblebees fed on a
honey/water mixture of osmotic concentration close to 2
osmol 1^{-1} produce frequent droplets of urine whose
osmotic concentration is as low as 40–50 mosmol 1^{-1}.

UPTAKE OF WATER VAPOUR BY TERRESTRIAL INSECTS

It was argued above that in flying insects the water
produced in flight could play a very valuable part in the
overall water balance of these insects. If this argument
is sound, then flightless insects would be disqualified
from enjoying this advantage. When Edney (1977) drew up a
list of all the insects known to be able to absorb water
vapour from the air, he noted that they were all
wingless. He supposed that this was nothing more than
coincidence, but it is now at least reasonable to suggest

that, since wingless insects could not as easily generate water by intense muscular activity as can insects able to fly, there may well have been more selection pressure on them to develop alternative water-gaining systems.

DIURESIS IN AQUATIC INSECTS

Fresh-water insects face a continuous osmotic influx of water. The evidence suggests that the excretory system reacts to this by the continuous production of urine. The Malpighian tubules constantly secrete fluid, under the control, it is presumed, of a diuretic hormone. Since Malpighian tubule fluid is iso-osmotic with the haemolymph, there must be extensive reabsorption of ions in the hindgut, and this is known to occur (Ramsay, 1950). There is evidence that tubule fluid production is controlled. In larvae of _Aedes aegypti_ transferred from fresh-water to a more concentrated medium, water elimination is reduced and the lumen of the tubules becomes filled with solid matter; evidently, the high rate of fluid production had slackened (Wigglesworth, 1953). Nothing is known of any hormonal control of the hindgut.

CONCLUSIONS

Diuresis in insects, in the sense of fast elimination of fluid from the body, is well known in blood-sucking insects, newly emerged flying adult insects and in insects living in fresh water. It now seems probable that this list must be extended to include insects in flight, where water production by metabolism may exceed the evaporative losses. In many cases it is known that underlying such diureses is the release into the haemolymph of a diuretic hormone which much accelerates the rate of fluid formation by the Malpighian tubules. In only a few cases is there evidence that hormones also affect the reabsorptive parts of the excretory system, though such action is probably essential, as the release unaltered of the fluid from the Malpighian tubules is not likely to be physiologically appropriate.

REFERENCES

Alexander, R. McN. (1971). Size and Shape. Arnold, London.

Altman, G. (1956) Die Regulation des Wasserhaushaltes der Honigbiene. Insectes soc. 3, 33-40.

Berridge, M.J. (1975) The physiology of excretion in the cotton stainer, Dysdercus fasciatus Signoret. IV. Hormonal control of excretion. J. exp. Biol. 44, 553-566.

Bertsch, A. (1984) Foraging in male bumblebees (Bombus lucorum L.): maximizing energy or minimizing water load? Oecologia 62, 325-336.

Bradley, T.J. & Phillips, J.E. (1977) The location and mechanism of hyperosmotic fluid secretion in the rectum of the saline-water mosquito, Aedes taeniorhynchus. J. exp. Biol. 66, 111-126.

Cazal, M. & Girardie, A. (1968). Controle humoral de l'equilibre hydrique chez Locusta migratoria migratorioides. J. Insect Physiol. 14, 655-668.

Cockbain, A.J. (1961) Water relationships of Aphis fabae during tethered flight. J. exp. Biol. 38, 175-180.

Dores, R.M., Dallmann, S.H. & Herman, W.S. (1979) The regulation of post-eclosion and post-feeding diuresis in the monarch butterfly, Danaus plexippus. J. Insect Physiol. 25, 895-901.

Edney, E.B. (1977) Water Balance in Land Arthropods. Springer, Heidelberg.

Gee, J.D. (1976) Active transport of sodium by the Malpighian tubules of the tsetse fly, Glossina morsitans. J. exp. Biol. 64, 357-368.

Gringorten, J.L. & Friend, W.G. (1979) Haemolymph volume changes in Rhodnius prolixus during flight. J. exp. Biol. 83, 325-333.

Kestler, P. (1985) Respiration and respiratory water loss. In Environmental Physiology and Biochemistry (ed. K.H. Hoffmann). Springer, Heidelberg.

Knowles, G. (1976) The action of the excretory apparatus of Calliphora vomitoria in handling injected sugar solution. J. exp. Biol. 64, 131-140.

Maddrell, S.H.P. (1962) A diuretic hormone in Rhodnius prolixus Stal. Nature, Lond. 194, 605-606.

Maddrell, S.H.P. (1980) The control of water relations in insects. In Insect Biology in the Future (eds. D.S. Smith & M. Locke). Academic Press, New York.

Maddrell, S.H.P. (1981) The functional design of the insect excretory system. J. exp. Biol. 90, 1-15.

Maddrell, S.H.P. (1982) Insects: small size and osmoregulation. In A Companion to Animal Physiology (eds. C.R. Taylor, K. Johansen & L. Bolis). Cambridge University Press, Cambridge.

Maddrell, S.H.P. and Phillips, J.E. (1978) Induction of sulphate transport and hormonal control of fluid secretion by Malpighian tubules of larvae of the mosquiteo, Aedes taeniorhynchus. J. exp. Biol. 75, 133-145.

Mills, R.R. (1967) Hormonal control of excretion in the American cockroach. I. Release of a diuretic hormone from the terminal abdominal ganglion. J. exp. Biol. 46, 35-41.

Mordue, W. (1969) Hormonal control of Malpighian tubules and rectal function in the desert locust Schistocerca gregaria. J. Insect Physiol. 15, 273-285.

Nicolson, S.W. (1976) The hormonal control of diuresis in the Cabbage white butterfly, Pieris brassicae. J. exp. Biol. 65, 565-575.

Nicolson, S.W. and Hanrahan, S.A. (1986) Diuresis in a desert beetle? Hormonal control of the Malpighian tubules of Onymacris plana (Coleoptera, Tenebrionidae). J. Comp. Physiol. B 156, 407-413.

Nicolson, S.W. and Louw, G.N. (1982) Simultaneous measurement of evaporative water loss, oxygen consumption and thoracic temperature during flight in a carpenter bee. J. exp. Zool. 222, 287-296.

Nijhout, H.F. and Carrow, G.M. (1978) Diuresis after a blood-meal in female Anopheles freeborni. J. Insect Physiol. 24, 293-298.

Nunez, J.A. (1956) Untersuchungen uber die Regelung des Wasserhaushaltes bei Anisotarsus cupripennis Germ.. Z. vergl. Physiol. 38, 341-354.

O'Donnell, M.J., Maddrell, S.H.P. & Gardiner, B.O.C. (1984) Passage of solutes through walls of Malpighian tubules of Rhodnius by paracellular and transcellular routes. Am. J. Physiol. 264, R759-R769.

Pasedach-Poeverlein, K. (1941) Uber das "Spritzen" der Bienen und uber die Konzentrationsanderung ihres Honigblaseninhalts. Z. vergl. Physiol. 28, 197-210.

Phillips, J.E. (1964) Rectal absorption in the desert locust, Schistocerca gregaria Forskal. III. The nature of the excretory process. J. exp. Biol. 41, 69-80.

Phillips, J.E. (1969) Osmotic regulation and rectal absorption in the blowfly, Calliphora erythrocephala. Can. J. Zool. 47, 851-863.

Phillips, J., Spring, J., Hanrahan, J., Mordue, W. & Meredith, J. (1982) Hormonal control of salt reabsorption by the excretory system of an insect. Isolation of a new protein. In Neurosecretion: Molecules, Cells, Systems (eds. D.S. Farner & K. Lederis), Plenum, New York.

Pilcher, D.E.M. (1970) Hormonal control of the Malpighian tubules of the stick insect, Carausius morosus. J. exp. Biol. 52, 653–665.

Ramsay, J.A. (1950) Osmotic regulation in mosquito larvae. J. exp. Biol. 27, 145–157.

Ryerse, J.S. (1978) Developmental changes in Malpighian tubule fluid transport. J. Insect Physiol. 24, 315–319.

Schwartz, L.M. and Reynolds, S.E. (1979) Fluid transport in Calliphora Malpighian tubules: a diuretic hormone from the thoracic ganglion and abdominal nerves. J. Insect Physiol. 25, 847–854.

Weis-Fogh, T. (1964) Diffusion in insect wing muscle, the most active tissue known. J. exp. Biol. 41, 229–256.

Wigglesworth, V.B. (1953) The Principles of Insect Physiology, 5th edition. Methuen, London.

Williams, S.C. and Beyenbach, K.W. (1983) Differential effects of secretagogues on Na and K secretion in the Malpighian tubules of Aedes aegypti (L.). J. comp. Physiol. 149, 511–517.

CHEMISTRY OF SYNAPSES AND SYNAPTIC TRANSMISSION

IN THE NERVOUS SYSTEM OF INSECTS

HEINZ BREER

UNIVERSITY OSNABRÜCK

4500 OSNABRÜCK, WEST GERMANY

INTRODUCTION

Neurons are the elementary signalling units of the nervous system which differ from other cells in their
highly developed ability to generate signals and to communicate with another and with other target cells
rapidly, precisely and over long distances. The communication between neurons occurs at specialized structures,
the synapses, by means of precisely regulated release of
primary messenger molecules, which effectively react on
specialized regions of the postsynaptic cell, triggering
changes in the permeability and the potential of the
membranes either directly or via cellular second messenger molecules. Thus synaptic transmission is considered
a fundamental process in neuronal function and it is
reasonable to speculate that higher functions in the
central nervous system must involve specific structural
and molecular elements of synapses. Due to the strategi-
cal role, synapses and synaptic transmission are central
for understanding how the nervous system works and great
efforts are presently made to explore aspects of synaptic transmission in molecular detail.
In the central nervous system of insects, synapses are
located mainly in the core of the ganglia, the neuropil,
where neurites branch to form a complex network of neu-

ronal processes and synaptic contacts. In insects, research on synapses has mainly concentrated on the function
and structure of nerve contacts: intracellular recordings have given important informations of physiological
mechanisms, electron microscopy has provided wonderful
insight into the structural details of synapses. The
picture that has emerged for the microphysiology and mi-
croanatomy of synapses in insects resembles that found
in vertebrates: each synapse is formed by a specific association between a nerve terminal of the presynaptic
neuron and a receptive patch on some part of the postsynaptic neuron. The nerve terminal, as highly specialized
cellular compartment is continously supplied with nutrients and biochemically active molecules via axoplas-
mic transport. Since electrical and chemical activity of
neurons are interlinked at synapses, synaptic contacts
are considered as the sites where neuronal activity is
most vulnerable and open to interruption in a variety of
ways.
In views of the importance of synapses for nerve communication and processing of neuronal information and also
their role as potential targets sites for insecticide
action, progress in unravelling the molecular characteristics of chemical synapses and in understanding the
dynamics of synaptic junctions in biochemical and molecular terms, is a major aim for basic insect neurobiology and is at the same time of immediate practical
importance. However, our knowledge about neurochemistry
of insect synapses, and about the molecular processes
and functional elements of nerve contacts in the insect
CNS, is still very limited and has lagged behind that of
other animal groups.

Table 1   Distribution of ACh, GABA, ChAT, AChE, and GAD
          in insect and mouse brain.

	ACh[*] nmol/g	GABA[**] umol/g	ChAT[*] nmol/min/mg	AChE[*] nmol/min/mg	GAD[**] nmol/min/mg
Locust (Head ganglia)	762	15.20	120.0	780.0	12.5
mouse (cortex)	15	2.4	2.0	50.0	1.5

[*]Breer, 1981
[**]Breer and Heilgenberg, 1985

There is general agreement that excitatory synaptic
action, which is effective in generating the discharge
of impulses, and inhibitory synaptic action, which tends
to prevent generation of impulses, are the two basic modes govern the neuronal activity. Comparing the concentrations of putative excitatory and inhibitory neurotransmitter revealed acetylcholine (ACh) and γ-amino-
butyric acid (GABA) as major candidates (Table 1). Therefore a main challenge for insect neurobiology at present time is to understand the distribution of cholinergic as well as gabaeric pathways and the molecular mechanisms of its synaptic terminals in the nervous system
of insects.

Transmitters and Related Enzymes

Both transmitters are produced in simple, one-step synthesis catalyzed by specific enzymes: Choline acetyl
transferase (ChAT; EC 2.3.1.6, M_r = 67.000) catalyses
the acetylation of choline by acetyl CoA, whereas the
pyridoxalphosphate-dependent enzyme glutamatic acid decarboxylase (GAD; EC 4.1.1.15, M_r = 85.000) synthesizes
GABA. High activities of both enzymes have been detected
in insect ganglia, and antibodies raised to purified
ChAT and GAD-preparations either from insect or vertebrate nervous tissue, have been used with striking success for histochemical localization of cholinergic
(Salvaterra, 1985; Sattelle et al., 1986; Buchner et
al., 1986) and gabaergic neurons (Fig.1), in the nervous
system of insects at the light and electron-microscopic
level.
Whereas ACh is inactivated by acetylcholinesterase
(AChE, EC 3.1.1.7), the primary inactivation of GABA is
achieved by high affinity reuptake in gabaeric neurons
and in surrounding glia cells where it is transaminate
by GABA-T (EC 2.6.1.19). Although it is now known that
AChE is not exclusively associated with cholinergic
neurons, the well established technique of Koelle (1963)
for histochemical detection of AChE has provided valuable information for mapping cholinergic systems and
tracts in insect ganglia which has now to be confirmed
using more specific markers.

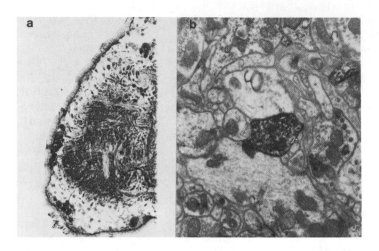

Fig.1 Light (a) and electron-microscopic (b) localization
of GAD-immunoreactive elements in the 2. thoracic
ganglion of locust.

Synaptosomes: in-vitro-Systems for Synapse Research

Biochemical and molecular analysis of further essential
elements of synapses, in particular, the dynamic
processes, like transport and release of transmitters
require as essential prerequisite the isolation of intact nerve terminals. The discovery, more than 20 years
ago, that during homogenizing nervous tissue under conditions of moderate liquid shear in isotonic sucrose solution, nerve terminals are sheared off from their axons,
reseal and can be isolated as discrete subcellular
particles, synaptosomes (Gray and Whittaker, 1962;
Whittaker, 1984), is now considered a landmark in neurochemistry and has opened the possibility to study
aspects of synapses and synaptic transmission in the
test tube.
Unfortunately, the conditions established for isolating
synaptosomes from vertebrate brain appeared to be not
applicable to insect nervous tissue and specific adaptations of tissue treatment and separation conditions are

obviously indispensible prerequisites for suitable approaches to isolate intact nerve endings from insect
ganglia. A new microscale flotation procedure for subcellular fractionation, notably for isolating the nerve
terminals of insect nervous tissue, has been designed
and employed to locust ganglia (Breer and Jeserich,
1980; Breer, 1981; Breer and Knipper, 1985a). The flotation procedure appears to be most suitable, since insect
do not have myelin, and therefore its light membrane
fragments do not contaminate the floating synaptosomal
fractions; furthermore, in a continous microgradient the
sensitive synaptosomes can be kept under constant physi-
cochemical conditions (osmolarity, pH) throughout the
whole procedure and the homogeneous distribution of the
P_2-fraction minimizes artifacts caused by bulk
sedimentation. Sectioning of the isolated ganglia prior
to homogenization destroys the peripheral connective
tissue, thus exposing the centroganglionic neuropil.

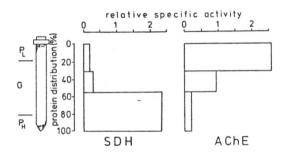

Fig.2 Graphic representation of the distribution of
      marker enzymes in the flotation gradient.

Detailed biochemical and morphological analysis (Breer
and Knipper, 1985a; Breer, 1986) have revealed that the
pellicle fraction was highly enriched in synaptosomal
markers (Fig.2) and contained numerous synaptosomal profiles with synaptic vesicles mostly of the electrolucent
type (Fig.3), thus characterizing the pellicle as a typical synaptosomal fraction.

The metabolic properties of insect synaptosomes indicate that they are viable particles, representing an autonomous, active miniature nerve cell, which maintained and generated a transmembrane potential of about 60 mV (Breer, 1986); furthermore, a clear enrichment and high specific activities of cholinergic and gabaergic activities are considered as an indication that high proportions of cholinergic and gabaergic nerve endings are accumulated in the pellicle fraction. Synaptosomes thus be considered as suitable in vitro models for studying relevant elements and processes of synapses in insects.

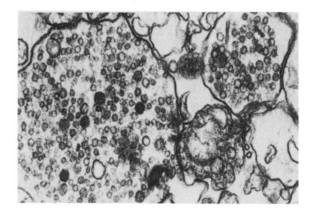

Fig.3 Electromicrograph of a synaptosomal fraction
      derived from head ganglia of locust by the
      flotation procedure.

Transport Systems for Neurotransmitters

Specific transport systems in the plasma membranes of nerve terminals appear to be particular important for presynaptic function; they apparently evolved to "inactivate" the neurotransmitters by rapid removal from the synaptic space to the interior of nerve terminals and evidences suggest that each neuron carries only the transport system for the neurotransmitter it releases. A high affinity uptake system for choline is considered to be unique for cholinergic nerve endings it seems to be the rate limiting, regulatory step for the synthesis of ACh and thus is of particular importance for cholinergic synapses (Kuhar and Murrin, 1978). Isolated nerve endings from locust, accumulated exogenous choline in a linear, temperature-sensitive process; and concentration-dependent experiments showed saturation, indicating that choline was taken up by a carrier-mediated process. Analysis of the transport kinetics revealed the existence of a low-and a high-affinity system; the latter seems to be very important in insects as well, since most of the choline accumulated via this high-affinity pathway was immediately converted to ACh (Breer, 1982). The high-affinity choline transport was very efficiently inhibited by hemicholinium-3 and a K_i value of 100 nM was obtained. However, the basic mechanisms of the carrier-mediated choline translocation are still not understood; this may be due to the fact that an adequate interpretation of transport data concerning aspects of energetics and regulation has always been rather difficult because of the intracellular metabolic activities and the sequestration in intact nerve endings. More recently, there has been significant progress in exploring synaptosomal transport processes by following a shift from the complex synaptosomes to membrane vesicles obtained after osmotic lysis of synaptosomes. Such "synaptosomal ghosts" have proved to be extremely useful for studying neurotransmitter transport in isolation from other processes; but unfortunately, it was found that the transport activity for choline was lost following osmotic lysis in mammalian preparations (Meyer and Cooper, 1982; Kanner, 1983). Synaptosomal ghosts derived from the highly cholinergic insect nervous tissue, however, loaded with K^+ ions and suspended in NaCl were capable of accumulating choline via a high-affinity carrier-mediated process (Fig.4);the accumulation of choline inside the vesicles was energized by ion gradients only, showing that ATP or ATPase are not directly involved (Breer, 1983).

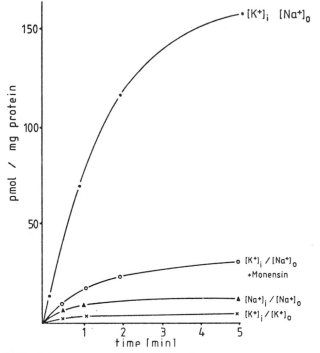

Fig.4 Choline accumulation by "synaptosomal ghosts" endowed with different ion gradients. Choline is only transported in the presence of an appropriate sodium gradient.

Manipulation of the ion composition in the intra- and extravesicular medium should elucidate whether the choline transport is chemiosmotically coupled, i.e. whether the accumulation of choline is driven by ion gradients. Substitution of Na^+ ions always caused a blockade of choline uptake; Na^+ could not be replaced by any other monovalent cation. Vesicles with sodium inside as well as outside, or treated with specific Na^+ ionophores, did not accumulate any choline, indicating that a Na^+ gradient $[Na^+]_o > [Na^+]_i$ must be considered as the main driving force. Accordingly, the transport rate was proportional to the Na^+ gradient, resulting in a sigmoidal relationship between choline accumulation and the sodium gradient. Analysis of the data gave a Hill coefficient of about 2, suggesting a stoichiometry of 2 Na^+ per molecule of choline transported (Breer and Lueken, 1983). Similar experiments concerning the anion specificity have shown that a chloride ion gradient is important as well; however, only one Cl^- appears to be involved in choline accumulation. Thus, possibly a complex of 2 Na^+ : 1 Cl^- : 1 choline is translocated. This points to an electrogenic transport process, i.e. net charges cross the membrane and the membrane potential should affect choline transport. In fact we found that valinomycin, a K^+ ionophore which increases the potential, enhanced choline uptake, whereas CCCP, an H^+ ionophore, which destroys the membrane potential, reduced choline uptake. These results indicate that the active transport of choline is driven by artificially imposed ion gradients as the sole energy source. Thus in the intact system the function of the Na^+K^+-ATPase is merely to convert the energy of cellular ATP into osmotic energy, e.g. electrochemical Na^+ and K^+ ion gradients; subsequently the electrochemical sodium gradient is utilized via cotransport, catalyzed by tightly coupled Na^+/Ch symporters to drive the accumulation of choline.

For the process of GABA accumulation by insect nerve terminals apparently very similar ionic conditions are required; the GABA uptake is sodium- and chloride-dependent as well (Gordon et al. 1982; Breer and Heilgenberg, 1985).

Studies on the efflux of choline from membrane vesicles have shown that the ion dependency for influx and efflux is quite similar, suggesting that the carrier is symmetrical with respect to the membrane and that the vectoral transportation under physiological conditions is caused by ion gradients and electrical potential across the membrane (Breer and Knipper, 1985b); this view is further supported by transactivating effects of choline, showing that the carrier obviously must be returned to the outside for influx to continue. Furthermore, the data on sodium dependence of influx and efflux support the concept that the sodium-choline cotransport system involves an ordered binding of choline first and then of two Na^+-ions on the carrier before translocating across the membrane.

By modulating the involved ion gradients there are multiple ways of regulating the transport process. However, the choline transporter may be regulated by other mechanisms as well. Very recently it was found that extrasynaptosomal ATP might be a modulator, an inhibitor of the choline carrier. Extracellular ATP inhibited the choline transport into synaptosomes at submillimolar concentrations. The finding is of particular interest, since ATP might be released together with ACh at cholinergic synapses in insects, as has already been shown for other cholinergic nerve endings (Zimmermann, 1982). This view is supported by the finding that the Na^+-dependent transport of GABA was not at all affected by addition of nucleotides. The effect seems to be very specific for ATP; neither ADP, AMP and adenosine nor GTP or non-hydrolyzed ATP analogues, like AMP-PNP, had any comparable effect; the high specificity for ATP suggests that a kinase reaction might be involved.

Due to its very efficient blocking of choline transport (see above), hemicholinium-3 is considered as a specific inhibitor of the high-affinity choline carrier; this compound may thus greatly aid the molecular identification of the choline transporter, and may possibly play a comparable role as phlorizin in the characterization of the glucose carrier (Semenza et al., 1984). Using radiolabelled hemicholinium-3, the number of binding sites for this ligand, tentatively the number of choline carriers, in various tissues have been estimated. The number of binding sites was found to be quite low in mouse cerebellum (about 4 fmol/mg), but concentrations as high as 120 fmol mg were determined for the ganglia of the locust. Interestingly, the distribution of binding sites for hemicholinium corresponds fairly well to the concentration of ACh and the rate of choline uptake in the different tissues (Table 2). Photoaffinity labelling experiments using labelled hemicholinium or appropriate derivatives should assist in the identification of polypeptides which are part of the choline carrier.

Table 2   Comparison of acetylcholine, choline uptake
          and hemicholinium binding sites.

	ACh nmol/mg	Ch uptake pmol/min/mg	HC-3 binding fmol/mg
Mouse			
Cerebellum	0.17	--	4
Cortex	0.3	15	25
Locust			
Ganglia	6.8	116	120

Release of Neurotransmitters in vitro

The release of ACh and GABA has been studied using a small perfusion technique (Breer and Knipper, 1985a) which prevents reuptake and allows a continous measurement of released material. Isotonic perfusion medium caused only little efflux; however, the addition of depolarizing agents, like veratridine or elevated potassium concentrations, induced a considerable release of transmitter (Fig.5). The evoked release was demonstrated to be essentially dependent on extracellular Ca^{2+}, in the millimolar range; Mg^{2+} ions antagonized the effect of calcium, suggesting that the ACh release from synaptosomes proceeds by a process of depolarization-secretion coupling (Breer and Knipper, 1984). Recently it was found that components in the venom of black widow spider had a very pronounced potential to induce transmitter release in locust synaptosmes suggesting that there are specific sites of action in insect synaptic membranes (Knipper et al., 1986).

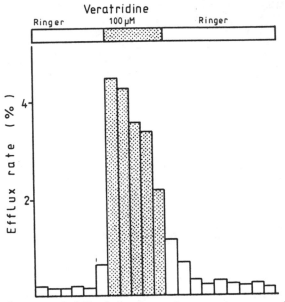

Fig.5 Release of ACh from locust synaptosomes in a
      micro-superfusion system.

Release experiments in the presence of ACh and AChE inhibitors gave a significantly diminished rate of
release, suggesting that ACh can influence its own release process, possibly via feedback mechanisms. When
the effect of various cholinergic agents were analyzed
it became clear that muscarinic agonists, like oxotre-
morine, inhibited the release, whereas muscarinic antagonists enhanced the evoked transmitter release. Nicotinic ligands, in contrast, did not show any significant
effect (Table 3).

Table 3 Effects of cholinergic ligands on the evoked
         release of acetylcholine

Compound	Percentage of control $\pm$ SD
Control	100
Muscarine (10 uM)	90.5 $\pm$ 1.2
Oxotremorine (1 uM)	87.4 $\pm$ 3.0
Atropine (1 uM)	112.1 $\pm$ 4.4
Nicotine (10 uM)	104.0 $\pm$ 8.5
Tubocurarine (100 uM)	98.2 $\pm$ 3.5

Thus, these results may be considered as support for the
concept that a feedback regulation of ACh release is
exerted via muscarinic autoreceptors located on cholinergic axon terminals. Furthermore, these results may
represent the first evidence for a physiological role of
muscarinic binding sites in insect nervous tissue; suggesting a presynaptic role of the muscarinic receptors.
Recent evidence from vertebrate pharmacology indicates
that there are two subclasses of muscarinic acetylcholine receptors (mAChRs), depending upon their affinity for
the antagonist pirenzipine and their coupling to a different second messenger system. M_1-type: high affinity
for pirenzipine, positively linked to phosphatidyl inositol turnover; M_2-type: low affinity for pirenzipine,
negatively linked to adenylate cyclase (Hammer and
Giachetti, 1982). Binding studies on pericarya and synaptosomes from locust revealed that the mAChR in cell
bodies are mainly of the M_1-type whereas the synaptosomal membranes contain mAChR of the M_2-subtype (Knipper
and Breer, 1986); consequently we propose that the modulation of ACh-release via presynaptic muscarinic autoreceptors is brought about by inhibiting the adenylate
cyclase and thus reducing the cAMP-level. Studies employing permeable cAMP-analog and specific drugs are in
progress to verify this concept.

Receptors for Neurotransmitters

The transduction of signals between neurons is mediated
through postsynaptic receptors; the chemical characterization of the receptor molecules is therefore of fundamental importance in understanding synaptic transmission
in insects. High-affinity, sodium-independent binding
sites for GABA, which are supposed to be GABA-receptors,
have been detected in insect nervous tissue (Breer and
Heilgenberg, 1985; Lummis and Sattelle, 1985; Lunt et
al., 1985).
From the earliest days of research on cholinergic synapses is thas been recognized that two categories of receptors for ACh exist, nicotinic and muscarinic AChRs,
which can now qualitatively be differentiated and quantitatively estimated using specific radiolabelled
ligands, notably α-bungarotoxin (BGTX) and quinuclidi-
nyl benzilate (QNB). When monitoring the number of puta-

tive receptors for ACh with specific ligands it was clear that there is a much higher concentration of cholinergic receptors in insects than in vertebrates; but is
was even more interesting to find that the relative levels of binding sites are reversed: predominantly QNB-
binding sites (mAChR) in the vertebrate brain but very
high concentrations of ⍺-toxin binding sites, putative
nicotinic receptors in insect ganglia (Fig.6).

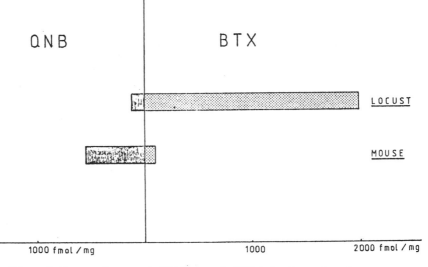

Fig. 6 Comparison of BGTX- and QNB-binding in the
        central nervous system of locust and mouse.

The preponderance of nicotinic binding sites and physiological evidences indicating that these sites are functional acetylcholine receptors in insect nervous tissue
(Sattelle et al., 1984) emphasize the particular importance of neuronal nicotinic acetylcholine receptors in
insects. In view of the enormous progress in identifying
the peripheral nicotinic AChR in vertebrates (McCarthy
et al., 1986) it was of particular interest to explore
the molecular properties of the insect AChR. This approach not only would extend our knowledge in insect neurochemistry but furthermore may offer some clues for understanding the molecular evolution of the receptors for
ACh.
ACh-receptors, as probed by BGTX binding, have been solubilized from locust neuronal membrane preparations

using mild detergents and subjected to analytical density gradient certrifugation. The binding activity was recovered in a single peak between the marker enzymes catalase and phosphatase which corresponds to a sedimentation coefficient of about 10S; a value which characteri-
zes the receptor as macromolecule similar to the vertebrate receptor (Breer et al., 1985).
The receptors were further purified by means of affinity
chromatography using ⍺-toxin immobilized covalently on
Sepharose 4B; the receptor protein was biospecificly
eluted with cholinergic ligands. Analysis of the eluted
proteins by polyacrylamid-microelectrophoresis gave a
single band corresponding to a molecular weight of about
300.000 in a gradient gel under non-denaturing
conditions, which confirms the 10S-value; under denaturing conditions in a SDS-gel, however, the same protein
migrated again as a single band now corresponding to a
molecular weight of about 65.000 (Fig.7).

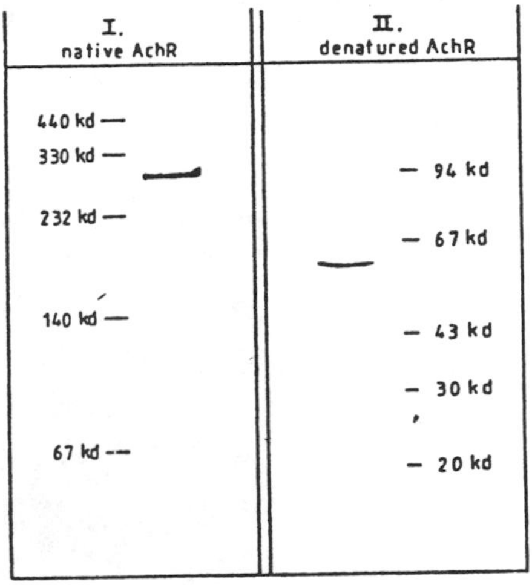

I. native AchR	II. denatured AchR
440 kd —	
330 kd — ▄▄	— 94 kd
232 kd —	— 67 kd
140 kd —	— 43 kd
	— 30 kd
67 kd –—	— 20 kd

Fig.7 Separation of native and denatured polypeptides
       of locust AChR in polyacrylamid electrophoresis.

This result suggests, that the native receptor apparently represents an oligomer of 4-5 identical or very similar polypeptides, indicating that the molecular organization would be significantly different from the vertebrate receptor. But before drawing such a conclusion it

was necessary to evaluate whether the isolated toxin
binding protein represents a functional AChR.
In a first approach immunohistochemical techniques have
been used for a topochemical localization. Monospecific
antisera against the affinity purified protein were raised in rabbits and used to incubate cryostat sections
from insect ganglia. The antigenic sites were visualized
by the peroxidase-antiperoxidase-technique. In the head
and thoracic ganglia of various insect species very distinct areas of the neuropil were found to be labelled;
many zones which are known to be very rich in neuronal
connections and several which are supposed to have many
cholinergic synapses. No labelling was found over the
fibers, the peripheral sheets and cell-layers (Fig.8).
Thus the appearance of the antigenic sites in immunocytochemical preparations would be consistent with a location at neuronal synapses.

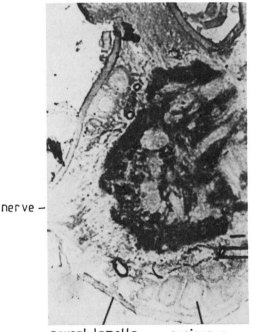

nerve —

neural lamella   pericarya

Fig.8 Immunocytochemical localization of AChR in the
      2. thoracic ganglion of locust.

Secondly, the purified receptor subunits were analyzed
for crossreactivity with monoclonal antibodies against
the Torpedo receptor. In immunoblots it was demonstrated
that a few monoclonals significantly crossreacted with
the insect receptor polypeptides which points to some
molecular relatedness of both receptor types. The immunological approaches thus provide further evidences that
the purified protein is in fact a constituent of the
AChR in the nervous system of insects.
The ultimate proof, however, that the isolated polypeptides represent the functional AChR and not only ligand
binding components, can only be achieved by reconstitu-
ting the protein in artificial lipid membranes, e.g.
planar lipid bilayers. To approach this goal, the purified insect receptor protein has been incorporated into
liposomes and these proteoliposomes subsequently fused
into a planar lipid bilayer; the conductance of this bilayer was then analyzed under voltage clamp conditions
and at high time resolution, which allows to resolve
single channel events. Without agonists in the aqueous
solution no conductance was observed; however, when agonists were added the appearance of specific channel
fluctuations was recorded.

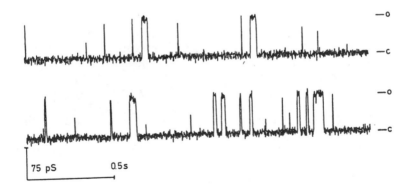

Fig.9 Fluctuations of an AChR channel from locust
      nervous tissue reconstituted in planar lipid
      bilayer.

In the presence of only very low concentrations, e.g.
0.5 µM suberyldicholine single channels with a conductance of about 75 pS and with a mean lifetime of a few
milliseconds were determined. The channel was found to
be selectively permeable for cations and activation was
blocked by cholinergic antagonists like d-tubocurarine
(Hanke and Breer, 1986a; Hanke and Breer, 1986b).
Thus, for the first time a neuronal ACh-controlled ion
channel, a functional neuronal ACh receptor, has been
reconstituted. The nicotinic ACh receptor in the nervous
system of insects obviously represents an oligomeric
membrane protein composed of identical or very similar
subunits and is thus different from the peripheral vertebrate receptor. The oligomeric structure of this receptor type is of particular interest in view of the receptor evolution, since based on immunological crossreactivity between subunits and the aminoacid sequence homology of the polypeptides forming the vertebrate receptor, is has been suggested that the four genes encoding
the subunits must have originated from a common and ancestral gene via duplication (Raftery et al., 1980; Noda
et al., 1984); thus the ancestral AChR supposedly has
been a homooligomeric complex from which the recent he-
terooligomeric vertebrate receptor has evolved. Whether
the insect receptor can be considered as a prototype of
such an ancient form is unclear and can only be decided
on the basis of the complete aminoacid sequence.
As a first approach the purified receptor protein was
subjected to gas phase microsequencing; unfortunately,
it was found that the amino-terminus was blocked and
therefore not accessible for sequencing; however, fragment of the receptor polypeptide was applied to the sequencer and a readily identifiable single sequence was
found: Ala-Val-Pro-Leu-Ile-Gly-Arg (Zenssen et al.,
1986). If this short sequence was compared with the
known sequences of the vertebrate receptor subunits, it
was observed that an almost identical sequence is present in all identified α-subunits (both Torpedo
species, chick, mouse, calf, human) and in the δ-sub-
unit of Torpedo (Boulter et al., 1985). This particular
amino acid stretch is located in the transmembrane region II/III; it is part of a region in the primary
structure, which was recently identified as most striking example of regional sequence conservation; a conservation which seems to be characteristic for membrane

proteins with multiple transmembrane segments (Boulter et al., 1985). This consideration points to the possibility that insect neuronal receptors display a similar transmembrane organization as the vertebrate receptor subunits.

To explore this aspect it will of course be necessary to unravel the complete amino acid sequence of the insect receptor polypeptides; this can only be achieved by employing techniques of molecular genetics. An essential early step before the powerful approaches of molecular genetics can be applied is the recognition of the mRNA coding for the receptor polypeptides. The translation of isolated mRNA in cell-free systems in combination with immunoprecipitaion using anti-(AChR)-antibodies represents a suitable assay for identification receptor specific mRNA. PolyA$^+$-RNA was isolated from nervous tissue of locust and translated in a rabbit reticulocyte lysate in the presence of ^{35}S-labelled methionine. Precipitation experiments followed by electrophoretic analysis indicated that the RNA-preparation from locust ganglia contained intact mRNA, capable to code even for high molecular weight polypeptides. To evaluate whether receptor protein have been produced as well, monospecific antiserum against the native AChR have been used for immunoprecipitation of specific polypeptides; it was found that 0.1 - 0.5% of the polypeptides could be separated, showing that antigenic sites of the native receptor already exist in the non-processed polypeptides and that the polyA$^+$-RNA fraction in fact contained receptor-specific mRNA. The in vitro translation system is, however, incapable to produce intact ligand binding sites, functional receptors; therefore as a next step, the Xenopus oocyte was used as in vivo translation system. The oocyte technique introduced by Gurdon as an efficient translation system for exogenous mRNA, capable to perform all post-translational modifications, has been adapted by Barnard and his colleagues for studying the expression of mRNA for receptors and ion channels (Barnard et al., 1984). When polyA$^+$-RNA isolated from the nervous tissue of young locusts were microinjected into oocytes after 1 day, α-bungarotoxin binding activity could be detected, depending linearly on the amount of applied RNA. Binding sites for α-toxin were found in the surface membrane of the oocytes, indicating that some of the AChR molecules are inserted in the membrane.

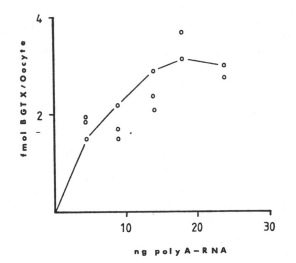

Fig.10 Expression of ∝-BGTX binding sites in
       Xenopus oocytes after microinjection of
       locust mRNA.

Immunoprecipitating of receptor polypeptides synthesized
in the presence of ^{35}S methionine, followed by electrophoresis and autoradiography revealed, that obviously a
homogenous population of polypeptides was isolated (M_r =
65.000), indicating that the in ovo synthesized constituents of the receptor have the same or a similar size as
the native subunits.
However, to verify that the expressed binding proteins
represent in fact functional AChRs it was necessary to
provide evidence for ACh-gated ion channel. This question was addressed using ion flux studies. When oocytes,
previously injected with insect mRNA were treated with
cholinergic agonists a significant influx of ^{86}Rb iso-
tops was induced, which was blocked by d-tubocurarine
(Breer and Benke, 1986). Thus, it can be concluded, that
after microinjection of mRNA from the nervous tissue of
insects, Xenopus oocytes produce ACh-controlled ion
channels. The oocytes can therefore be considered as an
excellent system to analyze the translation of specific
insect nucleic acid samples and possibly the expression
of interspecies receptor hybrids in forthcoming

molecular-genetic studies on the insect receptor.

Cloning of a Receptor-specific cDNA

A cDNA library was constructed by the procedure of Gubler and Hoffman, 1983, using polyA$^+$-RNA preparations from young locusts, probed for receptor specific mRNA in oocytes. The cDNA was cloned into the λ-gt11 expression vector which promotes synthesis of fusion proteins. An amplified λ-gt11 library was plated, the expression induced with IPTG and the fusion proteins transferred onto NC-filters. Approximately 10^5 recombinant phage plaque were assayed for the production of a β-galactosidase-AChR fusion protein that was reactive with monospecific anti(locust AChR)-antiserum. Eight plaques reacted with the antibodies and two were further plaque purified. Lysogens of the λ AChR-2 clone were made in E.coli Y1089 and grown under conditions to induce the synthesis of β-galactosidase fusion protein; a fusion protein sythesized by the AChR-2 lysogen was identified on Western blots using anti-β-galactosidase and anti-receptor-antibodies.

CONCLUSION

The results of recent experimental studies and the potential of new approaches suggest that synaptosomes, neuronal membranes and membrane proteins from insects are suitable systems for studying various molecular facets of synaptic transmissions in arthropods; these approaches may significantly contribute to explore the molecular architecture of the insect nervous system and to unravel molecular details of chemical signalling between nerve cells in general. Comparative studies on molecular neurobiology of invertebrates may gain new information concerning the evolution and phylogenetic development of biochemical processes and molecular elements involved in synaptic transmission. More detailed knowledge of synaptic elements in the nervous system of insects is also considered as essential prerequisite for analyzing the mode of action of chemicals that may be considered as putative insecticides and furthermore, to detect critical sites in the nervous system of insects, which might represent new targets for insecticides. Finally, evaluating the molecular properties of neuronal elements in insects may gain some new insight in the basis of severe problems of applied entomology like insecticide resistence and thus, contribute to detect and determine more effective, more selective and safer insect control agents.

ACKNOWLEDGEMENTS

The author gratefully acknowledges the technical assistance of Miss G. Hinz and Mrs. U. Mädler and expresses his graditude to Prof. W. Lueken for encouragement during the course of this work and to Mrs. G. Moehrke for typing the manuscript. The experimental studies were supported by the Deutsche Forschungsgemeinschaft.

REFERENCES

Barnard, E.A., Beeson, D., Bilbe, G., Brown, D.A., Constanti, A., Houamed, K. and Smart, T.G. (1984) A system for the translation of receptor mRNA and the study of the assembley of the functional receptors. J. Recept. Res. **4**, 681.

Boulter, J. Luyten, W., Evans, K., Mason, P., Ballivet, M., Goldman, D., Stengelin, S., Martin, G., Heinemann, S. and Patrick, J. (1985) Isolation of a clone coding for the α-subunit of a mouse acetylcholine receptor. J. Neurosci. **5**, 2545.

Breer, H. (1981) Characterization of synaptosomes from the central nervous system of insects. Neurochem. Int. **3**, 155-163.

Breer, H. (1982) Uptake of N-[Me-^{3}H]-choline by synaptosomes from the central nervous system of Lucusta migratoria. J. Neurobiol. **13**, 107-117.

Breer, H. (1983a) Choline transport by synaptosomal membrane vesicles isolated from nervous tissue. FEBS Lett. **153**, 345-348.

Breer, H. (1986) Synaptosomes systems for studying insect neurochemistry. In: Ford, N.G., Usherwood, P.N.R., Reay, R.C. and Lunt, G.G. (eds.) Neuropharmacology and Neurobiology, Ellis Horwood Books, pp. 384-413.

Breer, H. and Benke, D. (1986) Messenger RNA from insect
    nervous tissue induces expression of neuronal
    acetylcholine receptors in Xenopus oocytes. Molecular
    Brain Research, -in press-
Breer, H. and Heilgenberg, H. (1985) Neurochemical
    studies on GABAergic elements in the central nervous
    system of Locusta migratoria. J. comp. Physiol. A.,
    157, 343-354.
Breer, H. and Jeserich, G. (1980) A microscale flotation
    technique for the isolation of synaptosomes from
    nervous tissue of Locusta migratoria. Insect Biochem.
    10, 457-463.
Breer, H. and Knipper, M. (1984) Characterization of
    acetylcholine release from insect synaptosomes. Insect
    Biochem. 14, 337-344.
Breer, H. and Knipper, M. (1985a) Synaptosomes and
    neuronal membranes from insects. In: Breer, H. &
    Miller, T.A. (eds.), Neurochemical Techniques in
    Insect Research. New York, Springer, pp. 125-154.
Breer, H. and Knipper, M. (1985b) Choline fluxes in
    synaptosomal membrane vesicles. Cellul. Molec.
    Neurobiol. 5, 285-296.
Breer, H. and Lueken, W. (1983) Transport of choline by
    membrane vesicles prepared from synaptosomes of insect
    nervous tissue. Neurochem. Int. 5, 713-720.
Breer, H., Kleene, R. and Hinz, G. (1985) Molecular forms
    and subunit structure of the acetylcholine receptor in
    the central nervous system of insects. J. Neuroscience
    5, 3386-3392.
Buchner, E., Buchner, S., Crawford, G.D., Mason, W.T.,
    Salvaterra, P.M. and Sattelle, D.B. (1986) Choline
    acetyltransferase-like immunoreactivity inthe brain of
    Drosophila melanogaster. Cell Tissue Res., -in press-
McCarthy, M.P., Earnest, J.P., Young, E.F., Choe, S. and
    Stroud, R.M. (1986) The molecular neurobiology of the
    acetylcholine receptor. Ann. Rev. Neurosci. 9,
    383-413.
Gray, E.G.and Whittaker, V.P. (1962) The isolation of
    nerve endings from brain: an electronmicroscope study
    of cell fragments derived by homogenisation and
    centrifugation. J. Anat. 96, 79-88.
Gubler, U. and Hoffmann, B.J. (1983) A simple and very
    efficient method for generating cDNA libraries. Gene,
    25, 263.

Hammer, R. and Giachetti, A. (1982) Muscarinic receptor
    subtypes: M1 and M2. Biochemical and functional
    characterization. Life Sci. **31**, 2971-2998.
Hanke, W. and Breer, H. (1986a) A neuronal acetylcholine
    receptor purified from the nervous tissue of insects
    reconstituted in planar lipid bilayers. Nature **321**,
    171-174.
Hanke, W. and Breer, H. (1986b) Characterization of
    channel properties of a neuronal acetylcholine
    receptor reconstituted in planar lipid bilayers.
    J. Gen. Physiol.
Kanner, B.I. (1983) Bioenergetics of neurotransmitter
    transport. Biochim. Biophys. Acta **726**, 293-316.
Knipper, M. and Breer, H. (1986) Subtypes of muscarinic
    aceytlcholine receptors in the nervous system of
    insects. -in press-
Knipper, M., Maleddie, L., Breer, H. and Meldolesi, J.
    (1986) Black widow spider venom-induced release of
    neurotransmitters from synaptosomes: a unique venom
    component (α-latrotoxin) is active on mammalian,
    multiple components on insect nerve terminals.
    Neuroscience, -in press-
Koelle, G.B. (1963) Cytological distributions and
    physiological functions of cholinesterases. In:
    Koelle, G.B. (ed.) Handbuch der experimentellen
    Pharmakologie, Vol.XV, Springer, Berlin, pp. 188-298.
Kuhar, M.J. and Murrin, L.C. (1978) Sodium-dependent
    high affinity choline uptake. J. Neurochem. **30**, 15-21.
Lummis, S.C.R. and Sattelle, D.B. (1985) Insect central
    nervous system γ-aminobutyric acid receptors.
    Neuroscience Letters **60**, 13-18.
Lunt, G.G., Robinson, T.N., Miller, T.A., Knowles, W.P.
    and Olsen, R.W. (1986) The identification of GABA
    receptor binding sites in insect ganglia. Neurochem.
    Int. **7**, 751-754.
Meyer, E.M. and Cooper, J.R. (1983) High affinity uptake
    and calcium-dependent release of ACh from
    proteoliposomes derived from rat cortical
    synaptosomes. J. Neurosci. **3**, 987-994.
Noda, M., Takahashi, H., Tanabe, T., Kikyotani, S.
    et al. (1983) Structural homology of Torpedo
    california acetylcholine receptor subunits. Nature,
    302, 528-532.

Raftery, M., Hunkapiller, M.W., Strader, C.D., Hood, L.E.
    (1980) Acetylcholine receptor: Complex of homologous
    subunits. Science, **208**, 1454-1457.
Salvaterra, P.M., Crawford, G.D., Klotz, J.L. and
    Ikeda, K. (1985) Production and use of monoclonal
    antibodies to biochemically defined insect neuronal
    antigens. In: Breer, H. & Miller, T.A. (eds.)
    Neurochemical techniques in insect research. Springer,
    Berlin, pp. 223-242.
Sattelle, D.B., Harrow, I.D., Hue, B., Pelhate, M.,
    Gepner, J. and Hall, L.M. (1983) α -Bungarotoxin
    blocks excitatory synaptic transmission between cercal
    sensory neurones and giant interneurone 2 of the
    cockroach Periplaneta americana. J. exp. Biol. **107**,
    473-489.
Sattelle, D.B., Ho, Y.W., Crawford, G.D., Salvaterra,
    P.M. and Mason, W.T. (1986) Immunocytochemical
    staining of central neurons in Periplaneta americana
    using monochlonal antibodies to choline acetyl-
    transferase. Tiss. Cel.. **18**, 51-61.
Semenza, G., Kessler, M., Hosang, M., Weber, J. and
    Schmidt, U. (1984) Biochemistry of Na^{+}-D-glucose
    cotransporter of the small intestinal brush border
    membrane. Biochim. Biophys. Acta **779**, 343-379.
Whittaker, V.P. (1984) The synaptosome. In: Lajtha, A.
    (ed.), Handbook of Neurochemistry, 2nd., Vol.7. New
    York, Plenum, pp. 41-69.
Zenssen, Hinz, G., Beyreuther, K. and Breer, H. (1986)
    Sequence homology between the peripheral vertebrate
    receptor and the neuronal insect receptor. -in press-
Zimmermann, H. (1982) Insight into the functional role of
    cholinergic vesicles: In: Klein, R.L., Lagercrantz, H.
    and Zimmermann, H. (eds.), Neurotransmitter Vesicles.
    London, Academic Press, pp. 305-359.

BIOGENIC AMINE RECEPTORS AND THEIR MODE OF ACTION IN

INSECTS

PETER D. EVANS.

A.F.R.C. Unit of Insect Neurophysiology and Pharmacology,
Dept. of Zoology, University of Cambridge, Downing Street,
Cambridge, CB2 3EJ, U.K.

INTRODUCTION

Insect nervous systems contain a wide range of
biogenic amines that function as either neurotransmitters,
neuromodulators or neurohormones (Evans, 1980). The
structures of some of these biologically active amines are
shown in Fig. 1. In general amines such as dopamine,
serotonin (5-hydroxytryptamine, 5HT) and octopamine are
present in large quantities in insect nervous tissue and
we have a considerable amount of information about some of
the functional roles of the receptors for these amines in
the insect nervous system. Thus dopamine is likely to act
as the neurotransmitter in cockroach salivary glands (see
House, 1980) and may also have actions on gut and heart
muscle (see Evans, 1980). 5-HT is the well documented
neurohormonal activator of blowfly salivary glands
(Berridge, 1972; Trimmer, 1985) and may also act as a
neurotransmitter of central interneurones (Taghert and
Goodman, 1984) and some afferent and efferent neurones
(Tyrer et al., 1984). However, in the last ten years
studies on the functional role of the monophenolic
biogenic amine, octopamine have almost outpaced those on
other biogenic amines in insects (see Evans, 1985a).
Octopamine acts as a neuromodulator of neuromuscular
transmission and muscle contraction in skeletal muscle
(Evans and O'Shea, 1977, 1978; O'Shea and Evans, 1979;
Evans 1981; Evans and Siegler, 1982) and as a circulating
neurohormone controlling carbohydrate and lipid levels

117

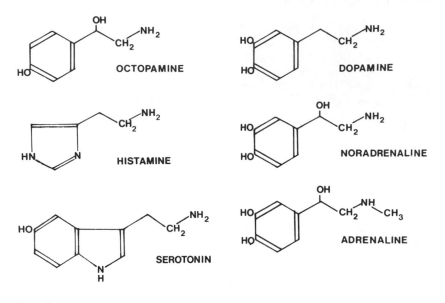

Fig.1.   The structures of some biogenic amines

(Matthews and Downer, 1974; Downer, 1979a,b; Orchard et al., 1981, 1982).   In addition it also acts as a true neurotransmitter in firefly light organs (Christensen and Carlson, 1981, 1982; Christensen et al., 1983) and in the release of peptide hormones from the corpora cardiaca (Orchard and Loughton, 1981; Orchard et al., 1983).   At present the function of noradrenaline in insects is enigmatic.   It is present in only low concentrations in the region of one tenth the concentration of dopamine present, but it is not known if the noradrenaline is present as a metabolic mistake or if a small number of specific noradrenergic neurones exist.   Another biogenic amine recently identified in the insect nervous system is histamine.   This is present in all regions of the nervous system with particularly large concentrations being found in the visual system, where it may function as the neurotransmitter of some insect photoreceptors (Elias and Evans, 1983).   In addition, insect nervous tissue can synthesize and metabolise histamine (Elias and Evans, 1983) and it is also concentrated against a concentration gradient by a specific set of glial cells in the visual system (Elias and Evans, 1984).

At present our knowledge of the pharmacology of biogenic amine receptors in the insect central nervous system, although increasing, is very limited. Much of it consists of information obtained from biochemical studies on nervous system homogenates or from physiological studies of bath application of amines to intact nervous systems. The information from such studies is hard to interpret meaningfully in terms of a functional role for specific biogenic amine receptors at identifiable target sites in the central nervous system. Thus the bulk of our detailed pharmacological knowledge of biogenic amine receptors in insects has been obtained from studies on receptors in peripheral target sites which are more accessible for study and in which fewer cell types are present to complicate the interpretation of the observed results. In this way a detailed knowledge of insect dopamine receptors has been obtained from cockroach salivary glands (House, 1980), of 5-HT receptors in blowfly salivary glands (Berridge, 1972) and of octopamine receptors on insect skeletal muscle (Evans, 1981).

Such pharmacological studies on insect peripheral target sites for biogenic amines have made it clear that in insects, as in the vertebrate nervous system (Snyder and Goodman, 1980), that multiple receptor types with different modes of action exist for various transmitters. Thus in blowfly salivary glands two distinct types of 5-HT receptor exist (Berridge and Heslop, 1981) and in locust muscle at least three (Evans, 1981) and maybe even more (see below) distinct types of octopamine receptors can be found. It is thus important for each biogenic amine receptor studied in a new insect preparation that the identity of any second messenger system used in that individual preparation be determined. At present a considerable array of different second messenger systems are being found to be associated with the activation of biogenic amine receptor systems in insects. However they can be subdivided into two general classes. First, those that use cyclic AMP as a second message and where membrane bound receptors on the outer face of the cell membrane control the interaction of the regulatory and catalytic subunits of adenylate cyclase, the enzyme that synthesises cyclic AMP (see Rodbell, 1984). The increased levels of cyclic AMP resulting from receptor activation cause the activation of specific cyclic AMP dependent protein kinases which bring about the phosphorylation of various target proteins (see Nestler and Greengard, 1984). It is the activation of the latter target proteins that bring about the diverse physiological responses of the cells to the applied biogenic amine. Second, are the complex set of interrelated second messengers that can be activated when specific receptors activate the breakdown of membrane phospholipids (see Berridge and Irvine, 1984). This second pathway involves a combination of second messengers including Ca^{2+} and two substances inositol trisphosphate (IP_3) and diacylglycerol (DG) generated by the breakdown of a specific membrane phospholipid. The IP_3 is released into the cytosol where it acts as a second messenger to release calcium from endoplasmic reticulum-derived stores. The DG remains in the membrane where it activates the enzyme protein kinase C_2 which can regulate various ionic mechanisms such as Ca^{2+}-dependent K^+ channels or the Na^+/H^+ exchanger.

In the rest of this paper I would like to concentrate on one specific example. Thus I will review the current status of the evidence for multiple pharmacologically distinct types of octopamine receptors on insect skeletal muscle and the evidence for their different modes of action.

OCTOPAMINE RECEPTORS ON LOCUST SKELETAL MUSCLE

The bulk of information on the pharmacology of octopamine receptors on insect skeletal muscle has been obtained on a single preparation, namely the extensor-tibiae muscle preparation from the locust hindleg. This preparation presents a very useful model system in which to study the pharmacology and mode of action of octopamine receptors on insect skeletal muscle (see Evans and Myers, 1986). The muscle is innervated by three physiologically identified motorneurones, a fast excitatory (FETi), a slow excitatory (SETi) and a branch of the common inhibitor (CI). In addition it is also innervated by an identified modulatory octopaminergic neurone (Evans and O'Shea, 1977, 1978) which has been designated DUMETi (Dorsal Unpaired Median neurone to the Extensor Tibiae) (Hoyle et al., 1974). The latter cell does not make specific neuromuscular contacts in the muscle, but rather its terminals end as blindly ending neurosecretory terminals between the muscle fibres (Hoyle et al., 1980). The extensor muscle itself is a large muscle and can provide large quantities of material for the biochemical analysis of second messengers (Evans, 1984a). Further, the muscle is highly differentiated such that different regions can be identified that contain exclusively, fast, slow or intermediate muscle fibre types (Hoyle 1978) and it is thus easy to study the actions of octopamine on the physiology and biochemical properties of regions of the muscle specifically containing these different types of muscle fibres (Evans, 1985b).

Octopamine has two separate actions on this muscle that are mediated by two distinct pharmacological classes of octopamine receptor (Evans, 1981). First, it can slow a myogenic rhythm of contraction and relaxation found in a proximal bundle of muscle fibres (Hoyle, 1975, Evans and O'Shea, 1978). The receptors mediating this action have been designated OCTOPAMINE$_1$ class receptors. Second, octopamine can modulated neuromuscular transmission and muscular contraction in the bulk of the non-myogenic fibres in the muscle. The receptors mediating these effects have been designated the OCTOPAMINE$_2$ class receptors. These can be further subdivided into the OCTOPAMINE$_{2A}$ receptors located presynaptically on the terminals of the slow motorneurone and when stimulated by octopamine released from the neurosecretory terminals of the DUMETi neurone (Morton and Evans, 1984), they potentiate both the spontaneous and neuronally evoked release of neurotransmitter. The second subtype, the OCTOPAMINE$_{2B}$, receptors are located postsynaptically, or extrasynaptically since the DUMETi neurone does not form morphologically distinct neuromuscular junctions, on the muscle fibres themselves. Stimulation of the latter class of receptors increases the rate of relaxation of both fast and slow excitatory motorneurone generated twitch tension (O'Shea and Evans, 1979), tetanic and 'catch' tension (Evans and Siegler, 1982) in this muscle.

OCTOPAMINE$_1$ receptors can be distinguished from OCTOPAMINE$_2$ receptors on the basis of both antagonist and agonist studies (Evans, 1981, see Tables 1A and 1B). Chlorpromazine and yohimbine are much better blockers of the former receptors than is metoclopramide, whereas the converse is true of the latter receptors. Also clonidine is a more potent agonist than naphazoline at OCTOPAMINE$_1$ receptors, while the converse is true for OCTOPAMINE$_2$ receptors. Similarly OCTOPAMINE$_{2A}$ receptors can be distinguished from OCTOPAMINE$_{2B}$ receptors on the basis of both agonist and antagonist studies (see Table 1A and 1B). Cyproheptadine, mianserin and metoclopramide are better

Table 1A Action of agonists on octopamine receptors.

Drug	Receptor Class		
	Octopamine$_1$ EC_{50} (M)	Octopamine$_{2A}$ EC_{50} (M)	Octopamine$_{2B}$ EC_{50} (M)
Clonidine	6.8×10^{-10}	6.4×10^{-6}	2.0×10^{-5}
Naphazoline	1.2×10^{-8}	1.3×10^{-8}	2.2×10^{-7}
Tolazoline	1.5×10^{-9}	3.2×10^{-6}	6.0×10^{-7}

EC_{50} for Octopamine$_1$ receptors is the concentration of a 5 min pulse of the drug required to reduce the frequency of the myogenic rhythm by 50%. EC_{50} for Octopamine$_{2A}$ and Octopamine$_{2B}$ receptors is concentration of a 30s pulse of drug required to produce 50% of maximal response to octopamine in SETi twitch amplitude and relaxation rate, respectively (SETi fired at 1Hz). For further details of experiments and additional drugs see Evans, P.D. (1981)

Table 1B Action of antagonists on octopamine receptors.

Drug	Receptor Class		
	Octopamine$_1$ EC_{50} (M)	Octopamine$_{2A}$ EC_{50} (M)	Octopamine$_{2B}$ EC_{50} (M)
Metoclopro-mide	–	1.0×10^{-6}	9.5×10^{-6}
Yohimbine	2.8×10^{-7}	–	–
Chlorpro-mazine	2.6×10^{-8}	1.6×10^{-4}	7.0×10^{-5}
Cyprohep-tadine	3.7×10^{-8}	2.2×10^{-6}	5.1×10^{-5}
Mianserin	4.5×10^{-6}	1.2×10^{-6}	2.0×10^{-5}

EC_{50} for Octopamine$_1$ receptors is concentration of drug required to reduce response of myogenic rhythm to a 5 min pulse of 10^{-7} M DL-octopamine by 50%. EC_{50} for Octopamine$_{2A}$ and Octopamine$_{2B}$ receptors is concentration of a drug required to reduce response of SETi-induced twitch amplitude and relaxation rate, respectively, to a 30s pulse of 10^{-6}M DL-octopamine by 50% (SETi fired at 1Hz). For further details of experiments and additional drugs see Evans, P.D. (1981)

blockers of 2A than 2B receptors, whilst the converse is true for chlorpromazine. In addition naphazoline is a much better agonist than tolazoline at 2A receptors and tolazoline is much better than clonidine at 2B receptors. An important point to emphasise is that this receptor classification was evolved on the basis of pharmacological differences between the receptor subclasses. This is preferable to any classification based purely on a functional or mode of action basis since it avoids situations where pharmacologically identical receptors are classified differently if they are coupled to different transducing mechanisms for mediating their actions. However, in the case of the locust octopamine receptors the pharmacologically distinct OCTOPAMINE$_1$ and OCTOPAMINE$_2$ class receptors do in fact also have different modes of action (Evans, 1981, 1984a,b,c)

The drugs that are effective at distinguishing the different classes of octopamine receptor in locust skeletal muscle turn out to be a mixture of agents that have previously been described to be active at vertebrate α-adrenergic receptors, dopamine receptors, 5-HT receptors and histamine receptors (Evans, 1981, 1984a,b, 1985a,b). Thus at present there are no readily available specific octopaminergic agents available for such studies. However, the pharmacological differences described above suggest that it should be possible to synthesise such specific agonists and antagonists that would be capable of distinguishing octopamine receptors from other biogenic amine receptors and also of distinguishing between the different pharmacological classes of octopamine receptor.

The above conclusion is supported by recent evidence from studies on the actions of different octopamine isomers on the octopamine receptor subtypes on locust skeletal muscle. Octopamine occurs as three different positional isomeric forms i.e. para, meta and ortho, each of which occurs as the (+) and (−) enantiomorphs. Each of these six isomeric forms of octopamine has been synthesised and characterized by Prof. J. Midgley (University of Strathclyde, Glasgow) and we have examined their effects on the different subclasses of octopamine receptor in the locust extensor-tibiae muscle (Evans, Thonoor and Midgley, in preparation). Fig. 2 shows the actions of some of the isomers on the OCTOPAMINE$_{2B}$ receptors mediating the increase in relaxation rate of slow motorneurone induced twitch tension. It can be seen that the most effective form is the (−)para isomer which

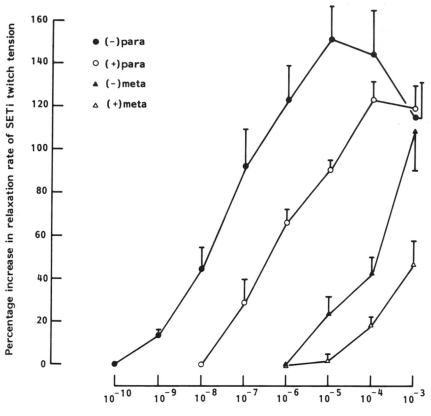

Fig.2   Dose-response curve for the action of various octopamine isomers on the maximal relaxation rate of slow motorneurone (SETi) twitch tension in the extensor-tibiae muscle of the locust hindleg. SETi was fired at a frequency of 1Hz and each of the isomers was introduced into the superfusate for a period of 30s. Each point represents the mean of at least three determinations and the bars represent standard errors.

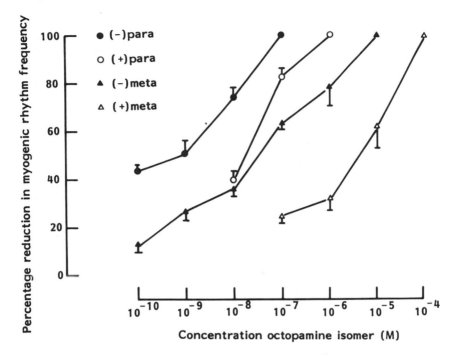

Fig.3.   Dose-response curve for the action of various octopamine isomers on the slowing of the myogenic rhythm in the extensor-tibiae muscle of the locust hindleg. The results are expressed as the percentage reduction in rhythm frequency at various isomer concentrations. The isomers were introduced into the muscle superfusate for a period of 5 min. Each point represents the mean of at least three determinations and the bars represent standard errors.

has a threshold of between 10^{-9} and 10^{-10}M and an ED_{50} of ~5 x 10^{-8}M. The (+) para isomer was around an order of magnitude less effective with an ED_{50} of~5 x 10^{-7}M. The (-) meta and (+) meta isomers were poor agonists that never maximally activated the receptors at concentrations below 10^{-3}M and needed between 10^{-6} and 10^{-5}M concentrations to demonstrate any significant effect. In contrast when the same isomers were tested on the OCTOPAMINE$_1$ receptors responsible for the slowing of the myogenic rhythm the meta isomers were relatively much more active. Fig. 3 shows that at the $OCTOPAMINE_1$ receptors that (-) para octopamine was active causing a 50% reduction in the rhythm frequency at $10^{-9}M$. Again the (+) para isomer was about an order of magnitude less effective. However, on this receptor both the (-) meta and (+) meta isomers were capable of causing a 100% reduction in rhythm frequency, with the (-) meta isomer causing a 50% reduction in frequency at $3 \times 10^{-8}M$ and with a threshold for an observable slowing of the rhythm occuring between 10^{-10} and $10^{-11}M$. Thus the (-) meta isomer of octopamine may well provide an important lead compound in the development of agonists and antagonists that are highly selective for the $OCTOPAMINE_1$ class receptors.

Recently a set of compounds, namely some substituted phenyliminoimidazolidines have been suggested to be selective agonists for $OCTOPAMINE_2$ receptors that bring about their actions by the activation of adenylate cyclase (Nathanson, 1985a,b). The structures of two of these derivatives, NC5 and NC7, are given in Fig.4 along with that, for comparison, of clonidine a related agonist which was shown previously to be active on octopamine receptors in the locust (Evans, 1981). Fig.5 shows that as expected NC5 and NC7 are effective agonists of the $OCTOPAMINE_{2B}$ receptors mediating the increase in relaxation rate of slow motorneurone induced twitch tension. They also activate the $OCTOPAMINE_{2A}$ receptors in this preparation (not shown). In both cases the compounds have thresholds between 10^{-9} and $10^{-10}M$ and NC5 is slightly more effective than NC7 especially at higher concentrations. However, if we look at their actions on the $OCTOPAMINE_1$ receptors mediating the slowing of the myogenic rhythm (Fig.6) it can be seen that both NC7 and NC5 are active, with NC7 being the most effective causing a 50% reduction in the rhythm frequency at $10^{-10}M$. Thus the phenylimino-imidazolidines are not specific for $OCTOPAMINE_2$ receptors in the locust.

A considerable amount of evidence indicates that the $OCTOPAMINE_{2A}$ and $OCTOPAMINE_{2B}$ receptors in locust skeletal muscle mediate their actions via increased cyclic AMP levels generated by the activation of adenylate cyclase (Evans, 1984a,b). All the physiological actions of these receptors can be mimicked by elevating cyclic AMP levels by mechanisms that bypass the receptor activation process. Thus they can be mimicked by the addition of the phos-

PHENYLIMINOIMIDAZOLIDINES

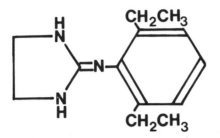

NC 5

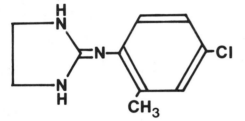

NC 7

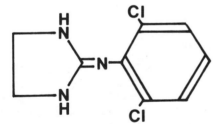

CLONIDINE

Fig.4.   The structures of some phenyliminoimidazolidine
derivatives.

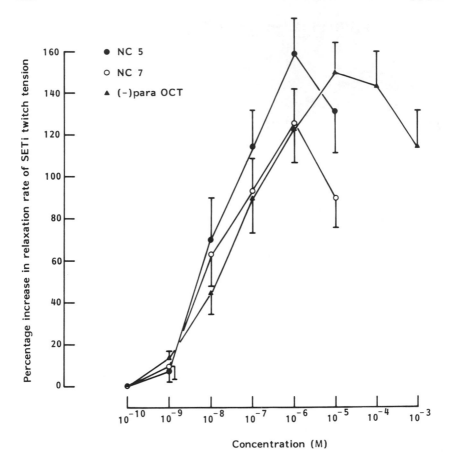

Fig.5.   Dose-response   curve   for   the   action   of
phenyliminoimidazolidine   derivatives   on   the   maximal
relaxation rate of slow motorneurone (SETi) twitch tension
in the extensor-tibiae muscle of the locust hindleg. SETi
was fired at a frequency of 1Hz and each of the isomers
was introduced into the superfusate for a period of 30s.
Each point represents the mean of at least three
determinations and the bars represent standard errors.

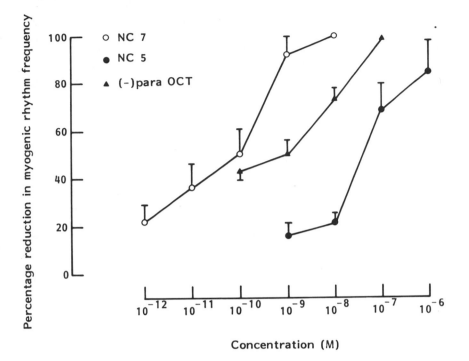

Fig.6. Dose-response curve for the actions of phenyliminoimidazolidine derivatives on the slowing of the myogenic rhythm in the extensor-tibiae muscle of the locust hindleg. The results are expressed as the percentage reduction in rhythm frequency at various concentrations of the derivatives. The derivatives were introduced into the muscle superfusate for a period of 5 min. Each point represents the mean of at least three determinations and the bars represent standard errors.

phodiesterase inhibitor isobutylmethylxanthine (IBMX) to the preparation. IBMX can also potentiate the physiological actions of octopamine. In addition these effects are mimicked by addition of forskolin, the specific diterpene activator of adenylate cyclase and also by the addition of highly permeable and slowly metabolised analogues of cyclic AMP, such as 8 chlorophenylthio cyclic AMP. Further evidence comes from the fact that short

pulses of octopamine introduced into the muscle
superfusate increase cyclic AMP levels, but not cyclic GMP
levels, in a dose dependent way. The pharmacological
profile and the time courses of the biochemical changes in
cyclic AMP level mirror those of the physiological
responses to octopamine.

The dose-response curve for the octopamine mediated
increase in cyclic AMP levels in the extensor-tibiae
muscle is unusual in that the rising phase of the sigmoid
curve extends over more than 2 log units of concentration
before entering the linear portion (Fig.7). This suggests
that there may be more than one component to the response.
This can be seen more clearly by replotting the data on a
log-log plot (Fig.8). This clearly reveals that there are
two proportionally related components to the response.
Each component has an initial linear functional slope of
unity where a ten fold increase in octopamine
concentration produces a ten fold change in cyclic AMP
accumulation. Thus in these regions of the curve there is
no cooperativity between agonist molecules. The existence
of these two linear components to the curve joined by a
non-linear section could be explained in two ways. It
could mean that the preparation has a single receptor type
that increases cyclic AMP levels and that at higher
agonist concentrations it may undergo some form of agonist
induced configurational change that alters its affinity
for octopamine. An alternative explanation is that there
are two distinct independent receptor sites involved.
However, the relationship between cyclic AMP levels and
physiological responses is complex since the increases in
cyclic AMP levels could be confined to a functionally
distinct subcompartment of muscle fibres within the
muscle. A direct test of the latter hypothesis reveals
that this is indeed the case (Fig.9 Evans, 1985b) with the
regions of the muscle that contain the largest proportions
of slow and intermediate muscle fibre types (regions, e,f
and 135c,d) exhibiting the largest octopamine dependent
increases in cyclic AMP levels. An examination of the
dose- responsiveness of the cyclic AMP levels to
octopamine in the seven different regions of the muscle
again provides evidence for the existence of more than one
component in the response. In all the regions of the
muscle the main component of the response is one that has
a half maximal effect between 10^{-4}M and 10^{-5}M and a
maximum in the region of 10^{-3}M octopamine. However a
second smaller effect is evident in some of the regions of

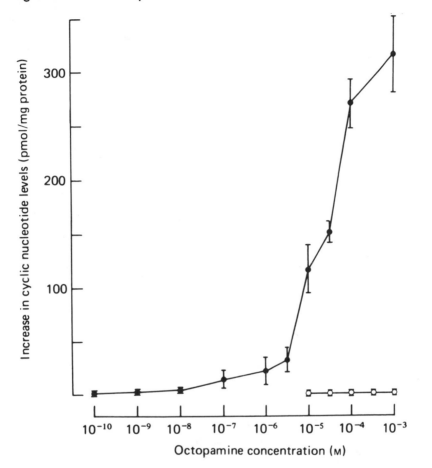

Fig.7.   Dose-response   curve   for   the   action   of
DL-octopamine on cyclic nucleotide levels in the extensor-
tibiae muscle.   The semi-log plot shows the effects on
cyclic AMP levels (●) and cyclic GMP levels (o).   The
results are expressed as the increase in cyclic nucleotide
levels in pmol mg^{-1} protein in the experimental muscle
above that found in the contralateral control muscle.
Both experimental and control muscles were pre-incubated
for 10 min in 10^{-4}M isobutylmethylxanthine (IBMX) before
the exposure of the experimental muscle to octopamine plus
IBMX for 10 min, and the control to a further 10 min
incubation in IBMX.   Each value is the mean of four
determinations and the bars represent standard errors of
the mean.

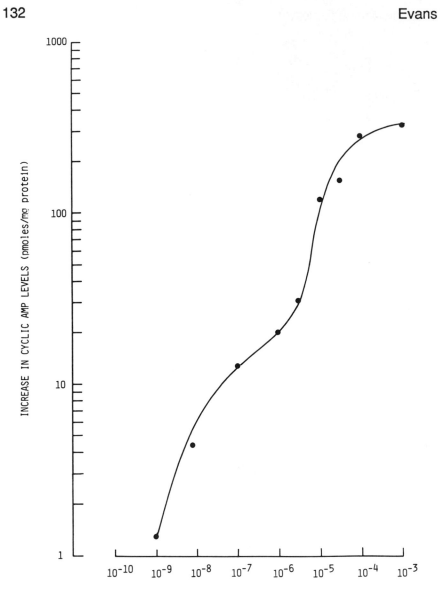

Fig.8.   A log-log plot of the data shown in Fig. 7 for the effects on cyclic AMP levels.

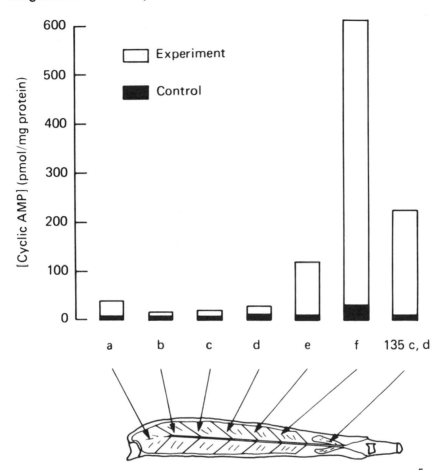

Fig.9. The effect of a 10min exposure to 10^{-5}M DL-octopamine on cyclic AMP levels in different regions of the extensor-tibiae muscle of the locust hindleg. The results are expressed as pmol cyclic AMP mg^{-1} protein for each of the blocks of muscle fibres from a single muscle. Open bars show the levels in the experimental muscle and filled bars the levels in the contralateral control muscle. Both experimental and control muscles were pre-incubated for 10 min in 10^{-4}M isobutylmethylxanthine (IBMX) for 10 min and the control for a further 10 min incubation in IBMX. The diagram indicates the location of the blocks of muscle assayed, proximal is to the left and distal to the right.

the muscle, particularly in regions d, e, f and 135c,d, which produces a shoulder on the dose-response curve at $10^{-6}M$ (Evans 1985b). This data supports the hypothesis, that the two components to the dose-response curve for the whole muscle represent two distinct receptor sites. At present it is not possible to identify definitely how the two biochemically identified receptor sites relate to the OCTOPAMINE$_{2A}$ and OCTOPAMINE$_{2B}$ receptors identified pharmacologically. However, it seems unlikely that the OCTOPAMINE$_{2A}$ sites on the presynaptic terminals of the slow motorneurone could account for a substantial part of the cyclic AMP accumulation measured in the whole muscle. It thus seems very likely that the higher affinity component of the cyclic AMP accumulation will represent the actions of the OCTOPAMINE$_{2B}$ receptors, since they both have the same thresholds and peak in the same regions of the dose-response curve. It further seems likely that the lower affinity component of the cyclic AMP increases could represent an octopamine mediate increase in cyclic AMP for which we have, as yet, not identified any corresponding physiological effect. Indeed it could even be related to a purely biochemical effect, such as changes in carbohydrate metabolism, that do not have any directly corresponding physiological responses. The idea that the two components of the cyclic AMP increase mediated by octopamine are generated by independent receptor systems receives further support from biochemical studies on the actions of NC5 and NC7 on the extensor-tibiae muscle (Fig.10). These compounds give dose-response curves for cyclic AMP accumulation that superimpose upon the higher affinity component of the octopamine curve, but lack any second component corresponding to the lower affinity octopamine component. Thus NC5 and NC7 although not specific agonists of OCTOPAMINE$_2$ receptors, since they activate OCTOPAMINE$_1$ receptors in the locust, can differentiate between the receptors mediating the two components of the octopamine sensitive increase in cyclic AMP levels in the extensor-tibiae muscle. These compounds have also been suggested to be able to distinguish differences in the OCTOPAMINE$_2$ receptors mediating increases in cyclic AMP levels in a number of other insect preparations including firefly light organ, cockroach nerve cord and tobacco hornworm nerve cord (Nathanson, 1985a,b).

At present the exact mode of action of the OCTOPAMINE$_1$ receptors mediating the slowing of the

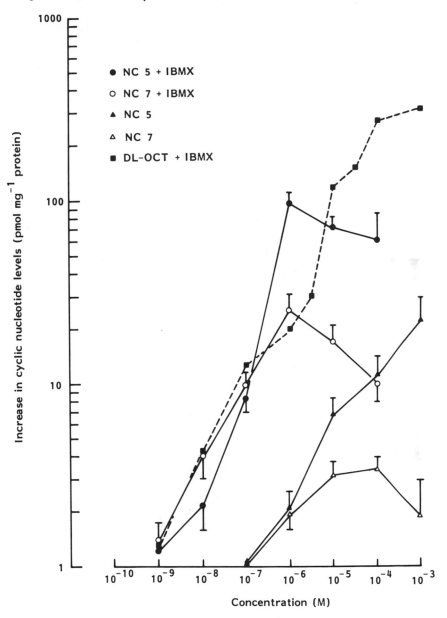

Fig.10.   A log–log dose response curve for the action of phenyliminoimidazoline derivatives on cyclic AMP levels in the extensor-tibiae muscle.   Details as for Fig.7.

myogenic rhythm remains to be elucidated. However, the bulk of the evidence available to date suggests that they do not mediate their actions via increases in cyclic AMP levels (Evans, 1984c). The frequency of the myogenic rhythm is determined by the actions of a small number of electrically coupled pacemaker fibres. Since these fibres are difficult to differentiate from the much larger number of follower fibres present, it is difficult to perform biochemical experiments on these fibres. Thus we have to rely on experiments where we manipulate the cyclic AMP levels by mechanisms that bypass the receptor activation stage. Thus IBMX increases the frequency of the myogenic rhythm, the opposite effect to octopamine which slows the rhythm. Further IBMX potentiates the actions of agents such as proctolin and 5-HT that accelerate the rhythm but has no actions on the octopamine mediated slowing of the rhythm. Further, forskolin also increases the frequency of the rhythm and its effects are again potentiated by IBMX. One of the few manipulations found to mimick the slowly action of octopamine on this rhythm is application of the Ca^{++} ionophore A23187. However, since extracellular Ca^{++} levels do not affect the action of octopamine on this system it is unlikely that octopamine mediates its effects by increasing the permeability of the preparation to extracellular calcium. Rather, it is more likely that octopamine mediates its actions by a mechanism involving the release of Ca^{++} from intracellular stores perhaps by a mechanism involving receptor activated breakdown of phospholipids (see Evans, 1984c).

CONCLUSION

Physiological, pharmacological and biochemical studies on the actions of octopamine on the locust extensor-tibiae muscle have lead to a classification of the octopamine receptors present into OCTOPAMINE$_2$ receptors which mediate their actions via increases in cyclic AMP levels and into OCTOPAMINE$_1$ receptors which probably mediate their actions via a release of calcium from intracellular stores (Evans 1981, 1984a,b,c). This raises the obvious question about the generality of such a classification and its applicability to other preparations. In general the octopamine receptors in other preparations have not yet been studied in the same detail as those on the locust extensor-tibiae muscle.

However, in a few other preparations where more detailed studies have been made the receptors that increase cyclic AMP levels have been found to have close pharmacological similarities with the $OCTOPAMINE_2$ receptors in the locust (e.g. see Uzzan and Dudai, 1982, Hollingworth, et al., 1984; Morton, 1984; Lafon-Cazal and Bockaert, 1985). However, in several of these cases it has not been possible to decide if they are more like 2A or 2B receptors. Even in the locust the distinction into 2A and 2B subtypes relies on the differences in the relative potencies of the agent used rather than clear cut specific blocking or agonist actions. Thus it is likely that such subtle variations could well be lost within the general differences between $OCTOPAMINE_2$ receptors found in different insect species. As yet no other preparation, apart from the locust, has been found to contain $OCTOPAMINE_1$ type receptors mediating their action via mechanisms that do not involve increases in cyclic AMP levels. Such preparations are urgently needed before the extent of the generality of the proposed classification scheme in the locust can be fully assessed.

Biochemical studies on octopamine mediated changes in cyclic AMP levels in the extensor tibiae muscle have revealed an additional receptor site that is pharmacologically similar to the $OCTOPAMINE_2$ receptors but which does not have any apparent physiological correlate at present. It is possible that this represents a receptor coupled specifically to a biochemical response in the muscle, such as the control of carbohydrate reserves. However, its discovery emphasises the point that the physiological relevance of various proposed octopamine receptor subtypes in any preparation can not be assessed from studies in which only changes in cyclic AMP levels are measured.

Perhaps the single most important problem in the study of octopamine receptors in insects at the present time is the development of specific agonists and antagonists that will first, distinguish octopamine receptors from the receptors of other biogenic amines and second distinguish between the different classes of octopamine receptors. Such compounds are urgently needed if we are to begin to expand our knowledge of the functional roles of octopamine receptors from the easily accessible peripheral systems studied to date, to those more complex systems that are as yet hidden away in the complexities of the insect central nervous system.

REFERENCES

Berridge, M.J. (1972). The mode of action of 5-hydroxytryptamine. J. exp. Biol. 56, 311-321.

Berridge, M.J. and Heslop, J.P. (1981). Separate 5-hydroxytryptamine receptors on the salivary gland of the blowfly are linked to the generation of either cyclic adenosine 3',5'-monophosphate or calcium signals. Br. J. Pharmacol. 73729-7389.

Berridge, M.J. and Irvine, R.F. (1984). Inositol trisphosphate, a novel second messenger in cellular signal transduction. Nature (Lond) 312, 315-321.

Christensen, T.A. and Carlson, A.D. (1981). Symmetrically organized dorsal unpaired median DUM neurones and flash control in the male firefly, Photuris versicolar. J. Exp. Biol. 93, 133-147.

Christensen, T.A. and Carlson, A.D. (1982). The neurophysiology of larval firefly luminescence direct activation through four bifurcating DUM neurons. J. comp. Physiol. 148, 503-514.

Christensen, T.A., Sherman, T.G., McCaman, R.E. and Carlson, A.D. (1983). Presence of octopamine in firefly photomotor neurons. Neuroscience 9, 183-189.

Downer, R.G.H. (1979). Trehalose production in isolated fat body of the American cockroach, Periplaneta americana. Comp Biochem Physiol 62C, 31-34.

Downer, R.G.H. (1979). Induction of hypertrehalosemia by excitation in Periplaneta americana. J. Insect Physiol. 25, 59-63.

Elias, M.S. and Evans, P.D. (1983). Histamine in the insect nervous system distribution, synthesis and metabolism. J. Neurochem. 41, 562-568.

Elias, M.S. and Evans, P.D. (1984). Autoradiographic localization of ^{3}H-histamine accumulation by the visual system of the locust. Cell Tissue Res. 238, 105-112.

Evans, P.D. (1980). Biogenic amines in the insect nervous system. Adv. Insect Physiol. 15, 317-473.

Evans, P.D. (1981). Multiple receptor types for octopamine in the locust J. Physiol. 318, 99-122.

Evans, P.D. (1984a). A modulatory octopaminergic neurone increases cyclic nucleotide levels in locust skeletal muscle. J. Physiol. (Lond.). 348, 307-324.

Evans, P.D. (1984b). The role of cyclic nucleotides and calcium in the mediation of the modulatory effects of octopamine on locust skeletal muscle. J. Physiol. (Lond.). 348, 325-340.

Evans, P.D. (1984c). Studies on the mode of action of octopamine, 5-hydroxytryptamine and proctolin on a myogenic rhythm in the locust. J. exp. Biol. 110, 231-251.

Evans, P.D. (1985a). Octopamine. In Comprehensive Insect Biochemistry, Physiology and Pharmacology. Eds. G.A. Kerkut and L. Gilbert pp499-530, Pergamon Press, Oxford.

Evans, P.D. (1985b). Regional differences in responsiveness to octopamine within a locust skeletal muscle. J. Physiol.(Lond.) 366, 331-341.

Evans, P.D. and Myers, C.M. (1986). Peptidergic and aminergic modulation of insect skeletal muscle. J. exp. Biol. 124, 143-176.

Evans, P.D. and O'Shea, M. (1977). An octopaminergic neurone modulates neuromuscular transmission in the locust. Nature, Lond. 270, 257-259.

Evans, P.D. and O'Shea, M. (1978). The identification of an octopaminergic neurone and the modulation of a myogenic rhythm in the locust. J. exp. Biol. 73, 235-260.

Evans, P.D. and Siegler, M.V.S. (1982). Octopamine mediated relaxation of maintained and catch tension in locust skeletal muscle. J. Physiol. 324, 93-112.

Hollingworth, R.M., Johnstone, E.M. and Wright, N. (1984). Aspects of the biochemistry and toxicology of octopamine in Arthropods. American Chemical Society Symposium. 225, 103-125.

House, C.R. (1980). Physiology of invertebrate salivary glands. Biol Rev. 55, 417-473.

Hoyle, G. (1975). Evidence that insect dorsal unpaired median (DUM) neurons are octopaminergic. J. exp. Zool. 193, 425-431.

Hoyle, G. (1978). Distributions of nerve and muscle fibre types in locust jumping muscle. J. exp. Biol. 73, 205-233.

Hoyle, G., Colquhoun, W. and Williams, M. (1980). Fine structure of an octopaminergic neuron and its terminals. J. Neurobiol. 11, 103-126.

Hoyle, G., Dagan, D., Moberly, B. and Colquhoun, W. (1974). Dorsal unpaired median insect neurons make neurosecretory endings on skeletal muscle. J. exp. Zool. 187, 159-165.

Lafon-Cazal, M. and Bockaert, J. (1985). Pharmacological characterization of octopamine -sensitive adenylate cyclase in the flight muscle of Locusta migratoria L. European J. Pharmacol. 119, 53-59.

Matthews, J.R. and Downer, R.G.H. (1974). Origin of trehalose in stress-induced hyperglycaemia in the american cockroach Periplaneta americana. Can. J. Zool. 52, 1005-1010.

Morton, D.B. (1984). Pharmacology of the octopamine stimulated adenylate cyclase of the locust and tick CNS. Comp. Biochem. Physiol. 78c, ·153-158.

Morton, D.B. and Evans, P.D. (1984). Octopamine release from an identified neurone in the locust. J. exp. Biol. 113, 269-287.

Nathanson, J.A. (1985a). Characterization of octopamine sensitive adenylate cyclase: Elucidation of a class of potent and selective octopamine -2 receptor agonists with toxic effects in insects. Proc. Natl. Acad. Sci. U.S.A. 82, 599-603.

Nathanson, J.A. (1985b). Phenyliminoimidazolidines. Characterization of a class of potent agonists of octopamine-sensitive adenylate cyclase and their use in understanding the pharmacology of octopamine receptors. Molecular Pharmacol. 28, 254-268.

Nestler, E.J. and Greengard, P. (1984). Neuron-specific phosphoproteins in mammalian brain. Adv. Cyclic Nuc. Protein Phosphoryl. Res. 17, 483-488.

Orchard, I., Carlisle, J.A., Loughton, B.G., Gole, J.W.D. and Downer, R.G.H. (1982). In vitro studies on the effects of octopamine on locust fat body. Gen. comp. Endocrinol. 48, 7-13.

Orchard, I., Gole, J.W.D. and Downer, R.G.H. (1983). Pharmacology of aminergic receptors mediating an elevation in cyclic AMP and release of hormone from locust neurosecretory cells. Brain Res. 288, 349-353.

Orchard, I. and Loughton, B.G. (1981). Is octopamine a transmitter mediating hormone release in insects? J. Neurobiol. 12, 143-153.

Orchard, I., Loughton, B.G. and Webb, R.A. (1981). Octopamine and short-term hyperlipaemia in the locust. Gen. Comp. Endocrinol. 45, 175-180.

O'Shea, M. and Evans, P.D. (1979). Potentiation of neuromuscular transmission by an octopaminergic neurone in the locust. J. exp. Biol. 79, 169-190.

Rodbell, M. (1984). Structure-function problems with the adenylate cyclase system. Adv. Cyclic. Nuc. Protein Phosphoryl. Res. 17, 207-214.

Snyder, S.H. and Goodman, R.R. (1980). Multiple neurotransmitter receptors. J. Neurochem. 35, 5-15.

Taghert, P.H. and Goodman, C.S. (1984). Cell determination and differentiation of identified serotonin-immunoreactive neurons in the grasshopper embryo. J. Neurosci. 4, 989–1000.

Trimmer, B.A. (1985). Serotonin and the control of salivation in the blowfly, Calliphora. J. exp. Biol. 114, 307–328.

Tyrer, N.M., Turner, J.D. and Altman, J.S. (1984). Identifiable neurons in the locust central nervous system that react with antibodies to serotonin. J. Comp. Neurol. 227, 313–330.

Uzzan, A. and Dudai, Y. (1982). Aminergic receptors in Drosophila melanogaster: responsiveness of adenylate cyclase to putative neurotransmitters. J. Neurochem. 38, 1542–1550.

EMBRYONIC FORMATION OF A SIMPLE NEUROSECRETORY NERVE IN

THE MOTH, MANDUCA SEXTA

Paul H. Taghert, Jeffrey N. Carr, John B. Wall
and Philip F. Copenhaver

Anat. & Neurobiol., Washington U. Med. School

660 S. Euclid, St. Louis, MO 63110.

INTRODUCTION

The development of the nervous system is marked by a striking diversity of cellular phenotypes: in many animals, individual neurons have unique identifiable properties. Equally remarkable, however, is the consistency of these diverse cellular morphologies and synaptic connections between conspecific animals. How this mixture ·of neuronal diversity and consistency is produced during embryonic development is a major challenge for developmental neurobiology. In recent years, much progress has been made in the analysis of neuronal determination and differentiation by utilizing the large embryos of certain insects (BASTIANI et al., 1985; THOMAS et al., 1984). In this chapter, we summarize recent observations we have made concerning the embryonic development of a simple neuroendocrine system in the tobacco hornworm, Manduca sexta.

Neurosecretory cells are specialized neurons which project axons into peripheral structures that are termed neurohaemal organs. Within these organs, the axons terminate with a profusion of varicosities and "blind endings" from which neurohormones are released into the blood (CARROW et al., 1984; COPENHAVER AND TRUMAN, 1986). In performing a cellular analysis of developing neurosecretory neurons our goals are to specify the rules

143

that underlie: (i) the stereotyped formation of axonal projections, (ii) the timing and induction of varicosity formation along terminal branches, and (iii) the basis of cell-specific expression of neurotransmitters at developmentally relevant times.

TRANSVERSE NERVES ARE SEGMENTAL NEUROHAEMAL STRUCTURES

Depending on the segment, each ganglion of the ventral nerve cord of Manduca consists of either ~700 or ~2000 neurons. In abdominal segments, there are three nerves that emanate from each ganglion to supply muscles and organs and to receive sensory information. The Dorsal (DN) and Ventral (VN) nerves are bilaterally paired; the third nerve is unpaired and is called the Transverse nerve (TN) (Fig. 2). The TN has two main functions in insects: (i) to supply the motor innervation to closer and/or opener muscles of the spiracles (e.g., LEWIS et al., 1973), and (ii) to serve as a point of release of various neurohormones. In the latter capacity, it is often termed part of the insect sympathetic nervous system (TRUMAN, 1973; TAGHERT and TRUMAN, 1981; see RAABE, 1982 for a review). The Peri-Visceral Organ (PVO) refers to thickened portions of the TN that lie on each side of the ganglion, the relative length of the PVO as a portion of the TN varies between different insect orders (evolutionary aspects are discussed by RAABE, 1982). While the PVO is obviously of major importance, it is clear that the neurohaemal function is distributed along the entire length of the nerve (Fig. 1).

In cross-sections, light and electron micrographs of the mature nerve reveal a pair of axons that run in an uninterrupted fashion through the TN: by reconstruction of serial images, these axons can be identified as the spiracle closer muscle motoneurons (SP MN's). The MN axons are surrounded by a thick glial sheath while the neurosecretory (NS) elements are associated with a more dispersed glial population (Fig. 1). More than one glial cell can be associated with an NS axon, and more than one type of NS axon (as judged by morphology of NS granules) can be associated with an ensheathing glial group (TAGHERT, unpub. results). NS elements are separate from the MN unit at all positions along the TN, although glial cell processes appear occasionally to be shared. Release

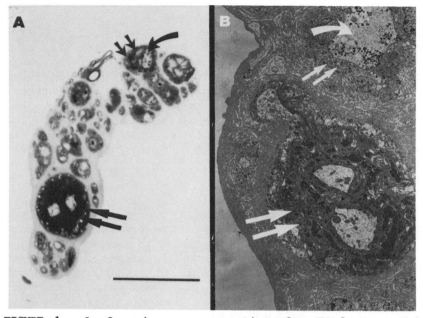

FIGURE 1.   A. One micron cross-section of a TN from an adult
moth.   Note large central cylinder (large double arrows) that is
composed   of   glial   cells   and   the   two   axons   of   the   mn's;
surrounding the cylinder are smaller inclusions (smaller double
arrows) that represent NS axons similarly surrounded by glia.
Scale = 25 μ.   B. Electron micrograph of a similar position along
the adult TN.   MN axons are prominent, as are nearby NS axons
that contain large dense cored neurosecretory granules.   Scale =
8μ.

of neurohormones is thought to occur at swellings along
the NS axons at which points, the glial processes are no
longer closely apposed (Raabe, 1982).

THE   TN   CONTAINS   AXON   PROJECTIONS   OF   BOTH   CENTRAL   AND PERIPHERAL NEURONS

Over the past 10 years, we have compiled a catalog of
all   the   neurons   that   project   axons   into   the   TN   of
abdominal   segments   in   Manduca.   A   certain   number
differentiate   during   embryogenesis,   but   the   number
increases   during   larval   life   and   again   during
metamorphosis.   By the end of embryogenesis, there are 2
motoneurons   (the   SP   MN's),   10   central   neurosecretory

neurons and 4 peripheral neurosecretory neurons. In
Figure 2 is shown their positions and cell-specific
trajectories. It is clear from this description that this
simple neuroendocrine effector organ is complex when
considered in fine detail. The different individual
neurons are specialized to secrete a variety of hormonal
substances, and further, they reach their end organ (the
TN) via diverse routes and at diverse times. The
following sections summarize our recent findings as
regards the embryonic development of this specialized set
of neurons.

EMBRYOGENESIS IN MANDUCA

Embryogenesis in Manduca is completed within
approximately 100 hr (at 25°C). The external morphology
of developing embryos can be used to accurately stage
animals to within 3 hours; we use the convention that was
adopted by BENTLEY et al. (1979) in staging grasshopper
embryogenesis and refer to Manduca embryonic development
in terms of percentages, where 100% is a fully completed
embryo at the hatching stage. Conveniently, 1% of
developmental time = 1 hour of real time. The timetable
of Manduca embryonic development was initiated by N.
Tublitz and M. Bate (per. comm.) and supplemented by our
own observations. In moths, as in grasshoppers, the
neurons of the segmental ganglia are the progeny of a
specialized class of cells termed neuroblasts (BATE, 1976;
TAGHERT et al., 1984; THOMAS et al., 1984; DOE and
GOODMAN, 1985). Neuronal progeny are first produced at
about the 25% stage and continue to be produced through
the late stages of embryogenesis. By 27%, the first
growth cones are produced by the first born neurons (BATE
and GRUNWALD, 1979; CARR and TAGHERT, unpublished
results).

The results presented here are derived from
observations of living embryos dissected out of the
eggshell and embryonic membranes, and viewed in a compound
microscope that is modified for Nomarski interference
contrast optics. In addition, whole embryos dissected as
above were stained as wholemounts using a variety of serum
and monoclonal antibodies. Finally, single cells in
living or fixed embryos were dye-filled with micro-
electrodes containing Lucifer Yellow, and processed

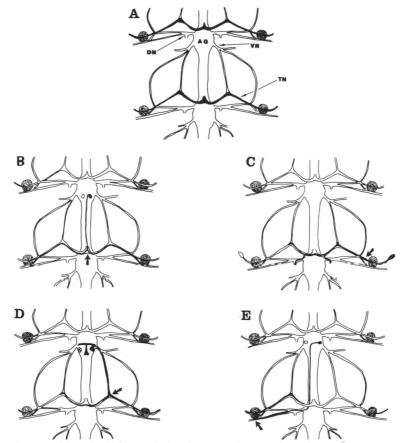

FIGURE 2. Axon pathways of TN-projecting neurons. Top: Diagram showing the arrangement of nerves and ganglia in two idealized abdominal segments. Middle and Bottom drawings represent the cell body positions and axon pathways of the various neurons that project to the TN. Middle Left: The SP MN's; Middle right: The peripheral NS, L1; Bottom left: NS's from the VN; Bottom Right: The fourth Bursicon-containing neuron. Arrows indicate the points at which the various neurons reach the TN. Pathways are indicated for cells only on one side. Contralateral homologues are indicated as open cell bodies.

with an anti-Lucifer Yellow antiserum (TAGHERT et al., 1982). Embryos are dissected by opening along the dorsal mid-line of the developing body; hence, dissected embryos typically are two-dimensional sheets of tissue, with the

segmental ganglia in the middle. In the next section we
provide a broad overview of TN formation from its
inception at ~30% of embryogenesis.

FORMATION OF THE TN

    The TN is created from the strap and the bridge.
Like the other nerves in this animal, the TN has a
characteristic   position,   shape,   and   trajectory.
Interestingly, much of the overall detail appears to be
specified by a collection of non-neural cells that form
just as the first neurons are born.   These cells are
likely to be mesodermal in origin; many of them
differentiate into glia that ensheath peripheral nerves.

    At about 27% of development, there occurs a group of
cells   that   lies   over   the   anterior   portion   of   each
segmental ganglion and is connected on either side to a
mass of largely undifferentiated mesoderm (Figs. 3A and
4A).   This collection of cells numbers about 50, and with
development to 34%, has changed its appearance.   It thins
out such that it is now only one to two cells wide and is
now inserted into the ectoderm at the segment border on
either   side   of   each   segmental   ganglion   (Fig.   3B   and
4B).   Moving distally from the insertion point, the group
forms   a   partial   ring   around   the   presumptive   spiracle
(Figs. 5 & 6).   At 39%, neuronal growth cones are first
beginning to reach these structures (Figs. 3C & 4C, also
see below); yet many of the features of the axonal
pathways they will form have already been put in place.
We call the group of cells over the ganglion the "strap"
and the group of cells around the forming spiracle the
"bridge".   In Figs. 3 and 5, these cells are highlighted
by virtue of antibody staining with Mab TN-1.

    By 43%, the strap has taken on an appearance more
similar to the mature TN (Fig. 3D), such that individual
strap cells now only occur at intervals along its
length.   In addition, the SP MN axons have reached the
origin of the TN and have branched to supply both sides
of the body.   By 61%, the TN is nearly fully formed (Fig.
3E and 4F):   all its component neurons have arrived onto
it and have nearly completed their morphogenesis (see
below).   In addition, cell-specific transmitters have
begun to be expressed (see below).

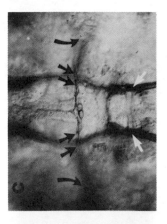

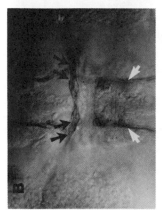

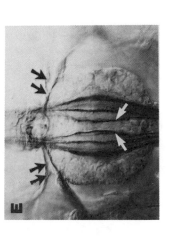

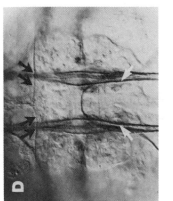

FIGURE 3. Photomicrographs of the developing TN as stained with monoclonal antibody (Mab) TN-1. A. 27%; B. 34%; C. 39%; D. 49%; E. 61%. Note the TN (double arrow) develops as a "strap" of cells at the anterior margin of the ganglion; curved arrow indicated site of strap insertion into the ectoderm. As a landmark, single arrows indicate a specific longitudinal axon pathway that is stained by Mab TN-1; magnificaton = x500.

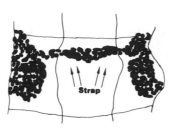

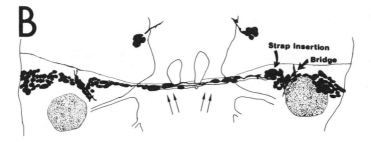

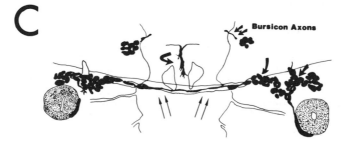

FIGURE 4.

E

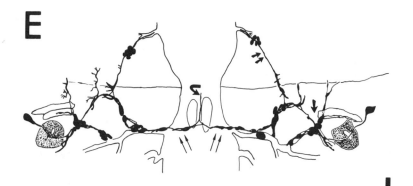

I

F

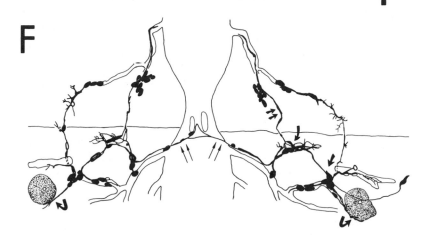

G

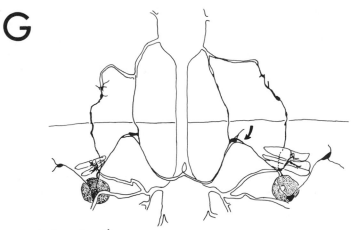

FIGURE 4 (continued)

FIGURE 4. Camera lucida drawings of the developing TN in abdominal segments. A. 32%; B. 37%; C. 43%; D. 49%; E. 57%; F. 61%; G. 90%. Scale Bar = 50 μ (A-F) and 100 μ (G). See text for further details.

This pattern of TN construction is outlined from camera lucida drawings in Figure 4, wherein is shown the developing strap and bridge at selected times in relation to the ganglion, and also to the point of arrival of certain neuronal growth cones. Next, we specifically discuss the development of the bridge and make an inference as to it role in TN formation.

The bridge serves as an interchange where many axons cross between diverse pathways. In addition to contributing to TN formation, the bridge also helps join other nerves. As diagrammed in Figs. 5 and 6, the bridge is a collecton of ~10-15 cells that were once continuous with the strap up to about the 39% time point. In mature larvae (Fig. 6F), this structure is an anastomosis between many convergent peripheral nerves. These include (i) the SP MN's from the strap insertion point to their muscle target, (ii) a nerve connection to the ventral nerve of the next anterior ganglion, (iii) the SP MN axon pathway to the closer muscle, (iv) a nerve connection of DN MN's to a muscle group that lies above the spiracle, and (v) a nerve connection to the DN.

Interestingly, as visualized with TN-1 immunostaining, many features of this future interchange of axon pathways is already in place by the configuraton of bridge cells and their elaborated processes (Fig. 6). We believe that most of these bridge cells later become glial cells (like the strap cells) because throughout development we observe TN-1 positive cells at these locations and they ultimately come to ensheath the axons that grow along them.

Hence we consider two major structures -- the strap and the bridge -- as providing much of the information necessary to build the TN according to its proper position and trajectory. Future experiments will test the hypothesis that these collections of future glial cells are required for proper TN formation by eliminating them prior to the arrival of specific neuronal growth cones. Although not our primary focus, we have also

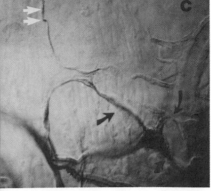

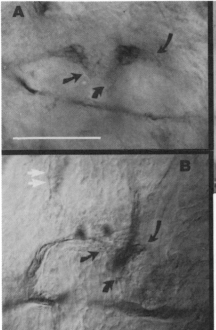

FIGURE 5. Photomicrographs of the developing bridge cells at three embryonic stages. The bridge becomes an anastomosis where five separate nerve pathways cross. Some are indicated by different arrow symbols. A. 34%; B. 39%; C. 57%. Scale = 40 μ (A,C); = 45 μ (B).

observed that cells similar in morphology and antigenic properties are placed at intervals along the pathways of future DN and VN nerve formation. However, in neither of these cases, is the complete pathway already in place, as appears to be the case for the TN. We now turn to consideration of the arrival and growth on the presumptive TN of its constituent neurons.

TN NEURONS ARRIVE AND GROW ONTO THE NERVE IN STEREOTYPED FASHION

There are basically three groups of neurons that project axons into the TN during embryonic development. (i) The SP MN's and (ii) the peripheral NS's arrive at approximately the same time but from different locations; (iii) a different group of NS's cells (these include the bursicon neurons) arrive slightly later and from a different trajectory. We have not yet analyzed the development of the fourth bursicon neuron (cf., Figure 2) and so it will not be discussed further. First, we will describe this pattern of neuronal differentiation in

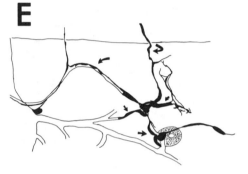

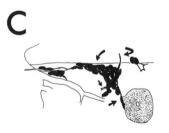

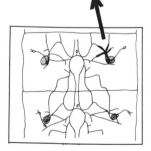

FIGURE 6. Camera lucida drawings of the bridge from TN-1 stained preparations. A. 39%; B. 41%; C. 43%; D. 57%; E. 90%. Arrows indicate specific nerve branches as in Figure 5. Inset at lower left indicates positon of the bridge within the segment; two segments are shown. Scale = 50 μ (A-D); = 67 μ (E).

terms of groups of axons growing roughly in concert to build the nerve proper. Next, we will consider the differentiation of a single neuron, the peripheral NS cell L1, in order to evaluate the mode and potential mechanisms that are utilized to elaborate a specific neurosecretory neuron.

The spiracular motoneurons. There are 2 SP motoneurons per abdominal ganglion, the cell bodies are located on the ventral surface, medial to the exit point of the ventral nerve. Within the ganglion, their axons travel medially within the neuropil, meet along the midline, then exit the ganglion posteriorly via what will become the median nerve (Fig. 2). In the abdominal ganglion, these events occur between embryonic stages 34-37%. By 39%, the growth cones of the two MN's are typically seen just approaching the strap (Fig. 4C). By 41%, they have traveled onto the strap and each has bifurcated to grow laterally along it. By 50%, they have reached the insertion point of the strap into the ectoderm, then, by travelling across the bridge cells, they reach their muscle target posterior to the spiracle by 57%.

The peripheral NS neurons. The 2 peripheral NS neurons that project into the TN are called L1 and L2. "L" refers to Link neurons: this nomenclature was initiated by FINLAYSON and OSBORNE (1969) in their studies of peripheral NS's that occur at link points between peripheral nerves in the stick insect Carausius and in the blowfly, Phormia. In Manduca, L1 and L2 are found around the spiracle in positions that can be slightly variable. We have first observed them at about 38% when they lie close to each other on the posterior surface of the trachae that are growing upwards from the spiracle. By 43% (Fig. 4D), L1 has grown its axon to the bridge and by 50% it has reached the level of the ganglion. More details concerning its pattern of growth are described in a following section. L2 first projects an axon into the DN back to the ganglion where it projects processes into the neuropil. Around 52%, it begins to send secondary axons along the TN: one peripherally towards the L1 cell body and one centrally, following the L1 axon along the TN where it grows up to the level of the ganglion.

Neurons from the VN. The majority of neurons that project to the TN do so via the ipsilateral VN of the next anterior ganglion. These include three of the four bursicon-containing neurons and two mid-line NS neurons (Fig. 2). During axonogenesis, they are highlighted at all times by TN-1 immunostaining (Fig. 4). These growth cones can be seen first exiting the ganglion by ~34%; at a distance halfway to the TN, they are consistently seen apposed to a group of TN-1 positive cells (future glia?) that lie along a presumptive oblique external muscle. Passing this group of cells the growth cones continue travelling posteriorly across the segment boundary until they reach the TN. The site at which they join the developing nerve is also a constant feature: this is the insertion point of the strap into the ectoderm.

These growth cones (typically one leads 2-3 others) reach this point by 47-49% (Figs. 4D and E) and there they undergo a morphological transformation. Whereas in transit, the growth cones were small thickened extensions of the axons, at the strap insertion point, they become considerably flattened, broad and display numerous far-reaching filopodia (Figs. 4 & 6). They continue in this state for at least 9-10% of development before they finally resume growth: each neuron now displays cell-specific directions of growth along the strap. The first growth cone grows laterally; the next two bifurcate and grow both centrally and laterally. What signals are bursicon cells using in making this navigational correction? We do not believe that the bursicon cells are fasciculating onto the motoneurons, but rather grow along the strap in parallel with them. These observations at the light microscope level need confirmation by electron microscopy; it would be in keeping with the confiquation of axons in the mature TN (Fig. 1) wherein the MN axons and those of the NS's are never in contact.

The L1 cell also displays a striking expression of filopodia in this same region at this same time but certain differences can be noted (see below). This specific point of connection is of interest because this area of intense lamellopodial and filopodial activity is the site of what later becomes the thickened PVO. It will be composed of numerous neuronal ramifications, glial cell projections and tracheal cell attachments (see below).

PRESUMPTIVE TRACHAEL CELLS ARE ALSO INVOLVED IN THE NORMAL DEVELOPMENT OF THE TN

During the period of time after the bursicon growth cones have reached the strap insertion point but before they have resumed growth (between 48 and 58%) a syncytium of three to four nuclei migrates into this position (Fig. 7A) from a lateral location near the spiracle. Its final position is usually along the insertion point of the intersegmental muscles; at this time the strap/TN is losing its connection with the ectoderm and hence its proximity to the syncytium varies (Fig. 7). Both the bursicon cell growth cones and the L1 axon (see section on L1 morphogenesis) show a strong tendency to adhere to this cell (Fig. 7). This tendency to adhere to the syncytium is manifested by the behavior of neuronal filopodia in showing selective preference to grow along its surface to the exclusion of most other surfaces and as well, in a tendency of the growth cones and of lamellopodial extentions of the axons to appear in close apposition to it. This tendency may last well into late embryonic stages. Post-embryonically the syncytium helps to form a tracheole that can be seen attached to the TN just near the attachment point with the bursicon axons from the VN.

The shape of the syncytium can be learned from dye-fills of pre-fixed embryos (Fig. 7A). Interestingly, we find that the syncytium is dye-coupled (Lucifer Yellow Dye MW = 450) to glial cells in the strap and bridge when the syncytium is filled in living embryos at this stage (Fig. 7B). There is some variability in the exact configuraton of bursicon axons, TN and syncytial cells: this is diagrammed by selected examples in Figure 7C, D, and E. We believe the time of arrival of the syncytium may vary. This fact coupled with the fact that the TN is losing its connection to the ectoderm (and therefore moving away from the syncytium) results in this morphological variability. It appears as if the decision of the bursicon axons to resume growing along the TN may not be contingent upon the arrival of the syncytium. If, however, it does arrive "in time", permanent growth of bursicon axons is seen along this cell in addition to growth along the TN. Hence, while most features of neuronal differentiation in the TN are strikingly stereo-

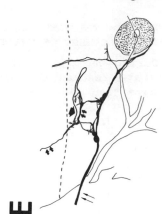

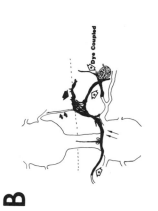

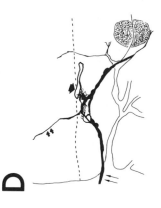

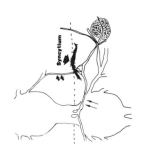

FIGURE 7.

FIGURE 7. Formation and configuration of the PVO. A. A syncytium that becomes a tracheoblast migrates along the segment boundary into the future PVO position. Lucifer Yellow dye fill of the syncytium (dark, wide arrow) at 56% of development following paraformaldehyde fixation to preclude dye-coupling. Small double arrow indicates the branch of the VN carrying the axons of the bursicon cells as they cross the segment boundary to join the TN at the future PVO. B. Lucifer Yellow dye-fill of the syncytium (dark, wide arrow) without pre-fixation: the syncytium is dye-coupled to all the glial cells of the strap (open arrows with asterisks) and the bridge. Note the specificity of coupling: glial cells in the DN are only coupled via the bridge. Same age as in (A). (C-E) Camera lucida drawings of this same nerve juncture to indicate the variable configuration of bursicon growth cones (double arrows) in relation to the strap. Bursicon growth cones grow along syncytium (wide arrow) in preference to the strap; all examples from 57% of development. See text for further details. Scale = 67 μ (A,B); = 50 μ (C,D,E).

typed, here we note this common source of variability in bursicon cell projection. Further elucidation of the exact role played by this syncytium must await its ablation in an experimental context.

MANY GLIAL CELLS ALONG THE STRAP ASSUME INDIVIDUAL IDENTITIES BY VIRTUE OF POSITION AND MORPHOLOGY

It was earlier mentioned that most cells of the early forming strap and bridge appear to differentiate into glial cells that ensheath the axons that grow along them. While it has long been known that individual neurons are unique and identifiable in this and many other animals (e.g., in insects: GOODMAN and BATE, 1981; TAGHERT and TRUMAN, 1982; COPENHAVER and TRUMAN, 1986), it was surprising to find that many of these differentiating glial cells also assume unique and identifiable properties. Single identified glial cells have previously been documented in the leech (KUFFLER and POTTER, 1966) and in the developing CNS of embryonic grasshoppers (M. BASTIANI and C. GOODMAN, per comm.).

We find at least 7 glial cells along the developing TN on each side of the segment that become identifiable

due to their position and/or morphology (data not
shown). G-1 usually lies just on either side of the
ganglion cell isolated from any other glial nuclei. G-2,
3 and 4 lie as a group just where the TN and DN separate
from each other, their positions with respect to one
another may vary. G-2 sends a short process that is TN-1
positive to the strap insertion point and insinuates
itself between the strap and the in-growing bursicon
axons. G-3 and 4 send short processes into the DN and
back up towards the ganglion respectively. G-5 and 6 are
two prominent glial cells whose nuclei lie close to or
just past the strap insertion point. Their morphology is
currently unknown. G-7 is the most dorsal member of the
bridge and it sends a short process back centrally
towards the strap insertion point.

This network of glial cells (both those that are
uniquely identifiable and those presently identifiable
only as a group) are extensively dye-coupled to one
another and as well to the presumptive trachael syncytial
cells that lie close to the strap insertion point (Fig.
7B). Possible roles that these glial cells may play
during the development of TN prior to mature glial
function are discussed later in this chapter.

CELL DIFFERENTIATION OF L1; AN IDENTIFIABLE PEPTIDERGIC NEURON OF THE TN

In the preceding sections we have described in
general and specific terms the construction of the TN in
abdominal segments of Manduca. Motoneuron and
neurosecretory neuron axons arrive and grow into a
previously elaborated framework of cells (the strap and
the bridge) to create a stereotypically formed nerve.
Here we consider the differentiation of a single member
of the neural complement in order to more carefully
examine the details of the developmental schedule that is
employed. We would like to know at a cellular and
ultimately at a molecular level how single neurons come
to assume mature and unique cellular phenotypes. We have
focussed on neuron L1 -- a peripheral neurosecretory
neuron. This choice was made for two specific reasons:
(i) a monoclonal antibody raised against the cardioactive
peptide of Aplysia specifically stains this and other
neurons in Manduca that are suspected to contain cardio-

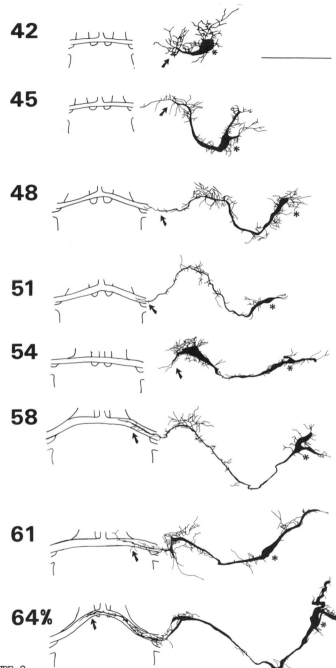

FIGURE 8.

FIGURE 8. Morphogenesis of the peripheral neuron L1. Camera
lucida drawings of NS cell L1 following Lucifer Yellow dye-fills
at the indicated stages of development. Note the stereotyped,
error-free pattern of growth. Note also that growth pauses
between 47 and ~60% of development at a position (arrow) just
ipsilateral to the mid-line. A profusion of filopodia
(sometimes including a large lamellopodium (e.g., 54%) is
produced at the future PVO location (cf. Figure 7). L1 is
interacting with either or both the syncytium and the bursicon
growth cones. Asterisks indicate cell bodies; scale bar = 50 μ.

active peptides (TUBLITZ and TRUMAN, 1985; COPENHAVER and
TAGHERT, unpub. results) and hence, it is a marker for
biochemical differentiation; and (ii) this neuron is
configured in the living animal such as to give the
experimenter many of the advantages of a neuron in tissue
culture. By this we mean that its processes are arrayed
in a 2-dimensional aspect and these processes are
likewise all accessible.

   Morphological differentiation of L1. A timeline of
morphological growth for neuron L1 is shown in Figure
8. These data are camera lucida drawings of the neuron
after Lucifer Yellow dye injection via a
microelectrode. L1 (and L2, another peripheral NS cell)
is first observed at about 38% at a position just
posterior to the developing trachae that grow out from
the spiracle. The initial contiguity of the L1 and L2
cell bodies allows the possibility that they may derive
from a single precursor cell in the underlying
ectoderm. Over time, the L1 cell body moves laterally
and anteriorly to a position closer to the dorsally-
positioned heart of the animal. It projects a growth
cone centrally around the trachae and towards the bridge
cells, which it reaches by 42%. Simultaneously, L1 grows
one and sometimes two distal growth cones: in the mature
state, this neuron is typically multipolar, as is L2,
with its distal processes close to or within the heart.
By 45%, the centrally-directed growth cone has reached
the strap insertion point and is following the strap
cells back towards the ganglion. By 47%, it has reached
the vicinity of the glial cell G-1. Here, the growth
cone typically pauses for at least 10% of development.
Between 62 and 64%, L1 growth resumes across the mid-line

in the TN, down the TN on the other side, now growing
distally.   By late stages (~90%), L1 has reached the
contralateral strap insertion point and here has
terminated growth.

Like the later arriving growth cones of the bursicon
neurons, the L1 axon exhibits a strong preference to
adhere to the trachael syncytial cells (Figs. 7 and 8),
although the syncytium appears on the scene long after
the L1 growth cone has passed through the strap insertion
point.   By contrast the SP MN's, which have passed
through this same point simultaneously but growing in the
opposite direction, do not exhibit this behavior.   The
mature L1 neuron displays numerous varicosities and
processes grown in parallel to the axon along the length
of the TN and adjoining nerves (Fig. 9).   These
structures probably represent the points of transmitter
(hormone) release -- neurohaemal sites.   These aspects of
its mature morphology are produced between 60 and 80% of
development, before the L1 neuron has completed its limit
of total axon growth.

Biochemical differentiation of L1.   Also occurring
during the pause in axon growth (47-60%) is the
initiation of transmitter expression (Fig. 9).   SCP-like
immunoreactivity is first expressed in L1 at ~50% along
its axon and cell body.   The L1 cell is immunoreactive in
all segments from 50-70% of development, but by 90%, only
thoracic L1 neurons reliably express this antigen;
abdominal neurons display reduced levels according to a
gradient from anterior to posterior.   By the 5th larval
instar, expression is seen only in thoracic segments,
despite the continued survival of the L1 neuron.   This
transient expression is reminiscent of the transcient
expression of Tyrosine Hydroxylase by certain peptidergic
cells in vertebrate embryos (TEITLEMAN et al., 1978).

DISCUSSION AND SUMMARY

In this chapter we have described the embryonic
development of a simple neuroendocrine effector organ,
the Transverse Nerve, in abdominal segments of <u>Manduca</u>.
These results suggest a pattern of stereotyped neuronal
differentiation that include specific cell interactions
between developing neurosecretory neurons, and between

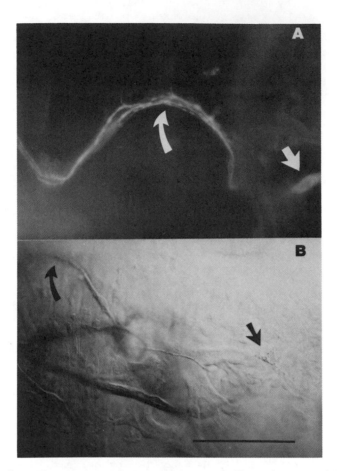

FIGURE 9. The mature L1 expresses a cardioactive peptide-like antigen and displays varicose and diffuse endings along the length of the TN.   A. Photomicrograph of the L1 cell body (straight arrow) and axon within the TN at 90% of embryonic development.   A Mab to Aplysia SCP was visualized with a FITC-coupled secondary antibody.   Note the varicose aspect to the stained processes of the cell, especially at the PVO (the former strap insertion point - curved arrow)   B. TN-1 stained 45% embryo to show the position of the L1 cell body and axon in relation to the developing TN.   Scale = 60 μ.

the neurons and identified glial and trachael precursors. Further, numerous features of TN trajectory and position appear to be "pre-figured" by the prior arrival and assemblage of non-neuronal cells into cellular pathways that are later taken by the growing axons of the TN neurons.

TN POSITION AND TRAJECTORY IS PREFIGURED BY NON-NEURONAL CELLS

The hypothesis that axonal pathways may be laid down before the arrival of neuronal growth cones by non-neuronal cells is one of many that have been posited to explain stereotyped nerve formation during development. Specifically, Singer et al. (1979) have proposed the "Blueprint Hypothesis" in which axon pathways are formed from pre-existing avenues of access that are afforded by "holes" in the neuroepithelium. In the past 6 years, studies in insect embryos have provided insights into the mechanisms of nerve formation and growth cone guidance in vivo (BASTIANI et al., 1985). These studies have emphasized the importance of cell surface interactions between the developing neuron and its local cellular environment. Sources of guidance that have been specifically implicated include single neurons and muscle pioneers (BENTLEY and CAUDY, 1983; BASTIANI et al., 1984; BALL et al., 1985) and as well, aspects of the overlying basement membrane and underlying epithelium (NARDI, 1982; BERLOT AND GOODMAN, 1983).

What is novel about the formation of axon pathways in the TN is the presence not of discontinuous, single landmark cells, or the use of a two-dimensional sheet of cells, but the assemblage of a continuous pathway composed of non-neuronal cells (the strap and the bridge) that prefigure the exact position and trajectory of the future TN. As emphasized earlier, this pattern is probably not generalizable to all other nerve pathways; rather, these results indicate a diversity of mechanisms that are utilized to generate specific neuronal connectivity.

GLIAL CELLS MAY SUBSTITUTE FOR THE TARGET CELLS THAT NEUROENDOCRINE NEURONS LACK

Developing neurons are dependent on target cells for a variety of developmental choices that are made. Depending on the specific example studied, these choices may include: survival (HAMBURGER, 1970), and the positioning of the terminal arbor (MURPHEY and LEMERE, 1984). In the case of neurosecretory neurons that lack closely apposed target cells, what are the mechanisms that regulate their viability, the positioning and number of transmitter release sites?

During the development of the pituitary in vertebrates (analogous to the TN in having a neurohaemal function), non-neural cells (pituicytes) appear to play a pivotal role in the differentiation of neurons (DELLMAN, 1981). Before neuronal growth cones arrive from the hypothalamus, pituicytes are present and in a quiescent mode. After the arrival of the neurons, these pituicytes switch to secretory mode. DELLMAN et al. (1981) interpret this process to implicate these non-neuronal cells in effecting the outgrowth or differentiation of the neurons. In the case of the developing TN, we have observed specific interactions between the growing axons of NS cells and developing glial cells that lie along the strap. These observations of normal events strongly imply a necessary flow of information from glia to neurons. We suspect that some or all of these glial cells may substitute as "fictional" targets for the developing NS cells that must produce blind, nonsynaptic varicose endings within the nerve.

For the purpose of designing specific experiments, we hypothesize three forms of information that may be utilized and which may be tested: (i) Axon guidance information. Within the assemblage of glial cells, there probably exists directional information that is utilized by the developing neurons as each takes cell-specific pathways along and within the strap. (ii) Induction of terminal growth. Here we posit the existence within the strap of signals to direct the growth cones to stop growing and to induce them to produce secondary and tertiary branches that carry terminal varicosities. Stop signals for insect motoneurons have been localized

previously to single cells (BALL et al., 1985). The
nature or source of signals that induce varicosity
formation are unknown. (iii) Regulation of
neurotransmitter expression. Transmitter expression has
been shown repeatedly to be a rather plastic phenotype in
vertebrates (BLACK, 1978; PATTERSON, 1978); in
invertebrates, specific evidence is lacking. In the case
of the TN neurons, L1 shows a segment-specific pattern of
SCP-like expression following a transient embryonic
period during which L1 neurons in all segments are
expressing. Perhaps, the signals that modulate
expression are to be found in the specific interactions
L1 has with glial cells in the strap or with the trachael
syncytium.

INDIVIDUAL PEPTIDERGIC CELLS DEVELOP ALONG STEREOTYPED SCHEDULES

The analysis of the development of the L1 neuron
argues that neuroendocrine neurons in insects
differentiate in a manner that is very similar to that
previously documented for motoneurons and interneurons
(RAPER et al., 1983; HO et al., 1985; TAGHERT and
GOODMAN, 1984). That is to say, morphological
development proceeds smoothly without evidence of error
in guidance, and with suggestions of specific cell-
surface interactions between the growth cone and certain
identified cells that are encountered along the way.
Further, the biochemical differentiation of L1 (as
indicated by the onset of its putative transmitter, a
cardioactive peptide-like substance) occurs after the
initiation but before the completion of axon outgrowth.
This schedule is similar to that seen in the cases of
serotonergic neurons in grasshopper embryos (TAGHERT and
GOODMAN, 1984), other peptidergic neurons in insects
KESHISHIAN and O'SHEA, 1985; COPENHAVER and TAGHERT,
unpub. results), and in peptidergic neurons in other
animals including vertebrates (e.g., PICKEL et al.,
1984). Particularly attractive features of the L1 neuron
for a mechanistic analysis of transmitter differentiation
include not only its stereotyped schedule, but also its
segment-specific expression, and the complete
accessibility of all its processes for experimental
manipulation (WALL and TAGHERT, expts. in progress).

PROSPECTS FOR A MOLECULAR ANALYSIS OF TN FORMATION

We have described the development of a peripheral nerve by analyzing the behavior of individual neurons and non-neural cells during embryogenesis. A recurring theme in these observations is the remarkably precise recognition that individual embryonic cells display as they explore their environment during their differentiation. For example, the L1 growth cone navigates smoothly along the cells of the bridge and strap insertion point, but it reliably pasues at or near the G-1 glial cell. Furthermore, this neuron (and other NS cells in the TN) later make varicose endings at non-random positions along the nerve (WALL and TAGHERT, unpub. results). These precise cellular events imply equally precise underlying molecular recognition, the study of which would greatly expand our understanding of neuronal differentiation. It must be conceded however, that these molecules, expressed by small numbers of neurons or non-neural cells, are likely to be extremely rare. One possible technical avenue of approach would be to utilize monoclonal antibodies.

The TN-1 Mab is an extremely useful histological stain in _Manduca_ in that it specifically and strongly labels subsets of neurons and non-neuronal cells in developing embryos. It was generated from a fusion that utilized adult TN as the immunogen and in which clones were selected by virtue of cell-specific expression of antigen within the CNS (TAGHERT et al., 1984). TN-1 is a cell surface antigen as indicated by its binding to antibody in living embryos (CARR and TAGHERT, unpub. results) and may be related to a family of cell-surface glycoproteins recently discovered in grasshopper embryos (C.S. GOODMAN, per. comm.). TN-1 has proven invaluable in this analysis of TN formation (CARR and TAGHERT, in prep.; WALL and TAGHERT, in prep.) and as well in describing the formation of the stomatogastric nervous system in _Manduca_ embryos (COPENHAVER and TAGHERT, in prep.). Other Mabs from this fusion hold equal promise in further elucidating the cellular interactions that underlie the formation of this nerve. For example, six other Mabs also show some cell specific staining of neuronal and non-neural elements of the developing TN. The prospect that some of these Mabs, which are presently useful for elucidating cellular events, are also indicating molecules that help mediate some of these processes will be addressed in the near future.

SUMMARY

Within this simple, anatomically distinct nerve -- the TN -- there occur many of the basic features of neuronal development. Cell-specific axon guidance, the halting of axon growth, the creation of secondary and tertiary branching patterns, the induction of release elements (varicosities) along axons. What makes this nerve attractive as a simple system in which to study these processes is that it represents a coherent and relatively homogeneous tissue; one that is not seriously contaminated by cells performing other functions (e.g., muscles, other neuron types, epithelium, etc.). Future studies will utilize this numerical and functional simplicity to further analyze the cellular and molecular events that underlie neuronal differentiation.

ACKNOWLEDGEMENTS We thank Boris Masinovsky, Stephen Kemp and A.O. Dennis Willows for a gift of the α-SCP$_B$ antibody and we thank our colleagues for sharing unpublished results. Joe Hayes and Susan Eads provided excellent photographic and secretarial assistance. We are grateful to acknowledge support of the studies described in this chapter by an NSF Pre-Doctoral Grant (JNC), an NIH Post-Doctoral Fellowship (PFC) and by funds from the McDonnell Center for Higher Brain Function (Washington Univ. School of Medicine) and a grant from the NIH #NS21749 to PHT who is a Sloan Fellow.

REFERENCES

Ball, E., R.K. Ho and C.S. Goodman, 1985, Development of neuromuscular specificity in the grasshopper embryo: guidance of motoneuron growth cones by muscle pioneers. J. Neurosci. 5, 1808-1819.

Bastiani, M.J., C.Q. Doe, S.L. Helfand and C.S. Goodman, 1985, Neuronal specificity and growth cone guidance in grasshopper and Drosophila embryos. Trends in Neurosci. 8, 257-266.

Bate, C.M., 1976, Embryogenesis of an insect nervous system. I. A map of the thoracic and abdominal neuroblasts in

Locusta migratoria. J. Emb. Exp. Morph. 35, 107-123.

Bate, C.M. and E. Grunwald, 1979, Embryogenesis of an insect nervous system. II. A second class of neuronal precursors and the origin of the intersegmental connectives. J. Emb. Morph. 61, 317-330.

Bentley, D., H. Keshishian, M. Shankland and A. Rorian-Raymond, 1979, Quantitative staging of embryonic development of the grasshopper. J. Emb. Exp. Morphol. 54, 47-74.

Bentley, D. and M. Caudy, 1983, Pioneer axons lose directed growth after selective killing of guidepost cells. Nature 304, 62-65.

Berlot, J. and C.S. Goodman, 1984, Guidance of peripheral pioneer neurons in the grasshopper: Adhesive hierarchy of epithelial and neuronal surfaces. Science 223, 493-496.

Black, I.B., 1978, Regulation of autonomic development. Ann. Rev. Neurosci. 1, 183-214.

Carrow, G.M., R.L. Calabrese and C.M. Williams, 1984, Architecture and physiology of cerebral neurosecretory cells. J. Neurosci. 4, 1034-1044.

Dellman, H.C., K. Sikora and M. Castel, 1981, Fine structure of the rat supraoptic nucleus and neural lobe during pre- and post-natal development. In: "Neurosecretion: Molecules, Cells and Systems", Eds., D. Farner and K. Lederis. Plenum Press, N.Y., pp. 177-186.

Doe, C.Q. and C.S. Goodman, 1985, Early events in insect neurogenesis. I. Development and segmental differences in the pattern of neuronal precursor cells. Dev. Biol. 111, 193-205.

Finlayson, L. and M.P. Osborne, 1969, Peripheral neurosecretory cells in the stick insect (Carausius morosus) and the flow-fly (Phormia terranovae). J. Insect. Physiol. 14, 1793-1811.

Goodman, C.S. and M. Bate, 1981, Neuronal development in the grasshopper. Trends in Neurosci. 4, 163-169.

Hamburger, V, 1975, Cell death in the development of the lateral motor column of the chick embryo. J. Comp. Neurol. 160, 535-546.

Keshishian, H. and M. O'Shea, 1985, The acquisition and expression of a peptidergic phenotype in the grasshopper embryo. J. Neurosci. 5, 1005-1015.

Kuffler, S.W. and D.D. Potter, 1964, Glia in the leech central nervous system: physiological properties and neuron-glia relationship. J. Neurophysiol. 27, 290-320.

Lewis, G.W., P.L. Miller and P.S. Mills, 1973, Neuromuscular mechanisms of abdominal pumping in the locust. J. Exp. Biol. 59, 149–168.

Murphy, R.K. and C.A. Lemere, 1984, Competition controls the growth of an identified axonal arborization. Science 224, 1352–1355.

Nardi, J., 1983, Neuronal pathfinding in developing wings of the moth, Manduca sexta. Dev. Biol. 95, 163–174.

Patterson, P.H., 1978, Environmental determination of autonomic neurotransmitter functions. Ann. Rev. Neurosci. 1, 1–17.

Pickel, V.M., K.K. Sumal and R.J. Miller, 1982, Early prenatal development of substance P- and enkephalin-containing neurons in the rat. J. Comp. Neurol. 210, 411–431.

Raabe, M., 1982, "Insect Neurohormones". Plenum Press, N.Y.

Raabe, M., N. Baudry, J.P. Grillot and A. Provensal, 1974, The perisympathetic organs of insects. In: "Neurosecretion -- The Final Common Pathway", Eds., F. Knowles and L. Vollrath. Springer-Verlag, N.Y., pp. 60–71.

Raper, J.A., M.J. Bastiani and C.S. Goodman, 1983, Pathfinding by neuronal growth cones in grasshopper embryos. I. Divergent choices made by growth cones of sibling neurons. J. Neurosci. 3, 20–30.

Singer, M., R.H. Nordlander and M. Egar, 1979, Axonal guidance during embryogenesis and regeneraton in the spinal cord of the newt. "The Blueprint Hypothesis" of neuronal pathway patterning. J. Comp. Neurol. 185, 1–22.

Taghert, P.H. and J.W. Truman, 1982, Identification of the bursicon-containing neurons in abdominal ganglia of the tobacco hornworm, Manduca sexta. J. Exp. Biol. 98, 385–401.

Taghert, P.H., M.J. Bastiani, R.K. Ho and C.S. Goodman, 1982, Guidance of pioneer growth cones: filopodial contacts and coupling revealed with an antibody to Lucifer Yellow. Dev. Biol. 94, 391–399.

Taghert, P.H., N.J. Tublitz, J.W. Truman and C.S. Goodman, 1983, Generation of monoclonal antibodies to the neurohaemal transverse nerve of the moth, Manduca sexta. Soc. Neurosci. 9, 256.

Taghert, P.H. and C.S. Goodman, 1984, Cell determination and differentiation of identified serotonin-immunoreactive neurons in the grasshopper embryo. J. Neurosci. 4: 989–1000.

Taghert, P.H., C.Q. Doe and C.S. Goodman, 1984, Cell determination and regulation of neuroblasts in grasshopper embryos. Nature 307, 163–165.

Teitleman, G., T.H. Joh and D.J. Reis, 1978, Transient expression of a noradrenergic phenotype in cells of the rat embryonic gut. Brain Res. 158, 229–234.

Truman, J.W., 1973, Physiology of insect ecdysis. III. Relationship between the hormonal control of eclosion and tanning in the tobacco hornworm, Manduca sexta. J. Exp. Biol. 57, 805–816.

Thomas, J.B., M.J. Bastiani, C.M. Bate and C.S. Goodman, 1984, From grasshopper to Drosophila: A common plan of neuronal development. Nature 310, 203–207.

Tublitz, H.J. and J.W. Truman, 1985, Insect cardioactive peptides. III. Identification of the cardioacceleratory peptides (CAPs)-containing neurons in the ventral nerve cord of the tobacco hawkmoth Manduca sexta. J. Exp. Biol. 116, 355–410.

Summaries of Recent Research

Neurochemistry

OVERVIEW: PROGRESS IN INSECT NEUROCHEMISTRY

L.L. Keeley

Dept. of Entomology, Texas A&M University

College Station, Texas 77843

ICINN-2 stands at a crossroads in the history of insect neurochemistry. Three years ago at the time of ICINN-1, there were only two structurally defined insect neuropeptides -- proctolin and the locust adipokinetic hormone, and the majority of neuroendocrine research was being performed using crude extracts of neurohemal organs. During the intervening three years, there have been nearly a dozen new hormone structures reported either completely or in part. At this meeting alone, there have been seven new hormone sequences described: four peptides that stimulate muscle contraction (leucokinins) (Cook et al.); one peptide that suppresses muscle contraction (leucomyosuppessin) (Holman et al.); a new hypertrehalosemic hormone (T. Hayes and Keeley) and the complete structure for one of the prothoracicotropic hormones (PTTH) of lepidoptera (Suzuki). In addition to these new hormones, other peptides were described as being near to structural definition. These include a diuretic hormone (Proux et al.), additional members of the adipokinetic-hypertrehalosemic hormone (AKH-HTH) family (Gade) and a pheromonotropic hormone (Jaffe et al.). Finally, other peptide factors are in the process of being isolated such as: allatotropins, allatostatins, other PTTHs, and a series of vertebrate-like hormones that have uncertain effects in the insect systems.

The description at these meetings of new hormone sequences and the promise of further structures to come, marks a critical turning point in the history of insect neurochemistry. The ability to work with chemically-defined peptides opens the way for future research to examine important questions about specific neurohormones such as their synthesis, secretion, circulating hemolymph titers and degradation. For example, there is no conclusive chemical evidence that the peptides so far studied and isolated actually are secreted and circulate. The development of immunoassays for characterized peptides will permit experiments to resolve the question of secretion and circulation. Furthermore, one of the first public reports on the existence and nature of precursor molecules for insect peptide hormones was presented for AKH at this meeting (O'Shea et al.). Another report describes a membrane-bound aminopeptidase that degrades proctolin (Isaac). Both of these studies were made possible by the availability of structurally-defined hormones. It is apparent that the continued definition of neuropeptides will facilitate research on the critical questions of insect neuroendocrinology.

One aspect of the chemistry of the neuropeptides that has become evident as a result of having defined sequences concerns the variety of peptides that regulate either a single physiological function or several closely-related functions. This was evident first with the AKH-HTH family and was extended at this meeting to include the group of myokinins present in the brain of cockroaches (Cook et al.). It is apparent that for any given physiological function of insects that is neuroendocrine regulated there may exist multiple natural forms (bioanalogs) for the peptidic hormones affecting that function. This is true both within a single species, where several structurally related but distinct forms of a hormone may exist, and between species. Within silkworm larvae, both large and small molecular weight forms exist for PTTH. Furthermore, at least three bioanalogs are known for the small PTTH (Suzuki). In the case of AKH-HTH, there are nine known bioanalogs that differ between insect species. There are three combinations reported for the AKH-HTHs: two bioanalogs may be present in the same species; different bioanalogs may be present in different species; or the same bioanalog may be present in several species. The reasons for this type of structural-functional redundancy among the

hormones is unclear at present. Comparative studies on
species-related activity for the various AKH-HTH bioanalogs
have resulted in proposed structure-activity relationships
for the AKH-HTH hormones (Goldsworthy and Wheeler).
Finally, in at least one case, two distinct physiological
effects are exhibited by either the same hormone or similar
hormones. Both PTTH and egg-development neurohormone
activities are found in the same tissue preparations, and
the two hormones appear to be either structurally related
or identical (Whisenton et al.).

New findings are elucidating the mode of action of
neuropeptides on their target cells. Second-messenger
systems continue to be indicated as the means by which
neurohormones regulate biochemical processes within target
tissues, but more than cyclic nucleotides serve as
intracellular messengers. Inositol phosphotides are
identified in insects as second messengers for both
hormones and neurotransmitter chemicals (Duggan and Lunt,
Evans). Of interest was the report by Birnbaum et al. that
the polyamines spermine and spermidine act as second
messengers in the insect nervous system. Casein kinase
activity was stimulated by spermine in the Manduca nervous
system but was insensitive to cyclic AMP. In contrast, the
protein kinases of frogs and rats were sensitive to the
cyclic nucleotides but not to the polyamines.

Little information exists concerning the receptor
molecules of insect membranes. Future work on membrane
receptors for the neuropeptides awaits the development of
tagged analogs. Preliminary studies on possible membrane
binding proteins were reported for the AKHs (D. Hayes et
al.). Considerably more is known about neurotransmitter
receptors than about neurohormone receptors. Octopamine
receptors are discriminated into types 1, 2a and 2b based
on differential sensitivities to agonists and antagonists
and on their modes of action (Evans). Type 1 receptors act
by releasing intracellular CA^{+2}; whereas, type 2 receptors
act via cyclic AMP. A new microflotation technique was
developed for isolating nerve termini (synaptosomes) from
insect nervous systems (Breer). Synaptosomes provide a
useful model for elucidating the structure of membrane
receptor proteins as well as for investigations into
neurotransmitter chemistry and the nature of synaptic
function in insects. The acetylcholine receptor protein
has been isolated and characterized from locust

synaptosomes. This receptor protein was reassembled into a planar lipid bilayer and showed specific channel responses when exposed to cholinergic agonists. The channels were selectively permeable to cations and were blocked by cholinergic antagonists. Structural analysis of the receptor proteins is being performed by molecular biology.

The relative youth of insect neuroendocrinology is evidenced by the notable paucity of reports on studies that employ the powerful methodologies of molecular biology. No neurohormone genes have been described to date. Methods for using molecular biology to isolate and characterize neuropeptides were described by both O'Shea et al. and Verhaert et al., and it is apparent that this is a potentially useful approach that will see wide applications in the future when suitable nucleic acid probes become available.

In closing, the neuropeptides have a significant potential for application to the problems of insect pest management. Neurohormones are the master regulators for the production of the ecdysteroids and juvenile hormones that regulate insect growth, metamorphosis and reproduction. Furthermore, neurohormones directly control specific processes such as eclosion behavior, cuticle tanning and pheromone production. Finally, neurohormones regulate homeostatic processes and general physiological functions such as: circulating metabolite levels, water and salt equilibria, basal metabolism, heartbeat rate, and muscle contraction. Disruption of nearly any of these processes would be debilitating to treated insects and could result in incapacitation or death. The wide variety of neurohormone structures combined with their potential insect- or species-specificity makes attractive the potential of disrupting the neuroendocrine balance as a means for insect control. Peptide hormones per se are not good candidates as insecticidal agents, but the natural hormones can be used as models for designing stable, nonpeptide, agonistic or antagonistic analogs that will mimic or inhibit the hormones at acceptor sites (tissue receptors, degradative enzymes). Other means of causing endocrine disruption may involve disturbing neurosecretion or infecting insects with genetically-engineered microorganisms bearing genes for either neurohormones or antihormones.

APPROACHES TO GENERATING MONOCLONAL ANTIBODIES TO THE PROTHORACICOTROPIC HORMONE IN MANDUCA SEXTA

M.A. O'Brien, T.R. Flanagan, N. Agui, H. Duve[2], A. Thorpe[2], G. Haughton[1], L. Arnold[1], E. J. Katahira and W.E. Bollenbacher

Department of Biology and [1]Department of Immunology, University of North Carolina, Chapel Hill, NC 27514 and [2] School of Biological Sciences, Queen Mary College, University of London, London E18, UK

The prothoracicotropic hormone (PTTH), the primary regulator of insect molting and metamorphosis, is a cerebral peptide which in at least two insects exists in multiple molecular forms. In the tobacco hornworm, Manduca sexta, two PTTHs (big ~28 kD and small ~7 kD) have been identified that activate larval and pupal prothoracic glands differentially (Bollenbacher et al., 1984). Similarly, in the silkworm, Bombyx mori, two comparable molecular weight classes of PTTH have been identified, with the small 4kD moiety apparently existing in three forms (Ishizaki et al., 1983; Nagasawa et al., 1984). The physiological significance of multiple forms is not known. In large part, the dearth of knowledge on PTTH has been a result of the difficulty encountered purifying the neurohormone.

In our laboratory, an on going project has been the generation of monoclonal antibodies to big PTTH from Manduca to obtain probes to investigate its chemistry and physiology. To date, several approaches have been taken, and these have included the use of different immunogens: 1) big PTTH isolated by conventional purification procedures; 2) big PTTH purified by HPLC; and 3) the prothoracicotropes (L-NSC III) themselves (Bollenbacher and Granger, 1985). In addition, different methods have been employed to screen for antibody production,

179

including: 1) an enzyme-linked immunosorbent assay
(ELISA); 2) immunofluorescence cytology of brain whole
mounts; and 3) horseradish peroxidase (HRP) immunocytology
of paraffin sections of brains.

Big PTTH isolated by conventional purification
procedures (Bollenbacher et al., 1984) was the immunogen
used initially. Serum from the mouse immunized with this
material was screened for antibody production with the
ELISA. Using highly purified big PTTH as the coating
antigen, a strong immune response was detected in the
serum, indicating the mouse was a good candidate for
monoclonal antibody production. The standard hybridoma
technique was employed to obtain clones and the ELISA was
used to screen them for antibody production. Supernatants
from expanded clones were screened with the ELISA using
increasingly purified big PTTH. One clone (6F10) remained
positive throughout the screening. Medium from a 6F10
subclone was then used in the ELISA to trace the big PTTH
activity eluting from a reverse phase HPLC column.
Concurrently, the HPLC fractions were screened for PTTH
activity with the in vitro prothoracic gland PTTH bioassay
(Bollenbacher and Granger, 1985). ELISA and PTTH positive
fractions were detected, and they were essentially the
same. However, attempts to block the activation of the
prothoracic glands by PTTH with the antibody were
unsuccessful, precluding confirmation that the antibody
was directed against big PTTH.

An immunocytological approach was taken as an
alternative means of ascertaining if the 6F10 monoclonal
antibody was specific for big PTTH. Immunofluorescence of
brain whole mounts revealed the antibody did not bind to
the prothoracicotropes (L-NSC III). This finding was
supported by an HRP histological method of localizing
antibody binding. The 6F10 antibody bound to a pair of
cells in each lobe of the Manduca brain smaller than the
prothoracicotropes and located medial to them. Thus it
appeared the 6F10 monoclonal antibody was not binding to
big PTTH. With this result, we concluded alternative
approaches had to be taken to generate monoclonal
antibodies to the big PTTH.

The second big PTTH immunogen used was isolated by
HPLC. It was more pure than the previous immunogen as
indicated by its higher specific activity. Serum from the

mouse injected with this PTTH was screened for antibody production by both the ELISA and immunocytological assays. While the ELISA yielded negative results, the immunocytological screens revealed staining of the L-NSC III, as well as the large medial NSC. Initial proof that the serum contained antibodies directed against big PTTH was provided by preabsorbing the serum with big PTTH before the immunocytological assay. Adjacent brain tissue sections incubated with either antiserum or antiserum plus big PTTH revealed that the staining of the L-NSC III was decreased considerably while medial cell staining was not, suggesting antibodies to big PTTH were present. Thus this mouse is being used for hybridoma production.

The third PTTH immunogen was a preparation of the prothoracicotropes. The cells were assayed for PTTH, demonstrating the presence of the neurohormone by its activation of the prothoracic glands in vitro, and the extract was injected into a mouse. Serum was screened immunocytologically, and specific immunofluorescence of the L-NSC III was obtained. In addition, the dendritic arbors and axons of the cells, inclusive of their terminals in the corpus allatum (the neurohemal organ for PTTH) (Agui et al., 1980), were strongly immunoreactive. This prothoracicotrope-specific staining was partially blocked by preabsorbing the serum with big PTTH. Demonstrating that this serum contained antibodies against big PTTH was accomplished by its ability to inhibit PTTH activation of the prothoracic glands in vitro. Therefore, this mouse was ideal for hybridoma production. To date, screening of the supernatants from hybrid clones derived from this mouse has revealed that 8% of the clones obtained produced antibodies specific for the L-NSC III. Monoclonal antibodies are now being generated, and once sufficient antibodies are obtained, clones recognizing big PTTH will be confirmed by: 1) blocking of immunocytochemical staining of the L-NSC III by preabsorption with PTTH; and 2) blocking of the in vitro biological activity of the neurohormone. Lastly, we hope to use western blot analysis as an additional method of verifying the anti-PTTH specificity of these monoclonal antibodies.

In summary, it appears that polyclonal antibodies to the Manduca big PTTH have been obtained using two

different immunogens; now monoclonal antibodies are being
generated. The success of using a preparation of isolated
NSC as an antigen for obtaining antibodies to the cells
themselves, and more specifically to their principal
protein product (a neurohormone), is a significant
accomplishment. It is conceivable that this approach
could be used to obtain antibodies to various insect
neurohormones without having to purify them first.

Acknowledgments: The authors thank Ms. Kathy Luchok
for preparing this manuscript and Ms. Debbie Sasser for
technical assistance. This research was supported by
grants NS-18791 and AM-31642 to W.E.B..

REFERENCES

Agui N., Bollenbacher W.E., Granger N.A. and Gilbert L.I.
(1980) Corpus allatum is release site for insect
prothoracicotropic hormone. Nature 285, 669–670.

Bollenbacher W.E. and Granger N.A. (1985) Endocrinology
of the prothoracicotropic hormone. In Comprehensive
Insect Physiology, Biochemistry and Pharmacology, Vol.
7, Endocrinology I (Ed. by G.A. Kerkut and L.I.
Gilbert), pp. 109–152, Pergamon Press, Oxford.

Bollenbacher W.E., Katahira E.J., O'Brien M.A., Gilbert
L.I., Thomas M.K., Agui N. and Baumhover A.H. (1984)
Insect prothoracicotropic hormone: Evidence for two
molecular forms. Science 224, 1243–1245.

Ishizaki H., Mizoguchi A., Fujishita M., Suzuki A., Moriya
I., O'Oka H., Kataoka H., Isogai A., Nagasawa H., Tamura
S. and Suzuki A. (1983) Species specificity of the
insect prothoracicotropic hormone (PTTH): The presence
of Bombyx- and Samia-specific PTTHs in the brain of
Bombyx mori. Develop. Growth and Differ. 25, 593–600.

Nagasawa H., Kataoka H., Isogai A., Tamura S., Suzuki A.,
Ishizaki H., Mizoguchi A., Fujiwara Y. and Suzuki A.
(1984) Amino-terminal amino acid sequence of the
silkworm prothoracicotropic hormone: Homology with
insulin. Science 226, 1344–1345.

STRUCTURE/ACTIVITY RELATIONSHIPS IN THE ADIPOKINETIC HORMONE/RED PIGMENT CONCENTRATING HORMONE FAMILY

Graham J. Goldsworthy and Colin H. Wheeler.

Birkbeck College, The University of London.

Malet Street, London, WC1E 7HX, U.K.

The structures of 3 locust adipokinetic hormones are known: the decapeptide adipokinetic hormone I is common to all species of locust investigated so far, but the octapeptide AKH-II's of Locusta (AKH-IIL) and Schistocerca (AKH-IIS) differ by a single amino acid (see Table 1). In addition to these locust peptides, structurally related but often functionally different peptides from other insect species have been sequenced. These include the adipokinetic peptide from Heliothis zea (Jaffe et al., 1986), which is identical to that in Manduca sexta (Ziegler et al., 1985); hypertrehalosaemic peptides I and II from Periplaneta americana (sequenced independently by various groups; see Goldsworthy et al., 1986a); and hypertrehalosaemic factor II from Carausius morosus (Gäde and Rinehart, 1986). Together with a crustacean peptide, Red Pigment Concentrating Hormone, these peptides apparently represent a group of closely related molecules, the AKH/RPCH family: they possess striking sequence and structural similarities, including blocked amino and carboxy terminals. We have compared the activities of these peptides in the locust hyperlipaemic assay to give an insight into the structural features necessary for the fit of the AKH molecule to its receptor on the fat body.

In earlier work on the structure/activity relationships of adipokinetic peptides, Stone et al. (1978) identified some key requirements for activity in the locust hyperlipaemic assay: at least 8 amino acid residues from

Table 1. Structure and dose response relationships for eight members of the AKH/RPCH family. The data refer to hyperlipaemic activity assayed in adult male Locusta migratoria (Goldsworthy et al., 1986b), except where indicated specifically.

Amino acid sequence	ED$_{50}$ pmol	ED$_{max}$ pmol	Maximum response %	Name of peptide
pGlu–Leu–Asn–Phe–Thr–Pro–Asn–Trp–Gly–Thr—NH$_2$	1.0	3	100	AKH-1
pGlu–Leu–Thr–Phe–Thr–Pro–Asn–Trp–Gly–Thr—NH$_2$	2	8	100	Carausius -2
pGlu–Leu–Thr–Phe–Thr–Ser–Ser–Trp–Gly—NH$_2$	10	40	45	Manduca & Heliothis AKH
pGlu–Leu–Asn–Phe–Ser–Pro–Gly–Trp—NH$_2$	4.6	20	100	RPCH
pGlu–Leu–Asn–Phe–Ser–Ala–Gly–Trp—NH$_2$	2	3	50–60	AKH-2 Locusta
pGlu–Leu–Asn–Phe–Ser–Thr–Gly–Trp—NH$_2$	12* / 2	25* / 5	50–60* / 90	AKH-2 Schistocerca
pGlu–Leu–Thr–Phe–Thr–Pro–Asn–Trp—NH$_2$	5	30	~90	Periplaneta -2
pGlu–Val–Asn–Phe–Ser–Pro–Asn–Trp—NH$_2$	5	20	~90	Periplaneta -1 Neurohormone D

*Assayed in Schistocerca

the N-terminus; a C-terminal threonine amide; and an N-terminal enantiomer of pyroglutamic acid. In addition, the sequence Pro-Asn-Trp (positions 6 to 8) is important for maximum activity of the decapeptide AKH-I, and this is circumstantial evidence in support of there being a β-bend in the molecule at this position.

Our recent studies (Goldsworthy et al., 1986a,b; Gäde et al., 1986) show that the naturally occurring members of the AKH/RPCH family fall into two categories: peptides (AKH-I, Carausius-II, RPCH, Periplaneta I and II) which, although varying in potency, elicit at sufficiently high doses the maximum possible adipokinetic response in locusts; and peptides (AKH-IIL and Manduca AKH) which, even at very large doses, elicit a truncated response (60-65% and 40-50% of the maximum respectively).

With the availability of synthetic AKH-IIS, we have re-investigated the dose-response relationships of both HPLC purified natural-AKH-IIS and synthetic-AKH-IIS. In both preparations the peptide elicits a full adipokinetic response in Locusta when injected in doses greater than 20pmol per locust. Our previous inability to demonstrate such a full response (Gäde et al., 1986; Goldsworthy et al., 1986b) with AKH-IIS may be attributed to lower-than-estimated recoveries of hormone during preparation (see also Siegert and Mordue, 1986); the dried AKH-IIS peptide is extremely difficult to re-dissolve! However, although AKH-IIS elicits a full response in Locusta, the response in Schistocerca to its own AKH-II at doses of up to 50pmol is only 50% of the maximum response: a result comparable with that in Locusta to AKH-IIL. This appears to suggest that the requirements for receptor fit of these octapeptides in Locusta are substantially different from those in Schistocerca.

Previously, we argued that the lack of proline at position 6 in AKH-IIL was the prime reason for the poor hyperlipaemic response elicited by this peptide. Our dose-response curve data obtained for Manduca AKH (Table 1) appeared to provide corroborative evidence for this hypothesis. However, the results obtained using AKH-IIS show these conclusions to have been premature; at least for octapeptides, proline (and therefore a β-bend?) may not be necessary to ensure complete receptor binding although it may still be important in determining the potency of AKH-I.

Acknowledgements - Original work described in this paper
was supported by the SERC (U.K.).   Synthetic Manduca AKH
was a gift from Professor R. Keller (Bonn).

REFERENCES

Gäde G. and Rinehart K.L., Jr (1986) Primary structure of
    the hypertrehalosaemic factor II from the corpus
    cardiacum of the Indian stick insect, Carausius morosus,
    determined by fast atom bombardment mass spectrometry.
    Biological Chemistry Hoppe-Seyler. In press.
Gäde G., Goldsworthy G.J., Schaffer, M.H., Cook J.C. and
    Rinehart K.L. (1986) Sequence analyses of adipokinetic
    hormones II from corpora cardiaca of Schistocerca nitans,
    Schistocerca gregaria,   and Locusta migratoria by fast
    atom bombardment mass spectrometry. Biochem. Biophys.
    Res. Comm. 134, 723-730.
Goldsworthy G.J., Mallison K., Wheeler C.H. and Gäde G.
    (1986a) Relative adipokinetic activities of members of
    the AKH/RPCH family. J. Insect Physiol. 32, 433-438.
Goldsworthy G.J., Mallison K. and Wheeler C.H. (1986b) The
    relative potencies of two known locust adipokinetic
    hormones. J. Insect Physiol. 32, 95-101.
Jaffe H., Raina A.K., Riley C.T., Fraser B.A., Holman G.M.,
    Wagner R.M., Ridgway R.L. and Hayes D.K. (1986) Isolation
    and primary structure of a peptide from the corpora
    cardiaca of Heliothis zea with adipokinetic activity.
    Biochem. Biophys. Res. Comm. 135, 622-628.
Siegert K.J. and Mordue W. (1986) Quantification of
    adipokinetic hormones I and II in the corpora cardiaca of
    Schistocerca gregaria   and Locusta migratoria.   Comp.
    Biochem. Physiol. 84A, 279-284.
Stone J.V., Mordue W., Broomfield C.E. and Hardy P.M.
    (1978) Structure-activity relationships for the
    lipid-mobilizing action of adipokinetic hormone.
    Synthesis and activity of a series of hormone analogues.
    Eur. J. Biochem. 89, 195-202.
Ziegler R., Eckart K., Schwarz H. and Keller R. (1985)
    Amino acid sequence of Manduca sexta adipokinetic hormone
    elucidated by combined fast atom bombardment (FAB)/tandem
    mass spectrometry. Biochem. Biophys. Res. Comm. 133,
    337-342.

LIPOPROTEIN/APOPROTEIN INTERACTIONS DURING ADIPOKINETIC HORMONE ACTION IN LOCUSTA.

Colin H. Wheeler and Graham J. Goldsworthy.

Birkbeck College, The University of London.

Malet Street, London, WC1E 7HX, U.K.

Adipokinetic hormones (AKH) I and II are released into the haemolymph during long-term flight in locusts and have several well defined sites of action: they increase the rate of release of diacylglycerols from the fat body and stimulate the uptake and utilisation of these lipids in the flight muscles (see Goldsworthy, 1983). Adipokinetic hormones also promote indirect effects such as haemolymph lipoprotein transformations (Goldsworthy et al., 1985). Lipids released from the fat body are transported as part of lipoproteins; in resting locusts the majority of haemolymph diacylglycerol is carried as part of lipoprotein Ayellow, but after hormone injection Ayellow becomes lipid-loaded and associates with non-lipid carrying apoproteins to become lipoprotein A$^+$, which carries 10 to 15X as much lipid as Ayellow (Mwangi and Goldsworthy, 1977; Van Der Horst et al., 1981a,b; Wheeler and Goldsworthy, 1983a,b).

Interestingly, similar lipoprotein transformations have been reported in the moth Manduca sexta: injection of locust AKH-I leads to conversion of the predominant haemolymph lipoprotein (density 1.11g/l) in the resting state to a lower density (1.06g/l), more highly lipid-loaded, form. In Manduca during low density lipoprotein formation, only one apoprotein, apolipophorin III, appears to associate with the lipoprotein (Shapiro and Law, 1983). Chino et al. (1986) have recently provided circumstantial support for earlier work of other groups

(Van Der Horst et al., 1981a,b; Wheeler and Goldsworthy,
1983a,b; Goldsworthy et al., 1985) which demonstrated the
association of apoproteins with locust lipoproteins during
AKH action. We have shown in our laboratory the
involvement of two apoproteins (molecular weights 20,000
and 16,0000) in lipoprotein A^+ formation. Polyacrylamide
gel electrophoresis (PAGE), under non-denaturing
conditions, of whole haemolymph from resting and
AKH-injected locusts provided strong evidence for the
involvement of two apoproteins in A^+ formation (Wheeler and
Goldsworthy, 1983a,b; Goldsworthy et al., 1985). Until
recently, pure preparations of these individual proteins,
CI and CII, were not available, but application of HP-IEC
(high performance ion exchange chromatography) on TSK
DEAE-5PW (Toya-Soda, Japan; in 0.05 M Tris, pH 8.5 with a
gradient from 0 to 0.25 M NaCl) has enabled their
purification, and provided a rapid method for following
their metabolism after injection into locusts.

Pure, tritium labelled (Wheeler and Goldsworthy,
1983b), CI and CII were injected into individual locusts
followed 20 min later by injection of AKH-I (2pmol).
Haemolymph samples taken before and 90 min after hormone
injection were diluted in buffer, filtered, and separated
directly by HP-IEC. Both CI and CII proteins associated
with lipoprotein A^+ during AKH action thus confirming the
association of both apoproteins with lipoprotein A^+.

What is the function of the binding of these
apoproteins to lipoprotein A^+? We have speculated
previously (Wheeler and Goldsworthy, 1983b) that
C_L-proteins could increase the lipid loading capacity of
lipoprotein A^+ particles and/or could regulate enzymes
necessary for lipid loading or unloading. Our experiments
support a role for C_L-proteins in regulating flight muscle
lipoprotein lipases which are responsible for lipid uptake
during flight (Wheeler et al., 1984); this does not,
however, preclude a role in lipoprotein stabilisation.
Flight muscle lipoprotein lipase in Locusta, which
preferentially hydrolyses lipids associated with
lipoprotein A^+ rather than Ayellow (Wheeler and
Goldsworthy, 1985a), is inhibited by C_L-proteins (Wheeler
and Goldsworthy, 1985b); each of the individual
HP-IEC-purified CI and CII inhibits the enzyme. Inhibition
is competitive at low concentrations but becomes mixed at

higher concentrations of C_L-proteins. Because these apoproteins can normally be found as part of the substrate of choice for the flight muscles, lipoprotein A^+, such inhibition is perhaps at first sight surprising. However it is the free (non-lipoprotein-bound) C_L-proteins that inhibit: when bound to lipoprotein A^+ they have no effect (or may even stimulate the enzyme). Hence flight muscle from resting animals has lower lipase activity than that from AKH-injected locusts (Wheeler and Goldsworthy, 1985b) in which the free C_L-protein concentration is reduced due to the formation of liporotein A^+.

The binding of C_L-proteins to A^+ is an indirect effect of AKH and appears to be part of a mechanism to control lipid uptake at the flight muscles. At the start of flight when carbohydrate utilisation predominates, excessive lipid uptake by the flight muscles would result in premature inhibition of carbohydrate oxidation at a time when there is insufficient lipid in the haemolymph to maintain flight (see Goldsworthy, 1983). However, at flight initiation when free C_L-protein concentrations are high (Goldsworthy et al., 1985), lipase activity in the flight muscle would be minimal (see Wheeler and Goldsworthy, 1985a); lipoprotein A^+ is also absent. Conversion of lipoprotein Ayellow to A^+ (with concomitant C_L-protein binding) not only produces the substrate of choice but also activates the lipase by relieving C_L-protein inhibition (Wheeler and Goldsworthy, 1985b). This mechanism appears to ensure that the uptake of lipid by the flight muscles is delayed until the quality and quantity of lipoprotein substrate in the haemolymph reaches optimum levels for prolonged flight.

Acknowledgements - Original work described in this paper was supported by the SERC (U.K.).

REFERENCES

Chino H., Downer R.G.H. and Takahashi K. (1986) Effect of adipokinetic hormone on the structure and properties of lipophorin in locusts. J. Lipid Res. 27, 21-29.

Goldsworthy G.J. (1983) The endocrine control of flight metabolism in locusts. In Advances in Insect Physiology, Vol. 17 (Eds. Berridge M.J., Treherne J.E. and Wigglesworth V.B.), pp. 149-204. Academic Press, New York.

Goldsworthy G.J., Miles C.M. and Wheeler C.H. (1985) Lipoprotein transformations during adipokinetic hormone action. Physiol. Entomol. 10, 151–164.

Mwangi R.W. and Goldsworthy G.J. (1977) Diglyceride-transporting lipoproteins in Locusta. J. comp. Physiol. B. 114, 177–190.

Shapiro J.P. and Law J.H. (1983) Locust adipokinetic hormone stimulates lipid mobilisation in Manduca sexta. Biochem. Biophys. Res. Comm. 115, 924–931.

Van Der Horst D.J., Stoppie P., Huybrechts R., De Loof A. and Beenakkers A.M.Th. (1981a) Immunological relationships between the diacylglycerol-transporting lipoproteins in the haemolymph of Locusta. Comp. Biochem. Physiol. 70B, 387–392.

Van Der Horst D.J., Van Doorn J.M., De Keijzer A.N. and Beenakkers A.M.Th. (1981b) Interconversions of diacylglycerol-transporting lipoproteins in the haemolymph of Locusta migratoria. Insect Biochem. 11, 717–723.

Wheeler C.H. and Goldsworthy G.J. (1983a) Qualitative and quantitative changes in Locusta haemolymph proteins and lipoproteins during ageing and adipokinetic hormone action. J. Insect Physiol. 29, 339–354.

Wheeler C.H. and Goldsworthy G.J. (1983b) Protein-lipoprotein interactions in the haemolymph of Locusta during the action of adipokinetic hormone: the role of C_L-proteins. J. Insect Physiol. 29, 349–354.

Wheeler C.H. and Goldsworthy G.J. (1985a) Specificity and localisation of lipoprotein lipase in the flight muscles of Locusta migratoria. Biol. Chem. Hoppe-Seyler 366, 1071–1077.

Wheeler C.H. and Goldsworthy G.J. (1985b) Lipid transport to the flight muscles in Locusta. In Insect Locomotion (Ed. Gewecke M. and Wendler G.), pp.126–135. Paul parey Press, Berlin.

Wheeler C.H., Van Der Horst D.J. and Beenakkers A.M.Th. (1984) Lipolytic activity in the flight muscles of Locusta migratoria measured with haemolymph lipoproteins as substrates. Insect Biochem. 14, 261–266.

PRESENCE AND PRELIMINARY CHARACTERISATION OF FACTORS
REGULATING CARBOHYDRATE AND LIPID METABOLISM ISOLATED FROM
THE CORPUS CARDIACUM OF THE EASTERN LUBBER GRASSHOPPER.

Gerd Gäde and Jeffrey H. Spring

Department of Biology, University of
Southwestern Louisiana, Lafayette, LA 70504.

There appears to be considerable conservation within
the arthropods of the group of small neuropeptides which
are known as the AKH/RPCH-family (Gäde, 1986) because of
their structural similarity to the locust adipokinetic
hormone I (AKH I, Stone et al., 1976) and the crustacean
red pigment-concentrating hormone (RPCH, Fernlund and
Josefsson, 1972). The adipokinetic hormones (AKH I and AKH
II) are stored in the corpus cardiacum (CC) of grasshoppers
and their hyperlipaemic action is thought to be associated
with the large increase in metabolic rate which accompanies sustained flight (e.g. Locusta, Schistocerca). We
therefore decided to study the flightless and slow-moving
lubber grasshopper, Romalea microptera, to determine
whether it had the ability to synthesise neuropeptides of
similar structure to AKH and, if so, what their intrinsic
function might be.

Initially, lubbers were injected with methanolic
extracts of their own CC (RCC).Haemolymph samples were
taken at the time of injection and again after 90 min, and
assayed for total carbohydrate (anthrone-positive material)
and total lipid (vanillin-positive material). RCC had no
effect on blood carbohydrates although it caused a slight
but significant increase in blood lipids at doses of 0.1
(from 11.5 to 14.2 mg/ml) and 1.0 RCC-equivalents (from
10.6 to 13.9 mg/ml). Synthetic AKH I (100 pmol) had no
effect on either blood lipids or carbohydrates. Insects

191

starved for 72 hours had greatly reduced levels of lipids and carbohydrates, but failed to respond to injections of either RCC (1.0 RCC-equivalents) or AKH I (100 pmol).

Methanolic extracts of RCC produced a strong hypertrehalosemic response in Periplaneta. The maximum increase in blood carbohydrate concentrations occurred with 0.004 RCC-equivalents and the minimum significant response with 0.0003 RCC-equivalents. Likewise, a large hyperlipaemic response occurred in Locusta with the maximum increase in blood lipids produced by 0.001 RCC-equivalents and the minimum significant increase by 0.0002 RCC-equivalents. For comparison, the minimum and maximum blood carbohydrate elevations produced in Periplaneta by injections of its own CC occurred with 0.005 and 0.05 CC-equivalents respectively. In Locusta, using its own CC, the minimum and maximum hyperlipaemic effects occurred with 0.001 and 0.10 CC-equivalents. Calculations based on these bioassays indicate that RCC must contain the equivalent of 2000-4000 pmol of AKH I compared to 100-200 pmol for Locusta CC.

Reversed phase-HPLC of methanolic extracts of RCC using the method of Gäde (1985) revealed two prominent absorbance peaks present in a 6:1 ratio (peak I: peak II). Both peaks showed hypertrehalosemic activity when injected into Periplaneta and hyperlipaemic activity when injected into Locusta. Comparison with commercially-available neuropeptides revealed that peak I eluted slightly before the cockroach hypertrehalosemic hormone I (MI) and peak II eluted slightly after. Co-chromatography with CC extracts from other insect species indicated that Romalea peak II and the adipokinetic factor from the cricket, Gryllus bimaculatus (Gäde and Scheid, 1986), are identical whereas Romalea peak I appears to be unique.

Despite the absence of a significant hypertrehalosemic effect in the lubber itself, we were recently able to show activation of fat body glycogen phosophorylase following injection of RCC. Phosphorylase activity in male Romalea was assayed in resting insects (control 1), or 20 minutes after the injection of distilled water (control 2), corpus cardiacum extract (0.1 RCC-equivalents), HPLC-purified peak I (0.25 RCC-equivalents) and peak II (0.25 RCC-equivalents). Phosphorylase activity was determined spectrophotometrically by monitoring the change in absorbance at 340 nm produced by the reduction of NADP during glycogen beakdown.

All values for active phosphorylase (in the absence of AMP) are given as a percentage of the total phosphorylase activity (in the presence of AMP). In both resting and water-injected lubbers, about 30% of the phosphorylase is in the active form. Phosphorylase is maximally activated (about 70% in the active form) following the injection of 0.1 RCC-equivalents. Similar results are obtained with injections of HPLC-purified peak I and peak II material, even though the peak II peptide is present in only about one-sixth the quantitiy of peak I.

In summary, we have shown that Romalea, although flightless and therefore incapable of the prolonged periods of rapid metabolism exhibited by migratory locusts, produces and stores large quantities of AKH-like material. Methanolic extracts of Romalea corpus cardiacum are very effective in elevating haemolymph carbohydrates in Periplaneta and lipids in Locusta, although they produce only a minor adipokinetic effect in the lubber itself. Using HPLC, the lubber bioactive material is eluted as two peaks (I and II) with about 85% present in peak I. In Romalea the major function of the peptides in peaks I and II appears to be the activation of glycogen phosphorylase in the fat body.

This work was supported by a grant (Ga 241/6-1) and a Heisenberg Fellowship (Ga 241/5-1) awarded by the Deutsche Forschungsgemeinschaft to GG and National Science Foundation grant DCB84-16829 awarded to JHS.

REFERENCES

Fernlund P. and Josefsson L. (1972) Crustacean color-change hormone: Amino acid sequence and chemical synthesis. Science, 177: 173-175.

Gäde G. (1985) Amino acid composition of cockroach hyper-trehalosaemic hormones. Z. Naturforsch., 40c: 42-46.

Gäde G. (1986) Relative hypertrehalosaemic activities of naturally occurring neuropeptides from the AKH/RPCH family. Z. Naturforsch., 41c: 315-320.

Gäde G. and Scheid M. (1986) A comparative study on the
    isolation of adipokinetic and hypertrehalosaemic factors
    from insect corpora cardiaca. Physiol. Entomol. 11: 145-
    157.

Stone J.V., Mordue W., Batley K.E. and Morris H.R. (1976)
    Structure of locust adipokinetic hormone, a neurohormone
    that regulates lipid utilization during flight. Nature,
    263: 207-211.

ISOLATION AND STRUCTURE OF THE HYPERTREHALOSEMIC HORMONE

FROM BLABERUS DISCOIDALIS COCKROACHES

Timothy K. Hayes and Larry L. Keeley

Texas A & M University

College Station, TX 77840

Trehalose is the major carbohydrate in most insects and its hemolymph concentrations are influenced by neuroendocrine factors from the corpora cardiaca (CC). In 1961, Steele first reported that CC extracts stimulated increased levels of hemolymph trehalose in the cockroach, Periplaneta americana (Steele, 1961). This phenomena was quickly confirmed in the tropical cockroach, Blaberus discoidalis (Bowers and Friedman, 1963). Since then, neuroendocrine effects were discovered in locusts, blowflies, stick insects, hornworm moths and others. Recently, two cardioaccelerator factors from P. americana were isolated and their structures confirmed by synthesis. These factors also have hyperglycemic activity (Scarborough et al., 1984). Further studies showed that these factors were the major hyperglycemic activity in CC extracts of P. americana. Each of these peptides are related to the structure of locust adipokinetic hormone (AKH1) and Crustacean pigment concentration hormone (RPCH). In this report, we describe the isolation and structure for the hypertrehalosemic hormone (HTH) from the cockroach, B. discoidalis, one of the first two insects in which this effect was identified.

HTH was tracked through the purification process with
the in vitro bioassay developed in our laboratory (Hayes
and Keeley, 1985). The assay uses fat body fragments
dissected from 24 h decapitated cockroaches and compares
the relative production of trehalose between two fragments
of the same fat body. Results are reported as activation
ratios (AR) that result from the ratio of trehalose
secretion for a fat body fragment exposed to a hormone
sample divided by the trehalose secretion of a fragment not
exposed to hormone. An AR value near one indicates no
activity and an AR above two indicates strong HTH activity.
The assay is specific for peptides that resemble HTH and is
not influenced by biogenic amines. The sugar secretion
stimulated by HTH is greater than 97% trehalose.

The bioassay was used to discover several fundamental
properties of HTH prior to isolation. HTH is small (1-2
kDa) and heat stable. HTH activity is sensitive to
chymotrypsin but not to aminopeptidase M, carboxypeptidase
A or trypsin. AKH1 and RPCH stimulate the HTH bioassay to
maximum sugar release. All of these properties indicate
that HTH is a member of the AKH family of peptides.

HTH was extracted from whole adult heads with ice cold
acidic acetone (acetone:0.1N HCl:thiodiglycol (TDG, 90: 10:
0.01). Acetone was driven off the extract with a stream of
nitrogen and the residual liquid was reconstituted in 0.1N
sodium phosphate (pH=7.2). The HTH activity was
concentrated on a C_{18}-cartridge, washed with water and
eluted with 40% acetonitrile (MeCN) in 5mM triflouroacetic
acid (TFA). The solvent was removed by vacuum
centrifugation and the sample stored for future use at
-20°C.

The relative hydrophobicity of HTH was determined by
reversed-phase TLC of the prepared extracts in conjunction
with standard hydrophobic peptides. TLCs were developed in
a solvent consisting of 35% MeCN in 5mM TFA. HTH extracts
and standard peptides were run in adjacent channels and the
channel containing the extract was scraped and extracted
for HTH by fractions defined by the mobility of the
standard peptides. HTH activity was localized between
glycylphenylalanylphenylalanine and triphenylalanine.
These standard peptides were used to develop optimal
solvent programs and define the elution range of HTH on
reversed-phase HPLC.

HTH was isolated from crude extracts by a rapid HPLC procedure. The crude extracts were fractionated by a reversed-phase step on a Biorad Hipore C-18 column with a separation gradient of 27 to 33% MeCN in 5mM TFA. HTH activity eluted in a single fraction at 30% MeCN in 5mM TFA. This fraction was collected in a polypropylene tube containing 100ug BSA and 10ul TDG. The HTH fraction was dried by vacuum centrifugation and redissolved in 200ul of 0.1N phosphate (pH=7.2) and fractionated on a Water's Protein-Pak I-60 column (equilibrated with the same buffer) at 0.5 mls/min. A peak (46 min) was observed to adsorb to the column beyond the salt volume (36 min). All other peaks eluted in the gel filtration fractionation range of 22 to 36 min. The fraction containing the adsorbed peak contained all of the HTH activity in the sample and was reinjected on the reversed-phase column under identical conditions as the first HPLC. Only one peak was observed for this fraction and the corresponding effluent contained HTH activity.

PTC-amino acid analysis was used to determine the amino acid composition of HTH. The sample was prepared for derivatization by vapor-phase hydrolysis (105°C, 20 h) with 6N HCl. The theoretical amino acid residues found in the HTH sample were Asx (1), Glx (1), Ser (1), Gly (2), Thr (1), Pro (1), Val (1), and Phe (1).

Gas-liquid phase sequence analysis was used to determine the primary structure of HTH. The HTH sample was prepared for analysis by removal of pyroglutamic acid from the N-terminus with pyroglutamyl amino peptidase. The analysis indicated the following sequence:

<Glu-Val-Asn-Phe-Ser-Pro-Gly-Trp-Gly-Thr-NH_2

Both the C-terminal amide and acid were prepared in our laboratory by solid-phase peptide synthesis. Only the amide coeluted with HTH on C_{18} and I-60 HPLC. Thus, the amide is assigned as HTH. Synthetic HTH_{10} had potent biological activity <u>in vitro</u> (ED_{50} around 10^{-10} Molar) and <u>in vivo</u>.

    This research was supported by NIH Grant # RO1
NS20137-02 to TKH. The authors wish to thank Mr. Daniel
Knight and Ms. Kathleen Thornton for their assistance in
the performance of the experiments.

                              References

Bowers, W. and Friedman, S. (1963) Mobilization of fat body
glycogen by an extract of corpus cardiacum. Nature (London)
198, 685.

Hayes T.K. and Keeley L.L. (1985) Properties of an in vitro
bioassay for hypertrehalosemic hormone of Blaberus
discoidalis cockroaches. Gen. Comp. Endocrinol. **57**,
246-256.

Scarborough R.M., Jamieson G.C., Kalish F., Kramer S.J.
McEnroe G.A., Miller C.A. and Schooley D.A. (1984) Proc.
Natl. Acad. Sci. USA **81**, 5575-5579.

Steele, J.E. (1961) Occurrence of a hyperglycemic factor in
the corpus cardiacum of an insect. Nature (London) **192**,
680-681.

COMPARATIVE ACTIONS OF A NEW SERIES OF MYOTROPIC

PEPTIDES ON THE VISCERAL MUSCLES OF AN INSECT

B. J. Cook, G. M. Holman and R. J. Nachman

USDA, ARS, VTERL

P.O. Drawer GE, College Station, TX 77841

For more than 30 years it has been known that the corpus cardiacum (CC) of insects contains substances that stimulate the contraction of visceral muscle (Cameron, 1953). Moreover, Brown (1965) demonstrated that at least two of these factors were peptides and in 1984 the structure of two cardioaccelerator peptides found in that neurodocrine organ were announced (Scarborough et al., 1984; Witten et al., 1984). Three years ago we presented evidence for at least six additional myotropic peptides in head extracts of the cockroach Leucophaea maderae. We now report the structure and comparative actions for members of a new family of cephalomyotropic peptides.

Hindguts isolated from the central nervous system of the cockroach were dissected and prepared for recording as previously described (Cook and Holman, 1978). Threshold concentrations were determined for both natural and synthetic peptides by adding known quantitites of each peptide to the chamber which contained the isolated hindgut. The threshold concentration was defined as that concentration of peptide required to evoke an observable change in the frequency and/or amplitude of hindgut contractions within 1 min. The procedures for the extraction and sep-pak purification of homogenates containing 3000 heads from adult cockroaches are described elsewhere (Holman et al, 1986). The four peptides were initially separated into two active fractions (52-54 min and 58-60 min) on a waters -Boundapak phenyl column by high-performance liquid-chromatography

(HPLC). Further separation and final purification of the peptides were achieved by the use of HPLC columns: 1) a Rainin Microsorb C1; 2) a Techsphere 3 C18; and 3) a Waters 1-125 Protein-Pac. The details of conditions and solvents have been described elsewhere (Holman et al 1986). The amino acid content of the peptides was determined with the Waters Picotag amino acid analysis system. The pure peptides were then sequentially degraded and the amino acids converted to the PTH-derivatives with a Model 470A protein sequencer (Applied Biosystems, Inc.). PTH-amino acids were identified by HPLC analysis with the Waters ALC-100.

HPLC analysis of the PTH-amino acids generated in each cycle by the protein sequencer yielded the following primary sequences:

Leucokinin I	Asp-Pro-Ala-Phe-Asn-Ser-Trp-Gly-NH$_2$
Leucokinin II	Asp-Pro-Gly-Phe-Ser-Ser-Trp-Gly-NH$_2$
Leucokinin III	Asp-Gln-Gly-Phe-Asn-Ser-Trp-Gly-NH$_2$
Leucokinin IV	Asp-Ala-Ser-Phe-His-Ser-Trp-Gly-NH$_2$

Analysis of the four synthetic peptide amides showed that in every case the retention time corresponded to the retention time of the appropriate natural product in all four systems. Thus structures were assigned for all four peptides as indicted above.

Threshold concentrations for the natural products and their synthetic counterparts are shown in Table 1. Each value $\vec{X}$ + S.D. was calculated from threshold measurements on five separate hindgut preparations.

Since near trace amounts of the leucokinins were found in head extracts (Table 2), any attempt to isolate and characterize these peptides from the CC would have required quite large numbers to be successful. Nevertheless preliminary distribution studies indicate that the leucokinins do predominate in extracts of both the brain and CC compared with other tissues in the head (Holman and Cook, unpublished).

To find an organism with four peptides of very similar structure, titre and biological activity could indeed be perplexing if one persists in the assumption that each peptide has only a single action in the organism. Obviously such is not the case because multiple sites and actions are of common occurrence with most neuropeptides of both verte- brates and invertebrates. Clearly chemical messengers often need the capacity to mobilize diverse cellular

Table 1. Threshold concentrations of natural and synthetic peptides leucokinins I, II, III & IV

Peptide		Threshold concentration $\bar{X} \pm SD$
L-I	Natural product	$2.1 \pm 0.48 \times 10^{-10}$ M
	Synthetic	$2.0 \pm 0.28 \times 10^{-10}$ M
L-II	Natural product	$1.8 \pm 0.37 \times 10^{-10}$ M
	Synthetic	$1.6 \pm 0.67 \times 10^{-10}$ M
L-III	Natural product	$9.4 \pm 2.3 \times 10^{-11}$ M
	Synthetic	$7.2 \pm 0.4 \times 10^{-11}$ M
L-IV	Natural product	$1.2 \pm 0.2 \times 10^{-10}$ M
	Synthetic	$1.4 \pm 0.5 \times 10^{-10}$ M

Table 2. Titres of leucokinins in the head of the cockroach, Leucophaea maderae

Peptide	Molecular weight	nm/head	ng/head
L-I	892	0.00048	0.43
L-II	851	0.00035	0.30
L-III	909	0.00022	0.20
L-IV	905	0.00023	0.21

responses to coordinate actions on the part of the whole organism. Thus the discovery of several peptides with the same site of action does not represent redundance but simply a need for that specific organ's response in a coor- dinated action directed by a given peptide. At the moment we have measured but one of perhaps many biological activi- ties that exist for these peptides. Only when all the diverse actions have been determined in a number of

physiological and biochemical systems of the insect will it
be possible to make firm statements about the principle
mode of action for a given peptide.

REFERENCES

Brown B. E. (1965) Pharmacologically active constituents of
    the cockroach corpus cardiacum: resolution and some
    characteristics. Gen. comp. Endocr. 5, 287-401
Cameron M. L. (1953) Secretion of an orthodiphenol in the
    corpus cardiacum of the insect. Nature, Lond. 172,
    349-350.
Cook B. J. and Holman G. M. (1978) Comparative pharmacolog-
    ical properties of muscle function in the foregut and the
    hindgut of the cockroach. Leucophaea maderae. Comp.
    Biochem. Physiol. 61C, 291-295.
Holman G. M., Cook B. J. and Wagner R. M. (1984) Isolation
    and partial chracterization of five myotropic peptides
    present in head extracts of the cockroach. Leucophaea
    maderae. Comp. Biochem. Physiol. 77C, 1-5.
Holman G. M., Cook B. J. and Nachman R. J. (1986) Isola-
    tion, primary structure and synthesis of two neuropep-
    tides from Leucophaea maderae: members of a new family
    of cephalomyotropins. Comp. Biochem. Physiol. In press.
Scarborough R. M., Jamieson G. C., Kalish F., Kramer S. J.,
    McEnroe G. A., Miller C. A. and Schooley D. A. (1984)
    Isolation and primary structure of two peptides with
    cardioacceleratory and hyperglycemic activity from the
    corpora cardiaca of Periplaneta americana. Proc. Natn.
    Acad. Sci. U.S.A. 81, 5575-5579.
Starratt A. M. and Brown B. E. (1975) Structure of the
    pentapeptide proctolin, a proposed neurotransmitter in
    insects. Life Sci. 17, 1253-1256.
Witten J. L, Schaffer M. H., O'Shea M., Cook J. Cl, Hemling
    M. E. and Rinehart K. L., Jr. (1984) Structures of two
    cockroach neuropeptides assigned by fast atom bombardment
    mass spectrometry. Biochem. Biophys. Res. Commun. 124,
    350-358.

LEUCOMYOSUPPRESSIN: A DECAPEPTIDE THAT INHIBITS HINDGUT CONTRACTILE ACTIVITY IN LEUCOPHAEA MADERAE

G. M. Holman, B. J. Cook and R. J. Nachman

USDA, ARS, VTERL

P.O. Drawer GE, College Station, TX 77841

During our initial isolation studies of several myo-tropic peptides from cockroach head extracts (Holman et al., 1984), we observed that one of the fractions isolated from the μ-Bondapak phenyl separation system contained a material that suppressed the spontaneous contractile activity of the isolated cockroach hindgut in a reversible manner. A similar inhibitory response was previously observed with a cockroach hindgut extract (Holman and Cook, 1970). This report describes the isolation, structural identification, synthesis, and biological activity of this inhibitory peptide present in head extracts.

Hindguts isolated from the central nervous system of the cockroach were dissected and prepared for recording as previously described (Cook and Holman, 1978). Threshold concentrations were determined for both natural and syn-thetic peptides by adding known quantities of each peptide to the chamber containing the isolated hindgut. The thres-hold concentration was defined as that concentration of peptide required to evoke an observable change in the frequency and/or amplitude of contractions within 1 min. The procedure for extraction and Sep-Pak purification of 3000-head homogenates is described elsewhere (Holman et al., 1986a). The inhibitory peptide was further purified by sequentially eluting the active fraction through four HPLC columns. These were, in order of sequence, Waters μ-Bondapak phenyl, Rainin C1, Techsphere 3 C18, and Waters I-125 Protein-Pac. The details of the HPLC conditions

and solvent systems are described elsewhere (Holman et al.,
1986b).

Micro-amino acid analysis with the Waters Picotag system showed that the inhibitory peptide was 98% pure and
contained equimolar amounts of glutamic acid, histidine,
arginine, and leucine; and twice that amount of aspartic
acid, valine, and phenylalanine. Based upon the phenylalanine value (the most accurate) we calculated that the 6000-
head extract contained 1.38 nanomoles of the peptide. Each
head, therefore, contributed 0.23 picomoles. The peptide
was named Leucomyosuppressin (LMS).

Incubtion of LMS with aminopeptidase M (Holman et al.,
1984) did not destroy the biological activity indicating
that LMS was blocked at the N-terminus. Presence of glutamic acid in the amino acid analysis suggested blocking
with pyroglutamic acid (pGlu). Incubation of LMS with
pyroglutamate aminopeptidase (Holman et al., 1986b, c) followed by HPLC isolation of the fragments yielded a material
with the same amino acid content as LMS, minus glutamic
acid. This truncated peptide was sequenced with the gas-
phase protein sequencer and a sequence was obtained. Two
peptides, based upon this sequence and having pGlu added at
the N-Terminus were synthesized; one with a free carboxyl
group at the C-terminus, and the second with a C-terminal
amide. Both peptides were evaluated in the HPLC purification systems and the amide was shown to have retention
times identical with natural LMS. Evaluation of both
materials in the bioassay system showed that the amide form
of LMS had a threshold concentration (7.8×10^{-10}M)
very similar to natural LMS (8.4×10^{-10}M) whereas the
acid was inactive at concentrations as high as 10^{-6}M.
We therefore assigned LMS the following structure.

pGlu-Asp-Val-Asp-His-Val-Phe-Leu-Arg-Phe-NH$_2$

The response of the isolated hindgut to LMS was dosage
dependent and reversible. At near threshold concentrations
both frequency and amplitude of contraction were initially
reduced. Although the frequency recovered to the pre-
exposure level within 2-3 min, the amplitude remained
depressed. With increasing concentrations of LMS, a drop
in tonus was also observed. Removal of LMS from the preparation resulted in an immediate recovery of tonus and frequency to pre-exposure levels, but the amplitude remained

somewhat depressed for 5 min (Fig. 1). The response of the
hindgut to LMS and three stimulatory neuropeptides (myotropins) is shown in Figure 2. The peptide concentrations

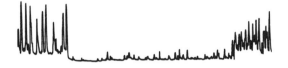

Fig. 1. Response of hindgut preparation to 2.5 X 10^{-9}M
LMS (up arrow). Inhibition was allowed to continue for 12
min followed by rinse of chamber and refill with fresh
saline at down arrow.

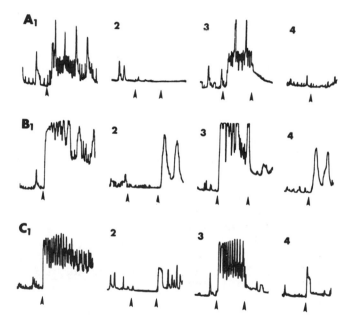

Fig. 2. Interaction of LMS (2 X 10^{-9}M) with insect
myotropins. Columns: (1) myotropin alone, (2) myotropin
after 3-min exposure to LMS, (3) LMS after 3-min exposure
to myotropin, and (4) LMS and myotropin added together.
Lines: (A) Leucopyrokinin (pGlu-Thr-Ser-Phe-Thr-Pro-Arg-
Leu-NH$_2$) 2 X 10^{-8}M., (B) Leucokinin II (Asp-Pro-Gly-
Phe-Ser-Ser-Trp-Gly-NH$_2$) 4 X 10^{-9}M. and (C) Proctolin (Arg-Tyr-Leu-Pro-Thr) 1 X 10^{-9}M.

were approximately 30X threshold in all cases. In all
three examples, the stimulatory effect was greatly reduced
or abolished by LMS , whether added before the myotropin
(A_2, B_2, C_2), after the myotropin (A_3, B_3, C_3),
or in conjunction with the myotropin (A_4, B_4, C_4).

Finally, the results of a preliminary extract of hindguts purified through the first HPLC fractionation suggest
that LMS is the inhibitory material in the hindgut. A larger extract is presently being processed in order to prove
or disprove this observation.

REFERENCES

Cook B.J. and Holman G. M. (1978) Comparative pharmacological properties of muscle function in the foregut and the
hindgut of the cockroach, Leucophaea maderae. Comp.
Biochem. Physiol. 61C, 291-295.

Holman G. M. and Cook B. J. (1970) Pharmacological properties of excitatory neurotransmission in the hindgut of
the cockroach, Leucophaea maderae. J. Insect Physiol.
16, 1891-1907.

Holman G. M., Cook B. J. and Nachman R. J. (1986a) Isolation, primary structure and synthesis of two neuropeptides from Leucophaea maderae: Members of a new family of
cephalomyotropins. Comp. Biochem. Physiol. (In Press).

Holman G. M., Cook B. J. and Nachman R. J. (1986b) Isolation, primary structure, and synthesis of Leucomyosuppressin, an insect neuropeptide that inhibits spontaneous
contractions of the isolated cockroach hindgut. Comp.
Biochem. Physiol. (In Press).

Holman G. M., Cook B. J. and Nachman R. J. (1986c) Primary
structure and synthesis of a blocked myotropic neuropeptide isolated from the cockroach, Leucophaea maderae.
Comp. Biochem. Physiol. (In Press.

Holman G. M., Cook B. J. and Wagner R. M. (1984) Isolation
and partial characterization of five myotropic peptides
present in head extracts of the cockroach, Leucophaea
Maderae. Comp. Biochem. Physiol. 77C, 1-5.

ENZYMATIC DEGRADATION OF PROCTOLIN IN THE CNS OF THE LOCUST, SCHISTOCERCA GREGARIA, BY MEMBRANE BOUND ENZYMES.

R. Elwyn Isaac

Department of Pure and Applied Zoology

University of Leeds, Leeds. LS2 9JT. U.K.

INTRODUCTION

Although a number of mechanisms may be involved in the termination of neuropeptide signals, recent studies in vertebrate neurochemistry have indicated an important role for extracellular membrane enzymes in the inactivation of neuropeptides (Turner et al., 1985). Whether similar membrane enzymes are also involved in the inactivation of insect neuropeptides is unclear. A study on the metabolism of proctolin (Arg-Tyr-Leu-Pro-Thre) by tissues of P. americana showed that this peptide was degraded primarily by two soluble enzyme activities which cleaved the Arg-Tyr and Tyr-Leu bonds (Quistad et al., 1984). Membrane fractions possessed only weak degradative activity towards proctolin.

The present paper describes two membrane-bound enzyme activities involved in the in vitro metabolism of proctolin by nervous tissue from Schistocerca gregaria. One of these activities, a proctolin aminopeptidase, was found to be highly localised in a synaptosomal membrane fraction.

The ability of nervous tissue to degrade proctolin was initially investigated by incubating proctolin (100μm) with a tissue homogenate at pH 7.3. Proctolin was degraded to yield Tyr-Leu-Pro-Thre and some tyrosine (Fig. 1a), indicating the importance of a neutral aminopeptidase in the metabolism of this peptide. Arg-Tyr was shown to be

a degradation product by adding the aminopeptidase
inhibitors, amastatin and bestatin, to the reaction
mixture. The inhibitors not only reduced the amount of
Tyr-Leu-Pro-Thre and tyrosine released but also allowed the
accumulation of Arg-Tyr (Fig. 1b).

Approximately 50% of the degradative activity was
localised in a 30,000g membrane fraction with a higher
specific activity (80pmol/h/μg of protein) than that found
in the supernatant (55pmol/h/μg of protein). At neutral pH
and a substrate concentration of 100μm, the supernatant
enzymes cleaved both the Arg-Tyr and Tyr-Leu bonds whereas
the membrane preparation appeared to be enriched in
aminopeptidase activity only. However, on lowering the
proctolin concentration employed in the assay to 0.1μM, it
was shown that neural membranes could catalyse the

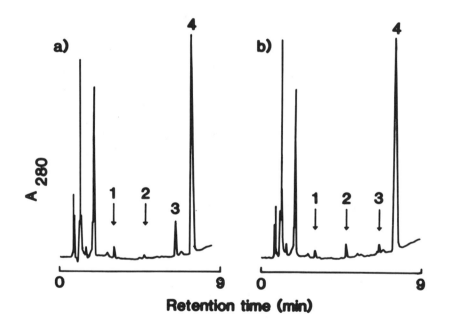

Fig. 1. H.p.l.c. separation of the degradation products
(1, tyrosine; 2, Arg-Tyr; 3, Tyr-Leu-Pro-Thre; 4,
proctolin) generated on incubation of proctolin with a
crude tissue preparation from S. gregaria nervous tissue in
the absence a) and presence b) of 0.1mM amastatin and 1mM
bestatin .

formation of the dipeptide, Arg-Tyr, at low substrate
levels. The formation of Arg-Tyr from [Tyr-3,5-³H]
proctolin was optimal at pH 7.0 and the enzyme involved
displayed a low Km of 0.3μM (Fig. 2a). The membrane-bound
aminopeptidase had a pH optimum of 7.0 and an apparent Km
of around 27μM (Fig. 2b). These properties are similar to
those of neuropeptide-degrading enzymes from vertebrate
nervous tissue.

    The 30,000g membrane preparation was further
fractionated into synaptosomal and mitochondrial fractions
which were assayed for acetylcholinesterase (AchE,
synaptosomal marker), succinate dehydrogenase (SDH,
mitochondrial marker) and the proctolin degrading enzymes
(A, aminopeptidase and B, Tyr-Leu hydrolysis).

    Proctolin aminopeptidase activity followed a similar
distribution pattern to that of acetylcholinesterase (Table
1) with 70-80% of the membrane-bound activity being
localised in the synaptosomal fraction. The high affinity
cleavage of the Tyr-Leu bond was predominantly associated
(60-70%) with the mitochondrial fraction. The latter
activity associated with the synaptosomal membranes was not
removed by osmotic shock treatment or by washing with 0.4M-
NaCl.

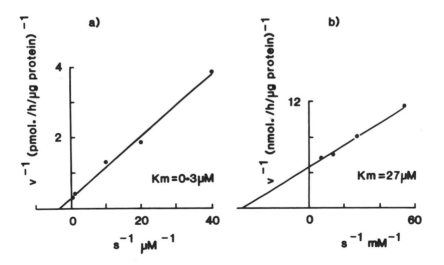

Fig. 2. Lineweaver-Burk plots for the hydrolysis of a) the
         Tyr-Leu bond and b) Arg-Tyr bond of proctolin

Table 1. Specific activities of proctolin degrading and
marker enzymes in synaptosomal and mitochondrial
membranes

Enzyme	Synaptosomal fraction	Mitochondrial fraction
AchE[a]	105 ± 18	28 ± 6
SDH[b]	140 ± 60	910 ± 250
A[b]	280 ± 40	95 ± 25
B[b]	1.08 ± 0.11	1.97 ± 0.56

a μmol/h/μg protein
b pmol/h/μg protein

The exact nature and subcellular location of the
membrane-bound enzymes involved in the hydrolysis of the
Tyr-Leu bond is not certain. However, it is clear that
locust synaptosomal membranes are enriched with an
aminopeptidase which is capable of degrading proctolin and
which may be involved in the in vivo inactivation of this
and other synaptically released neuropeptides.

REFERENCES

Quistad, G.B., Adams, M.E., Scarborough, R.M., Carney, R.L.
and Schooley, D.A. (1984). Life Sci. 34, 569-576.
Turner, A.J., Matsas, R. and Kenny, A.J. (1985).
Biochem. Pharmacol. 34, 1347-1356.

ENDOCRINE REGULATIONS OF PROTEIN SYNTHESIS BY THE IN VITRO

FAT BODY FROM BLABERUS DISCOIDALIS COCKROACHES.

L. L. Keeley, S. Sowa and T. K. Hayes.

Texas A&M University

College Station, TX 77843

Introduction

Juvenile hormone (JH) is the principal endocrine regulator of ovarian maturation in most insects. JH stimulates the fat body to synthesize vitellogenic proteins that are transported and deposited in the yolk of maturing oocytes. Neurohormones also affect vitellogenesis but their action is less well defined. In the tropical cockroach, Blaberus discoidalis, JH was essential for ovarian maturation, but the administration of corpora cardiaca (CC) extracts as a neuroendocrine source enhanced the amount of protein deposited in the ovaries in response to JH (Keeley and McKercher, 1985). The present investigations examined the interactions between JH and neurohormones as regulators of fat body protein synthesis and ovarian maturation in B. discoidalis.

The pattern of ovarian protein formation was determined for the first gonotrophic cycle. Ovarian maturation occurs in B. discoidalis independent of feeding or mating and starts on day 4 at 2 mg of total protein per ovary pair and is complete by day 22 with 140 mg of protein. Based on ovarian protein content, the 15-day-old female was chosen as the experimental animal for all subsequent studies. This age was selected because it was near the midpoint of ovarian maturation with 71 mg protein/ ovary pair.

211

Fat body protein synthesis was measured also during the first gonotrophic cycle. Protein synthesis was based on the rate of [^{14}C]leucine incorporation during a 9 h in vitro incubation in Reddy-Wyatt medium (1967) at 27°C in a 95:5 O_2:CO_2 atmosphere. The rate of synthesis for proteins secreted from the fat body was minimal on day 0 (4300 dpm/mg protein x h^{-1}) and increased until it reached a maximum on day 18 (51,000 dpm/mg protein x h^{-1}). The rate of protein synthesis declined for the remainder of the study period.

A dose-response was determined between JH-3 and the total ovarian protein content in females treated from adult emergence by either allatectomy (CA$^-$) + starvation or by decapitation (DC$^-$). JH-3 was administered by daily intrahemocoelic injection on days 2-14. Ovarian protein content was measured on day 15. Dose-responses were parallel over all JH doses for both experimental groups with the CA$^-$ females producing more ovarian protein than the DC$^-$ animals. Normal, 15-day levels of protein were obtained with CA$^-$ females at about 9 ug JH-3. Ovaries from DC$^-$ females contained 32% less protein at this dose of JH. This suggested that something associated with the head influenced ovarian protein formation in association with JH.

Further studies on ovarian maturation were performed with methoprene, a JH analog (JHA), which could be applied topically. A dose response for JHA and ovarian protein formation was determined by applying JHA topically to DC$^-$ females on days 3, 6, 9 and 12. Ovarian protein content was measured on day 15. Maximum ovarian protein was formed at a dose of 100 ug JHA.

In addition to JH deficiency, decapitation also causes neuroendocrine deficiency which may restrict the rate of yolk protein formation. The neuroendocrine effect was tested by stimulating yolk protein formation in females DC$^-$ on day 0 and treated with either CC extract, JHA or JHA + CC extract. JHA was administered topically on days 3, 6, 9 and 12 in a suboptimal (15 ug) dose so that protein synthetic systems would not be maximized and neuroendocrine effects might be observed. CC extract was administered as daily intrahemocoelic injections of 0.15 CC pair/10 ul Ringer on days 3-14. CC extract, alone, had no stimulatory effects on ovarian protein formation and resulting ovaries were equivalent to undeveloped ovaries from 0-day females. JHA, alone, resulted in 15-day ovaries that were only 74% mature. Treatments with JHA + CC extract produced ovarian protein contents comparable to those of normal, 15-day control animals.

Neuroendocrine extracts enhanced the action of JHA to stimulate ovarian protein formation. This effect was measured at the tissue level by determining the protein synthetic capacity of the fat body. Experimental animals were either normal animals or females DC^- on day 0. DC^- animals were treated on days 3-14 with the same endocrine regimens described in the previous experiment. The capacity for fat body synthesis of secreted proteins was measured in vitro at 15 days of age. Normal, 15-day females had a protein synthetic rate of 43,000 dpm/ mg protein x h^{-1}. Basal rates of synthesis were 6900 dpm/mg protein x h^{-1} for tissue from DC^-, 15-day-old females. Muscle extracts were used as tissue controls for CC extracts and did not affect the basal rate of protein synthesis in DC^- females. CC extracts, alone, elevated the synthesis of secreted proteins by 38% over the basal rate in DC^- females. JHA, alone, increased the basal rate by 2.7 times, but JHA + CC extract increased the synthetic rate of secreted protein by 4 times, to levels approaching those of normal, 15-day-old females. Treatments with highly purified cytochromogenic hormone (CGH) (same dosage and treatment regimen as CC extract) + JHA resulted in protein synthesis rates identical to animals receiving CC extracts. This suggests that the active factor in the CC extract may be CGH.

These results indicate that the CC contain a neurohormone that affects the general protein synthetic capacity of the fat body. CC extracts, by themselves, had only minor effects on the capacity of the fat body to synthesize export proteins and had no stimulatory effects on ovarian maturation. A 50% stimulatory dose of JHA tripled the rate of synthesis of export proteins and also stimulated partial ovarian maturation. This suggests that the fat body proteins synthesized in response to JHA were, at least in part, the vitellogenins. CC extracts combined with JHA resulted in greater rates of export protein synthesis than with JHA alone, and ovarian maturation reached normal, 15-day levels despite the fact that the JHA dose was only adequate to produce partial maturation. Since the CC extracts did not stimulate ovarian maturation by themselves, the data suggest that the CC extracts enhanced the general protein synthetic capacity of the fat body and the rate of vitellogenin synthesis. JHA stimulates vitellogenin synthesis but in the absence of the CC factors, the rate of vitellogenin synthesis was lower so that maturation required longer. When the CC extracts were administered in combination with suboptimal JHA, the protein synthetic capacity was elevated to normal and both the rate of vitellogenin synthesis and ovarian maturation reached normal levels within the 15-day observation period.

The data suggest that CGH may be the active factor in CC extracts that stimulates the protein synthetic capacity of the fat body. CGH stimulates heme synthesis for new cytochromes during mitochondriogenesis that occurs in conjunction with postimaginal fat body maturation (Keeley, 1981). Therefore, we speculate that CGH affects protein synthesis by determining the basal endergonic activity of adipocytes through their ATP-generating capacity. An increase in bioenergetic potential due to CGH effects would increase the biosynthetic capacity of the cells and elevate the rate of vitellogenin synthesis.

This research was supported by NSF Grant NO. 81-03277.

References

Keeley L.L. (1981) Neuroendocrine regulation of mitochondrial development and function in the insect fat body. In: Energy Metabolism in Insects, (Ed. Downer R.G.H.) 207-237, Plenum, NY.

Keeley L.L. and McKercher S.R. (1985) Endocrine regulations of ovarian maturation in the cockroach Blaberus discoidalis. Comp. Biochem. Physiol. 80A, 115-121.

Reddy S.R.R. and Wyatt G.R. (1967) Incorporation of uridine and leucine in vitro by cecropia silkmoth wing epidermis during diapause and development. J. Insect Physiol. 13, 981-994.

NEUROHORMONAL REGULATION OF PHEROMONE BIOSYNTHESIS IN

HELIOTHIS ZEA: EVIDENCE FOR MULTIPLE FORMS OF THE HORMONE

Ashok K. Raina[1], Howard Jaffe[2] and Richard L. Ridgway[2]

[1]Dept. Entomol., Univ. Maryland, College Park, MD
20742, and[2] Agric. Res. Serv., USDA, Beltsville, MD
20705, USA

Females of the corn earworm, Heliothis zea, produce the
sex pheromone only during the scotophase (Raina et al.,
1986). A brain factor has been reported to control
pheromone production in this species (Raina and Klun,
1984). The activity was found in both female and male
brains, even though the role of this factor in the males is
not yet known. The factor has recently been characterized
as a neuropeptide hormone originating in the suboesophageal
ganglion (SOG) and released through the corpora cardiaca
(CC) into the hemolymph to activate pheromone production
(Raina et al., in manuscript). The hormone has been
designated "pheromone-biosynthesis-activating neuropeptide"
or PBAN.

Isolation and purification of PBAN from brain-SOG-CC by
high-performance liquid chromotography (HPLC) yielded three
biologically active fractions (Jaffe et al., this volume).
All three activities were present in preprations of both
female and male brain-SOG-CCs. Initially, brain-SOG-CCs
were stored and homogenized in methanol:water:acetic
acid:thiodiglycol (90:9:1:0.1) solvent system, for the
isolation of PBAN. The total pheromonotropic activity
recovered after Method A- HPLC run (Jaffe et al. this
volume) was about 34% of the activity obtained by the
injection of fresh brain-SOG homogenate into a ligated
female (for assay technique see Raina and Klun, 1984).
There were three active fractions each contributing 33.1,
37.8 and 29.1% respectively to the total activity (Jaffe

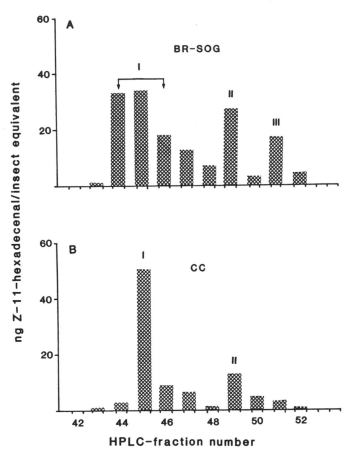

Fig. 1. Pheromonotropic activity expressed as ng of the
major component of H. zea pheromone produced by
ligated females following injection of HPLC-fractions
of homogenates of brain-suboesophageal ganglia (BR-SOG)
A and, corpora cardiaca (CC) B. Fractions representing
pheromone biosynthesis activating neuropeptide (PBAN)
I, II and III are shown.

et al. this volume). We designate these as PBAN-I,-II and
-III. Later, Bennett's buffer (5% formic acid v/v, 15%
trifluoroacetic acid v/v, 1% NaCl w/v in 1N HCl) was used as
a solvent system for the storage and homogenization of
brain-SOGs before HPLC fractionation. The recovery of the
activity was as high as 176% that of fresh tissue
homogenates (lower activity of fresh tissue homogenates may
be due to the action of endogenous proteolytic enzymes).
Once again three active fractions were isolated (Fig. 1A)
with 63.0, 23.8 and 13.2% activity corresponding to PBAN-I,
II and III respectively. However, when CC from females were
analyzed using this solvent system, most of the activity
(75.7%) corresponded to PBAN-I (Fig. 1B). There was no
distinct activity peak corresponding to PBAN-III. More
recently, brain-SOG-CCs after dissection, were instantly
frozen on dry ice before being put into Bennett's buffer and
stored at -90°C. Following this procedure, only two distinct
peaks of activity corresponding to PBAN-I and III with ca.
80% of the activity in the PBAN-I region were observed.

Significantly higher pheromonotropic activity was found
in CC from scotophase females compared to those from the
photophase females (Raina and Menn, in press). Likewise
pheromonotropic activity was present in the female hemolymph
only during scotophase. These results indicate that the CC
act as neurohaemal organs for the release of PBAN into the
hemolymph, and that the hormone is released only during the
scotophase. We speculate that PBAN-I may be the released
form of the hormone. PBAN-III, which is only seen in the
SOGs may represent the pro-PBAN. PBAN-II may be a
chemically or enzymatically altered form of PBAN; the
changes could be minor so that they effect only the HPLC-
elution times and not the apparent biological activity.
Confirmation of these speculations can be obtained only
after we determine the amino acid composition and sequence
of these different forms of PBAN.

ACKNOWLEDGEMENTS

This research was supported, in part, by Agric. Res.
Serv. Cooperative Grant No. 58-32U4-5-11. Scientific
Article No. A-4491, Contribution No. 7484 of the Maryland
Agricultural Experimental Station. Technical assistance of
Jeffrey Bray is acknowledged.

REFERENCES

Jaffe, H., Raina, A. K., and Hayes, D. K. (1987). HPLC
    isolation and purification of pheromone biosynthesis
    activating neuropeptide of Heliothis zea. Proc. Int.
    Conf. Insect Neurochemistry and Neurophysiology
    Aug. 4-6, 1986, College Park, MD. (Eds. Borkovec,
    A. B. and D. B. Gelman) in press.
Raina, A. K. and Klun, J. A. (1984). Brain factor control
    of sex pheromone production in the female corn
    earworm moth. Science 225, 531-533.
Raina, A. K., Klun, J. A., and Stadelbacher, E. A.
    (1986). Diel periodicity and effect of age and
    mating on female sex pheromone titer in Heliothis zea
    (Lepidoptera: Noctuidae). Ann. Entomol. Soc. Am. 79,
    128-131.
Raina, A. K., and Menn, J. J. (1987). Endocrine regulation
    of pheromone production in Lepidoptera. In Pheromone
    Biochemistry (Eds. Prestwich, G. D. and Blomquist, G.
    J.), Academic Press, Orlando, Florida, in press.
Raina, A. K., Jaffe, H., Klun, J. A., Ridgway, R. L. and
    Hayes, D. K. Characteristics of a neurohormone that
    controls sex pheromone production in Heliothis zea. J.
    Insect Physiol, in manuscript.

HPLC ISOLATION AND PURIFICATION OF PHEROMONE BIOSYNTHESIS ACTIVATING NEUROPEPTIDE OF HELIOTHIS ZEA

Howard Jaffe, Ashok K. Raina and D.K. Hayes

Livestock Insects Laboratory, ARS, USDA
Beltsville, MD 20705 and Department of
Entomology, University of Maryland,
College Park, MD 20742

A pheromone biosynthesis activating neuropeptide hormone (PBAN) of Heliothis zea has been recently reported by Raina and Klun (1984). This paper reports the isolation and purification of PBAN in addition to data on recovery of biological activity from various workups, molecular weight and amino acid analysis.

PBAN, like the adipokinetic hormone of H. zea was isolated and purified by sequential gradient elution using three reverse phase-high performance liquid chromatography (RP-HPLC) systems (Methods A-C, Jaffe et al., 1986) or by a combination of high performance-size exclusion chromatography (HP-SEC) and RP-HPLC (Methods B-C). Brain-subesophageal ganglion-corpora cardiaca (BR-SOG-CC) complexes of H. zea were dissected from adults and homogenized as previously described (Jaffe et al., 1986) using three different solvent systems. The three systems were (1) methanol:water:acetic acid:thiodiglycol (90:9:1:0.1) (2) water:acetonitrile:trifluoroacetic acid

Mention of a commercial product in this paper does not constitute an endorsement by the USDA.

(TFA) (60:40:0.1) and (3) 5% formic acid v/v, 15% TFA v/v,
1% NaCl w/v, 1N HCl (Bennett et al., 1978).

The resulting methanolic supernatant from solvent
system (1) was processed as previously described (Jaffe et
al., 1986) except that a Speed Vac™ (Savant) was used to
remove and/or concentrate the solvents. The filtered
sample was analyzed by HPLC method A (Supelcosil
LC-18DB/TFA).

The supernatant from solvent system (2) was filtered
through a Millex-HV filter (Millipore) and analyzed on 4
I-125 (Waters) HP-SEC columns eluted with 40% acetonitrile
containing 0.1% TFA at 1.0 ml/min. Effluent was monitored
at 214 nm. The columns were calibrated with 12 standards
ranging in mol. wt. from 556-67,000. A least-square plot
was drawn of log mol. wt. vs. retention time according to
published methods (Bennett et al., 1983).

The supernatant from solvent system (3) was adsorbed
onto a C-18 Sep-Pak™ (Waters) prepared by washing with
acetonitrile followed by 0.1% TFA. The Sep-Pak was washed
with 3 ml 0.1% TFA to remove the buffer and the peptide
fraction eluted with 2 ml 80% acetonitrile containing 0.1%
TFA (Bennett et al., 1978; Bohlen et al., 1980). The
resulting sample was concentrated in the Speed Vac, its
volume adjusted to ca. 500 µl with 0.1% TFA, and extracted
two or three times with ethyl acetate to remove lipids.
Residual ethyl acetate was removed under vacuum in the
Speed Vac and the sample filtered through a Millex HV
filter prior to analysis by Method A (Supelcosil
LC-18DB/TFA).

Elution profiles with superimposed PBAN activity of
BR-SOG-CC samples processed in solvent systems (1-3) are
shown in Fig. 1-3. In both Fig. 1 and 3, three distinct
peaks of PBAN activity (PBAN-I, II and III, Raina et al.,
this volume) were observed, with the first peak (PBAN-I)
clearly predominating in solvent system (3) workup. In the
HP-SEC workup (system 2), one major peak of PBAN activity
was observed corresponding to a mol. wt. of 4200.

Total recovery of PBAN activity for the solvent system
(1-3) workups expressed as a percentage of the activity of
one BR-SOG-CC homogenate in insect ringer is shown in

Table 1. Solvent system (3) gave the best recovery of biological activity, by far, with almost twice as much activity as the standard saline homogenate, probably because of the strong inhibition of enzymatic degradation by the low pH of this solvent medium.

PBAN-I was isolated and purified from fractions eluting between 33-35 min (Fig. 2) by the following sequence: HP-SEC ⟶ RP-HPLC (Method B) ⟶ RP-HPLC (Method C). PBAN-I was also purified from fractions eluting between 44-46 min resulting from the solvent system (3) workup by the following sequence: RP-HPLC (Method A) ⟶ RP-HPLC (Method B) ⟶ RP-HPLC (Method C). This purified PBAN-I (Fig. 4) was shown to be identical in chromatographic behavior to PBAN-I isolated by HP-SEC and somewhat purer probably because of the absence of a lipid extraction step in the HP-SEC workup. Amino acid analysis of purified PBAN-I by the pico-tag method (Waters) indicated the following amino acid composition: Ala (2), Asx (2), Gly (5), Glx (3), His (1), Ile (1), Pro (1), Ser (5), Thr (2), and Val (1) corresponding to a mol. wt. of 2126.

PBAN-I isolated by us from the BR-SOG-CC is identical in chromatographic behavior to PBAN isolated from a pure CC preparation (Raina et al., this volume), suggesting that PBAN-I is the actual released form of the hormone. The other active compounds observed in the solvent system (1) and (3) workups may be enzymatic or chemical degradation products of PBAN (e.g. Gln ⟶ Glu or Asn ⟶ Asp) or possibly the prohormone of PBAN which occurs in the SOG (Raina et al., unpublished) or possibly a dimeric form of PBAN, hence explaining the approximate doubling of the mol. wt. calculated from the amino acid analysis (2126) compared to the mol. wt. observed by HP-SEC (4200). All of these

Table 1. Recovery[a] of H. zea pheromonotropic activity after workup by solvent systems 1-3[b] and HPLC

Solvent System	% Recovery
1	33.5
2	39.4
3	176.0

[a]Expressed as a percentage of the activity of one BR-SOG-CC complex of H. zea homogenized in insect ringer.
[b]See text.

speculations await confirmation by determination of the primary structure of all three PBANs.

The authors gratefully acknowledge the assistance of E. Brown and S. Vohra in the maintenance and dissection of the insects and B. A. Fraser (FDA, Bethesda, MD) for the pico-tag amino acid analysis.

REFERENCES

Bennett, H. P. J., Hudson, A. M., Kelly, L., McMartin, C. and Purdon, G. E. (1978) A rapid method, using octadecasilyl-silica, for the extraction of certain peptides from tissues. Biochem. J. 175, 1139-114.

Bohlen, P., Castillo, F., Ling, N. and Guillemic, R. (1980) Purification of peptides: an efficient procedure for the separation of peptides from amino acids and salt. Int. J. Peptide Protein Res. 16, 306-310.

Jaffe, H., Raina, A. K., Riley, C. T., Fraser, B. A., Holman, G. M., Wagner, R. M., Ridgway, R. L. and Hayes, D. K. (1986) Isolation and primary structure of a peptide from the corpora cardiaca of Heliothis zea with adipokinetic activity. Biochem. Biophys. Res. Commun. 135, 622-628.

Raina, A. K. and Klun, J. A. (1984) Brain factor control of sex pheromone production in the female corn earworm moth. Science 225, 531-533.

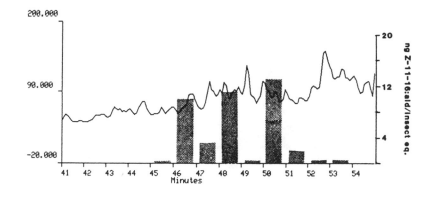

Fig. 1. Partial elution profile using HPLC Method A
obtained from ca. 550 BR-SOG-CC of H. zea processed using
solvent system (1) (0.220 A_{214} full scale) with
superimposed biological activity/insect eq. in ng
Z-11-16:aldehyde.

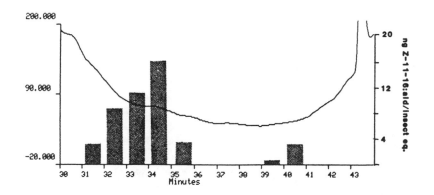

Fig. 2. Partial elution profile using HP-SEC obtained from
ca. 100 BR-SOG-CC of H. zea processed using solvent system
(2) (0.220 A_{214} full scale) with superimposed
biological activity/insect eq. in ng Z-11-16:aldehyde.

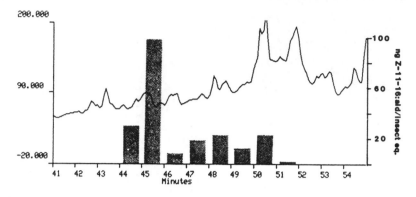

Fig. 3. Partial elution profile using HPLC Method A
obtained from ca. 400 BR-SOG-CC of H. zea processed using
solvent system (3) (0.220 A_{214} full scale) with
superimposed biological activity/insect eq. in ng
Z-11-16:aldehyde.

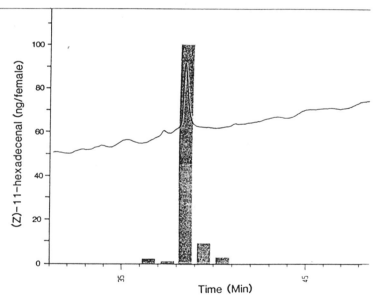

Fig. 4. Partial elution profile using HPLC Method C of
PBAN-I purified using solvent system (3) and the following
sequence: RP-HPLC (Method A) ⟶ RP-HPLC (Method B) ⟶
RP-HPLC (Method C) (0.030 A_{215} full scale).
Biological activity in ng Z-11-16:aldehyde for 1/80 of each
1-min fraction.

NOVEL NEUROPEPTIDES FROM THE CORPUS CARDIACUM OF A CRICKET AND A COCKROACH

Gerd Gäde

Institüt für Zoologie IV, Universität

Düsseldorf, D-4000 Düsseldorf 1, F.R.G.

The insect corpus cardiacum contains various neuropeptides known to influence energy metabolism, e.g. the so-called hypertrehalosaemic or phosphorylase-activating and the adipokinetic or hyperlipaemic peptides. These peptidergic neurosecretions belong to the AKH/RPCH-family, since they are all structurally related to the locust decapeptide adipokinetic hormone I (AKH I) and the crustacean red pigment-concentrating hormone (RPCH).

In 1984 Gäde screened extracts of corpora cardiaca (CC) from a wide range of insect species for their adipokinetic and hypertrehalosaemic activity in suitable test animals, migratory locusts and American cockroaches, respectively. Having established that the species under investigation contained such active compounds in their CC, he separated some of these extracts on reversed-phase high-performance liquid chromatography (HPLC), and compared the retention times of the active fractions with those of AKH I and RPCH. The powerful resolution of HPLC allowed discrimination between different peptides with adipokinetic and/or hyper-trehalosaemic activity and AKH I and RPCH in various species. In the meantime some of the peptides (locust adipokinetic hormones II, American cockroach hypertrehalo-saemic hormones I and II, stick insect hypertrehalosaemic factor II and moth adipokinetic hormone) have been isolated and the structures elucidated by different research groups. In the light of these recent findings, and our previous success at identifying new neuropeptides of the AKH/RPCH-family by a combination of bioassays and HPLC, the present study was undertaken to examine such peptides in some insect species not previously investigated.

It was found that methanolic extracts of CC (0.25 gland pairs) from the cricket Gryllus bimaculatus and the two cockroach species Nauphoeta cinerea and Leucophaea maderae had a hypertrehalosaemic effect when injected into the American cockroach (increase of 12, 18 and 26 mg/ml, respectively) and caused hyperlipaemia in acceptor migratory locusts (increase of 33, 8 and 14 mg/ml, respectively). In addition, we tested the ability of those CC extracts to increase lipids and/or carbohydrates in the donor species itself. It was not possible to perform these experiments in N. cinerea due to rapid clotting of its haemolymph. Injection of own CC extract into G. bimaculatus resulted in the elevation of blood lipids (16 mg/ml) when a high dose of one gland-equivalent was used; however, the haemolymph carbohydrate concentration was not affected. On the other hand, its own CC material (0.25 gland-equivalents) caused hypertrehalosaemia (17 mg/ml) in L. maderae, but the levels of blood lipids were unaffected.

HPLC was performed on a reversed-phase Nucleosil C-18 column. The aqueous solvent was 0.11% trifluoroacetic acid and the organic solvent was 0.1% trifluoroacetic acid in 60% acetonitrile. The solvents were applied as a linear gradient (25 to 80% in 45 min; flow rate 1ml/min) and the eluent was monitored at 210 nm. Methanolic CC extracts from both cockroach species showed numerous small absorbance peaks and one prominent one with a retention time of 23.6 min. Bioassays in American cockroaches and/or in L. maderae revealed that the hypertrehalosaemic activity was associated only with the material eluting at 23.6 min. From the known members of AKH/RPCH-peptide family (see introduction) only the hypertrehalosaemic hormone I (trade name: MI) of the American cockroach had a retention time similar to that of the CC from N. cinerea and L. maderae. However, the peptide material from the cockroaches under study eluted slightly ahead of MI (retention time: 24.6 min). A methanolic CC extract from the cricket showed three main absorbance peaks. Only the first absorbance peak, with a retention time of 25.1 min, caused hyperlipaemia in G. bimaculatus itself and hypertrehalosaemia in the American cockroach. The cricket adipokinetic material thus eluted just after MI and about 1.5 min later than the newly-found hypertrehalosaemic factors of N. cinerea and L. maderae.

In summary, in both of the cockroach species studied, N. cinerea and L. maderae, hypertrehalosaemic factors are found which have different retention times from any of the known members of the AKH/RPCH-family. The novel active cockroach material elutes close to, but just ahead of, the hypertrehalosaemic hormone I (MI) of the American cockroach and is clearly not identifical to it. A second new peptide can be purified from the CC of the cricket, G. bimaculatus, that has an adipokinetic effect in the donor species it- self. This novel active material elutes just after MI and, consequently, much later than the hypertrehalosaemic fac- tors of the other two cockroaches. The retention time of the G. bimaculatus adipokinetic factor is also different from any other known member of the AKH/RPCH-peptide family. Surprisingly, both cockroach species under study as well as the cricket species apparently contain only one active hypertrehalosaemic/adipokinetic factor, whereas locusts, American cockroaches and stick insects have been shown to contain two active peptides (see, for example, Gäde, 1986).

The author acknowledges financial support by a grant (Ga 241/6-1) and a Heisenberg Fellowship (Ga 241/5-1) awarded by the Deutsche Forschungsgemeinschaft.

REFERENCES

Gäde G. (1984) Adipokinetic and hyperglycaemic factors of different insect species: separation with high performance liquid chromatography. J. Insect Physiol. 30: 729-736.

Gäde G. (1986) Relative hypertrehalosaemic activities of naturally occurring neuropeptides from the AKH/RPCH family. Z. Naturforsch. 41c: 315-320.

CHARACTERIZATION OF VERTEBRATE PEPTIDE HORMONE-LIKE MATERIALS IN THE AMERI-

CAN COCKROACH: DIFFERENT METHODS EMPLOYED TO INVESTIGATE THESE SUBSTANCES

AND THEIR RESULTS

Peter Verhaert, Roger Huybrechts, Dominique Schols, Jozef
Van den Broeck, Arnold De Loof and Frans Vandesande

Zoological Institute of the University

Naamsestraat 59, B-3000 Leuven, Belgium

INTRODUCTION

During the last few years many investigations have reported the occurrence of substances in insects that resemble vertebrate (neuro)peptide hormones. Our particular contribution to this study concerns the exploration of
Periplaneta americana L. (Blattidae) for suchlike products. This paper will
critically comment three complementary approaches which, in our opinion, may
lead to the ultimate characterization of these substances.

THE PHYSIOLOGICAL APPROACH: BIOASSAYS

To detect homologies between vertebrate peptide hormones and invertebrate factors at the level of biological activity, we make use of bioassays.
Homogenates/extracts of interesting cockroach tissues are examined in well
established vertebrate peptide bioassays. E.g. by use of a mouse Leydig cell
and a frog skin melanophore bioassay we could demonstrate lutropin- and me-
lanotropin-like biological activity in respectively immunopositive cockroach
corpora cardiaca. Complementarily, vertebrate peptides can be investigated
for a potential activity in cockroach assays. In this regard we detected a
plain stimulatory effect of methionine-enkephalin and FMRFamide on roach gut
contraction. These results unequivocally show the binding of some factors
contained in blattarian neurohormonal tissue to vertebrate neuropeptide receptors and the binding of (vertebrate) peptides to cockroach receptors respectively. Of course they do not give any information about the possible role
or physiological function of the vertebrate peptide-like substances in the
insect itself, neither do they answer the fundamental question as to the
real structure of these chemicals. Both types of biological assays, however,
could indeed prove to be valuable as they could be very helpful in the extraction (and subsequent purification) of interesting active cockroach sub-

stances. Nevertheless we do not prefer these "classical" isolation methods. Even where they have already proved to be possible (the physiological functions of most of the recently discovered insect peptides (especially those of the vertebrate-type) still remain entirely unknown, which implies that a needful (bio)assay is not yet available), they often proved to be very laborious and not always quite successful (material loss or breakdown due to peptide instability, etc.); exeption made for some small peptides (e.g. AKH).

THE IMMUNOLGICAL APPROACH

The use of vertebrate peptide antibodies allows us to detect immunochemically similar materials in roach tissues. Apparently immunoperoxidase histochemistry, both single and multiple labelings, is an elegant method to achieve this (e.g. Verhaert and De Loof, 1985), proving to be fast, easy to perform and sensitive. At the present antisera to about 50 different neuropeptides (see Table 1), mainly all of vertebrate origin, have been demonstrated as labeling various cell bodies and their axonal processes, both in the blattarian nervous system and intestine. It is evident that one has to realize that immunological studies only reveal a relatedness at the antigenic determinant level of some tissue compound(s) to some antibodies present in the applied antiserum. Results thus obtained are therefore notoriously difficult to interpret and not always very informative as to the true identity of the detected molecules with respect to the immunizing antigen. This especially when antisera (i.e. polyclonal (i.e. a mixture of various) antibodies) that are raised against larger molecules such as holoprotein hormones (e.g. somatotropin, prolactin) containing a lot of antigenic sites are used. Notwithstanding this we believe that, directly in proportion to the degree of characterization of the antiserum, immunological results may have a substantial significance as a primary step in the identification process of "new" insect peptide-like chemicals. Of course the use of monoclonal antibodies (McAbs) would provide more valuable indications with respect to the nature of the products demonstrated. Therefore available vertebrate neuropeptide McAbs are being as well included in our tests. These probes, however, do not often appear to be very suitable for the immunocytochemical detection of the respective "tissue fixed" antigen (which in the insect will most probably not be identical to the vertebrate immunogen). Some positive results, however, have been obtained with monoclonal substance P (Verhaert and De Loof, 1985), human gonadotropin and somatotropin antibodies. A direct way to produce McAbs that recognize epitopes common to vertbrate peptide hormones and insect material has been elaborated in our laboratory and is currently being used. It principally consists of immunizing mice with vertebrate peptide antigens and screening the antibody containing hybridoma supernatants immunohistochemically on insect tissues. Furthermore, because of the theoretically unlimited quantities that can be produced, McAbs can provide a sufficient amount of efficient purification tools in immunoaffinity chromatography. An isolation procedure of interesting invertebrate neuropeptide-like materials based upon

immunizing antigen	NS	IC
hFSH (follicle stimulating hormone)	+	−
hLH (luteinizing hormone)	+	−
GnRH (gonadotropin releasing hormone)	+	+
hPRL (prolactin)	+	?
fPRL	+	0
hGH (growth hormone)	+	−
bGH	+	0
fGH	+	0
hGRF (growth hormone releasing factor)	+	0
SRIF (somatostatin)	+	0
SP (Substance P)	+	+
BB (Bombesin)/GRP (gastrin releasing peptide)	+	0
Gastrin	+	+
CCK (cholecystokinin)	+	+
INS (insulin)	+	+
Glucagon	+	?
VIP (vasoactive intestinal polypeptide)	+	0
PHI (peptide histidine isoleucine)	+	0
bPP (pancreatic polypeptide)	+	+
NPY (neuropeptide tyrosine)	−	0
FMRFa (phenylalanine-methionine-arginine-phenylalanine-amide)	+	+
VT (vasotocin)	+	0
AVP (arginine vasopressin)	+	+
OT (oxytocin)	+	?
IT (isotocin)	?	0
NPI (neurophysin I)	+	+
NPII (neurophysin II)	+	?
met-ENK (methionine-enkephalin)	+	+
leu-ENK (leucine-enkephalin)	+	+
α-END (α-endorphin)	+	+
β-END (β-endorphin)	−	0
α-MSH (α-melanocyte stimulating hormone)	+	+
β-MSH (β-melanocyte stimulating hormone)	?	0
γ₃-MSH (γ₃-melanocyte stimulating hormone)	+	0
ACTH (corticotropin) sequence 01-24	+	+
sequence 11-24	+	+
CRF (corticotropin releasing factor)	+	+
NT (neurotensin)	+	−
PTH (parathormone)	?	0
CGRP (calcitonin gene related peptide)	+	0
ANF (atrial natriuretic factor (a cardiodilatin sequence))	+	0
AKH (adipokinetic hormone)	+	0
TSH (thyroid stimulating hormone)	−	0
TRH (thyrotropin releasing hormone)	−	0

Table 1. List of peptide antisera yielding positive staining reactions in cockroach nervous system (NS) and intestinal epithelium cells (IC) (Abbreviations: h, human; f, fish (salmonid); b, bovine; +, clear immunopositive labeling; ?, weak or doubtful labeling; −, not detectable immunostaining; 0, not tested)

their immunochemical affinity towards vertebrate peptide antibodies does not need the availability of an appropriate bioassay, unlike the tedious and often frustrating conventional purification methods. Therefore it may indeed comprise a realistic and competitive alternative to these classical isolation procedures which, as stated above, are not held to be attractive by us. In this view our research group has produced, by immunization with crude insect tissue homogenates, a McAb to a factor contained in locust corpus cardiacum glandular lobes and also to roach brain and corpus cardiacum material. Further characterization of this antibody (which, as a direct consequence of the immunization procedure, may in fact well be directed against a non-peptide factor) is in progress. In brief, at least one sure and important conclusion may be drawn from all the present immunological data: the insect systems under investigation are astonishingly complex.

THE MOLECULAR BIOLOGICAL APPROACH

The introduction of <u>in vitro</u> recombinant DNA technology/genetic engineering in this research field originated in the idea that cloning of insect (cockroach) nucleic acids encoding for interesting peptides might allow huge amounts of pure insect material to be produced and might relatively easily lead to their identification (via nucleic acid sequence determination). A

difficulty to overcome is the problem of selection of the nucleic acid sequences of interest. The most direct way to achieve this is to construct a
cDNA library of various insect tissues in an expression vector system (λ-
phage derived genome) and to screen for vertebrate peptide-like gene products (e.g. immunochemically). A major problem that one faces here is the
rare occurrence of neuropeptide messengers (mRNAs) requiring a great expression efficiency of the cloning system. A screening method based on nucleic
acid hybridization with (radiolabeled or biotin-labeled) probes may comprise
a feasible alternative, particularly when one avails oneself of a common
non-expression cDNA library. If part of the amino acid sequence of the peptide looked for is known custom made synthetic oligonucleotide probes may be
used in STRINGENT HYBRIDIZATION tests, but, unfortunately this technology is
extremely expensive. To assess whether available vertebrate peptide hormone
cDNAs might be applied as probes in RELAXED/PERMISSIVE CROSS-HYBRIDIZATION
tests (which would be relatively cheap and rather fast), we examined if and
under which conditions certain of these vertebrate neuropeptide cDNAs could
specifically hybridize to various cockroach tissue RNA populations immobilized on nitrocellulose. Our preliminary results from dot blot hybridizations
of brain and midgut RNAs with radiolabeled cDNAs for vertebrate holoprotein
hormones (somatotropins and prolactins) are quite promising (Verhaert et al.,
1985), but tests with biotin marked insulin-, cholecystokinin-, neuropeptide
Y- and oxytocin-cDNAs revealed that this type of test might possess a considerable risk for false positivities. Concurrently, proper controls, i.e. nor-
thern blot analysis and in situ hybridizations are being elaborated at this
moment.

CONCLUSION

In our belief, the three diverse approaches here discussed are all quite
complementary one to another and together they undoubtedly will provide ca-
pital data for the identification of vertebrate-type peptides in the American
cockroach and in insects (invertebrates) in general. Especially molecular
cloning techniques (and probably also affinity chromatography with suitable
monoclonal antibodies) are considered to be very powerful to alleviate the
need for obtaining enough pure invertbrate peptide hormone-like material to
finally elucidate and completely determine their primary amino acid structure.

ACKNOWLEDGEMENTS

The authors are indebted to "Vlaamse Wetenschappelijke Stichting",

"Vlaamse Leergangen", the Belgian "Queen Elizabeth's Foundation", "N.F.W.O." and "I.W.O.N.L." for financial support. P.V. and J.V.B. are recipients of a "N.F.W.O.-Aspirant"-fellowship.

REFERENCES

Verhaert P., Cardoen J., Vanden Broeck J., De Loof A., Belayew A., Lecomte C. and Martial J. (1985) Vertebrate cDNAs to peptide hormones hybridize to insect RNA dot blots. Arch. Int. Physiol. Biochim. 93, B121-B123.

Verhaert P. and De Loof A. (1985) Substance P-like immunoreactivity in the central nervous system of the blattarian insect Periplaneta americana L. revealed by a monoclonal antibody. Histochemistry 83, 501-507.

NEUROCHEMICAL AND BEHAVIORAL EFFECTS OF BIOGENIC AMINE DEPLETERS IN THE BLOW FLY, PHORMIA REGINA

Brookhart, Gary L., Leland C. Sudlow, Robert S. Edgecomb, and Larry L. Murdock

Department of Entomology, Purdue University

West Lafayette, IN 47906

INTRODUCTION

Drugs can have profound effects on blow fly feeding behavior. Octopamine agonists cause flies to consume abnormally large amounts of aqueous sucrose (Long and Murdock, 1983). Amphetamine (AMP) and related drugs depress blow fly responsiveness to sucrose solutions (Long et al., 1986). Responsiveness of flies can be quantitatively estimated as the mean acceptance threshold (MAT), which is the log10 of the minimum concentration of sucrose to which the average fly in a test group will respond (Sudlow, 1985). Recent studies employing HPLC with electrochemical detection have revealed that AMP and reserpine deplete CNS aromatic biogenic amines, octopamine (OA), dopamine (DA), and 5-hydroxytryptamine (5-HT), from the blow fly (Brookhart et al., in press). The time course and degree of amine depletion correlated well with the drug-induced changes in MAT. In the present paper, the neurochemical and behavioral effects of selected drugs related to AMP and of tetrabenazine are examined to further test the hypothesis that depletion of biogenic amines from the CNS results in an elevated MAT.

PHENYLETHYLAMINE DRUGS

Fenfluramine (FEN) and AMP had similar effects on MAT at 30 min after injection (Fig. 1a). FEN and AMP depleted

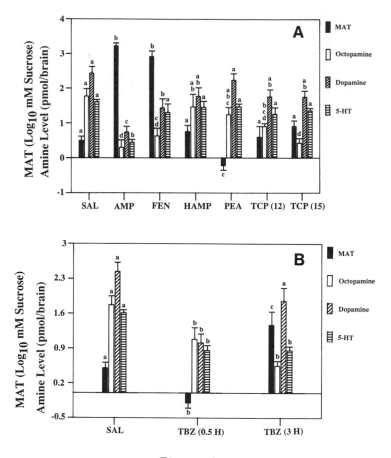

Figure 1

Effects of AMP congeners and tetrabenazine on MAT and brain
biogenic amines. Panel A depicts the effects of AMP (12 μ
g/fly), FEN (8 μg/fly), HAMP (12 μg/fly), PEA (12 μg/fly),
TCP (12 and 15 μg/fly), or saline. MAT and brain levels
of OA, DA and 5–HT were determined 30 min post–injection.
The effects of TBZ (10 μg/fly) were determined at 30 min
and 3 h (B) post–injection. The MAT values and brain
levels of amines are the average of at least three
determinations. T's at the top of each bar indicate ± SEM.
Significant differences among the means were determined by
Duncan's Multiple–Range Test (p < 0.05). For each
variable, bars with the same letter do not differ
significantly (p < 0.05).

brain OA to similar levels. However, unlike AMP, FEN did not affect brain DA or 5-HT. These results suggest that OA plays a key role in governing blow fly responsiveness to food stimuli.

MAT was not significantly affected in flies injected with p-hydroxyamphetamine (HAMP) (Fig. 1a). Brain levels of OA, DA, and 5-HT were also unaffected by this drug. These results are consistent with the amine depletion hypothesis.

Flies injected with β-phenylethylamine (PEA) exhibited a significant decrease in MAT, an effect opposite to that observed following treatment with AMP or FEN (Fig. 1a). Brain levels of OA, DA, and 5-HT were not affected by PEA. In vitro studies have demonstrated PEA to be a partial agonist of OA (cf. Bodnaryk, 1982). An agonist action may explain the increased responsiveness of PEA-treated flies.

Tranylcypromine (TCP) significantly depleted OA from the fly brain without affecting the MAT (Fig. 1a). Brain DA and 5-HT were not significantly affected. Increasing the dose of TCP still had no effect on MAT, even though OA was depleted to levels comparable to those found in the brains of AMP- and FEN-treated flies. Taken at face value, these results seem to contradict the amine depletion hypothesis. However, TCP may have another pharmacological action that counters the effect of OA depletion. Evidence for this is our recent preliminary observation that injection of fireflies (Photinus pyralis) with 10 μg TCP stimulated the light organs to light.

TETRABENAZINE

The effect of tetrabenazine (TBZ) on MAT was biphasic (Fig. 1b). Shortly after injection (30 min), the threshold of TBZ-treated flies was significantly lower than that of saline-injected flies. At this time, brain OA, DA, and 5-HT were significantly depleted. However, by 3 h the MAT of TBZ-treated flies was significantly elevated (Fig. 1b), although not to the extent observed in AMP-treated flies. At 3 h, brain OA was depleted to a level comparable to that observed in AMP-treated flies. The mode of action of TBZ is obviously complex. Its biphasic action is presumably the result of some combination of amine release, amine

depletion, and possibly agonist or antagonist action
(Reches et al., 1983). Octopamine release during the early
phase of TBZ action may partially explain threshold
reduction. Subsequent octopamine depletion may be
responsible for the elevation in threshold seen at later
times.

While the complex actions of TCP and TBZ are not
simply explained, the results of the present study are
generally compatible with the hypothesis that biogenic
amines play important roles in blow fly feeding behavior.
The neurochemical and behavioral effects of FEN (selective
depletion of OA concomitant with increased MAT) and PEA (OA
agonist leading to decreased MAT) are consistent with an
important role for octopamine in the regulation of blow fly
responsiveness to food stimuli.

REFERENCES

Bodnaryk R. P. (1982) Biogenic amine-sensitive adenylate
  cyclases in insects. Insect Biochem. 12, 1-6.

Brookhart G. L., Edgecomb R. S., and Murdock, L. L. (in
  press) Amphetamine and reserpine deplete brain biogenic
  amines and alter blow fly feeding behavior. J. Neurochem.

Long T. F., Edgecomb R. S., and Murdock, L. L. (1986)
  Effects of substituted phenylethylamines on blow fly
  feeding behavior. Comp. Biochem. Physiol. 83C, 201-209.

Long T.F. and Murdock L. L. (1983) Stimulation of blow fly
  feeding behavior by octopaminergic drugs. Proc. Natl.
  Acad. Sci. U.S.A. 80, 4159-4163.

Reches A., Burke R. E., Kuhn, C. M., Hassan M. N.,
  Jackson, V. R., and Fahn, S. (1983) TBZ, an
  amine-depleting drug, also blocks DA receptors in the rat
  brain. J. Pharmacol. Exp. Ther. 225, 515-521.

Sudlow L. C. (1985) Some factors which affect tarsal and
  labellar responsiveness in the black blowfly. Masters
  Thesis, Purdue University, West Lafayette, Indiana.

PROTECTIVE FUNCTION OF GANGLIOSIDES AGAINST THE DISRUPTIVE ACTION OF DDT AND DELTAMETHRIN

Joseph T. Chang & P. Gao

Department of Biology, Peking Univ.

Beijing, Peoples' Republic of China

The idea that gangliosides might influence ion permeability and flux in neural membranes was fostered by
earlier reports. They purported that gangliosides maintain
the excitability of nervous membrane in vitro and so offer
a protective function against the disruptive action of pro-
tamine and low temperature (Marks, 1959; McIlwain, 1961).
Though other findings cast doubt on the physiological significance of this ganglioside effect (Evans, 1967; Yogeewaran,
1973), in recent years, the idea of the role of gangliosides
in ion transport has been revived with evidence that gangliosides may influence surface potential.

Since most of the synthetic insecticides developed are
nerve poisons, and most of them (such as DDT and deltamethrin)
seem to be associated with ion permeability, it seems worth-
while to study the relationship of gangliosides and these
insecticides in order to see whether they have any bearing
on the mode of action of these insecticides.

Solution of 2.8×10^{-6}M DDT was perfused into cockroach
metapodium and spontaneous responses were recorded. Short
impulse trains appeared at 8 minutes in place of random
spikes, which were easy to recognize and subsequently called
the bursting time. Afterwards the trains increased in intensity and frequency, and repetitive discharges occurred.
Tetanic contraction of the muscle followed at 20 minutes.
At 40 minutes, tetanic contraction of the muscle was in high
frequency and persisted until 60 minutes, alternating with
blocking of nerve conduction. At 120 minutes, the block

was almost complete. When a solution of 2.8 x 10^{-6}M DDT
containing 6.6 x 10^{-7}M gangliosides was perfused into the
cockroach metapodium, trains did not appear until 30 minutes.
Repetitive discharge followed at various intervals. At 75
minutes, tetanic contraction of the muscle occurred. At 85
minutes, repetitive discharge decreased in frequency and
normal axon spikes were recorded. Normal spontaneous activity
reappeared at 120 minutes.

Similarly, 2.0 x 10^{-6}M deltamethrin solution was perfused into the cockroach metapodium. Nerve trains occurred
at 4 minutes, a complex high frequency burst of impulses
followed at 7 minutes and repetitive discharge came about at
12 minutes, lasting for 3 minutes and conduction was then
completely blocked. But when gangliosides were added (3.3
x 10^{-5}M) and perfused, nerve trains appeared at 42 minutes,
and the blocking did not occur until 54 minutes.

The results apparently showed that ganlgiosides offer
protection against the disruptive action of DDT and deltamethrin.

Early hypothesis on the possible function of gangliosides
implicated these lipids in special cation binding in transport and release mechanisms in biological membranes, particularly in the CNS (Triggle, 1971; Caputto et al, 1977).
Gangliosides have an affinity for divalent cations, especially
calcium, which binds more strongly to gangliosides than to
phospholipid because of the formation of chelate structure.

Calcium is vital to the proper functioning of all excitable tissues, and in particular, it has a stabilizing
effect on the nerve axon. High concentrations of external
calcium are well known to antagonize the DDT and pyrethroid
induced excitation (Gammon, 1980; Gordon & Welsh, 1948;
Matsumura & Narahashi, 1971). Conversely, reducing the extracellular calcium concentration around the arthropod nerve
axons elicit repetitive discharge reminiscent of DDT (O'Brien,
1966). Calcium is known to be intimately involved in conductance gating; specifically, extracellular calcium controls
the degree of coupling between membrane potential and Na^+
conductance.

There is proof that exogenous gangliosides can be taken
up by nerve membrane. In vitro incubations of cells in the
presence of exogenous gangliosides results in a saturated uptake of these cells (Keenan et al, 1974; Moss et al, 1976).

We therefore postulate that the protective function of
gangliosides against the two axonic poisons, DDT and deltamethrin, is due to their ability to bind with calcium. When
exogenous gangliosides are added, they are taken up by the

axonic membrane, and the binding of gangliosides to calcium
causes an increase of extracellular calcium concentration,
the result of which is the Na$^+$ gate becomes protected from
propping open by DDT or deltamethrin. It is now well
assumed that the disruptive action of DDT and deltamethrin
is the delayed closing of the Na$^+$ gate, due to the insertion
of the insecticide molecule into the Na$^+$ channel, resulting
in the increase of negative after-potential and repetitive
discharge. Gangliosides in this case antagonize the opening
of the Na$^+$ gate induced by DDT and deltamethrin, thus also
antagonizing their induced excitation.

Literature Cited

Caputto, R., Maccioni, A. H. R., & Caputto, B. L. (1977)
    Activation of deoxycholate adenosine triphosphate by
    ganglioside and asialoganglioside preparation. Biochem.
    Biophys. Res. Comm. 74: 1046.
Evans, W. H. & McIlwain (1967) Excitability and ion content
    of cerebral tissues treated with alkylation agents,
    tetanus toxin, or a neuraminidase. J. Neurochem. 14: 35.
Gammon, D. W. (1980) Pyrethroid resistance in a strain of
    Spodoptera littoralis is correlated with decreased sensi-
    tivity of CNS in vitro. Pestic. Biochem. Physiol. 13: 53.
Gordon, H. T. & Welsh, J. H. (1948) The role of ions in
    axon surface reactions to toxic organic compounds. J.
    Cell Comp. Physiol. 31: 395.
Keenan, T. W., Franke, W. W. & Wiegandt, H. (1974) Ganglio-
    side accumulation by transformed murine fibroblasts (3T3)
    cells and canine erythorcytes. Hoppe Seylers Z. Physiol.
    Chem. 355: 1543.
Marks, N. & McIlwain, H. (1959) Loss of excitability in
    isolated cerebral tissues, and its restoration by
    naturally occurring materials. Biochem. J. 73: 401.
Matsumura, F. & Narahashi, T. (1971) ATPase inhibition and
    electrophysiological change caused by DDT and related
    neuroactive agents in lobster nerves. Biochem. Pharmacol.
    20: 825.
McIlwain, H. (1961) Characterization of naturally occurring
    materials which restore excitability to isolated cerebral
    tissues. Biochem. J. 78: 24.
Moss, J., Fishman, P. H., Magniello, V. C., Vaughan, M. &
    Brady, R. O. (1976) Functional incorporation of ganglio-
    sides into intact cells. Proc. Natl. Acad. Sci. 73: 1034.
Narahashi, T. (1971) Effects of insecticides on excitable
    tissues. Ady Insect Physiol. 8: 1.

O'Brien, R. D.   (1966) Mode of action of insecticides.
   Ann. Rev. Ent. 11: 369.
Triggle, D. J.   (1971) Neurotransmitter-receptor
   interactions. Acad. Press N.Y.
Yogeeswaran, G., Murray, R. K., Pearson, M. J., Sanwal, B.
   D., McMorris, F. A. & Ruddle, F. H.   (1973) Glycosphin-
   golipids of clonal lines of mouse neuroblastoma and
   neuroblastoma X L cell hybrids. J. Biol. Chem. 248: 1231.

HYDROPHOBIC AND HYDROPHILIC COMPONENTS OF INSECT

ACETYLCHOLINESTERASE

Martine Arpagaus° and Jean-Pierre Toutant°°

°Laboratoire de Zoologie, ENS, 46 rue d'Ulm,
75005 Paris and °°Physiologie Animale, INRA,
9 place Viala, 34000 Montpellier, France.

A lot of studies have been devoted to the elucidation
of the molecular structure of Insect acetylcholinesterase
(AChE, EC 3117, see Eldefrawi, 1985 for a review). In
vertebrates, this enzyme presents a complex polymorphism
due first to the presence of globular and asymmetric
collagen-tailed forms. In addition, it is possible to
characterize, within a single globular form, molecular
variants which either interact with nondenaturing
detergents (hydrophobic component) or are detergent-
insensitive (hydrophilic component, see Massoulié and Bon,
1982). This property is of importance because it is
supposed to reflect the cellular localization of the
enzyme (intracellular, membrane-associated or secreted).
We studied the hydrophobic interactions and the
structure of AChE extracted from Pieris brassicae adult
heads, using centrifugation analyses on sucrose gradients
and nondenaturing electrophoreses (see Arpagaus and
Toutant, 1985).

Pieris heads AChE was extracted in a high-salt (1M
NaCl) medium containing 1% Triton X100, and analyzed in
parallel on high-salt sucrose gradients containing 1% TX100
or devoid of detergent. Figure 1 shows that AChE sedimented
as a single peak (6.5S) in the presence of TX100. When the
extract was centrifuged in the absence of detergent the
enzyme was resolved into two components: one which
aggregated (fraction A, 80% of total AChE activity), and

one which sedimented at 7.3S (fraction B). We observed a
similar pattern of AChE activity in dissected brains of
Pieris: the two variants of AChE are therefore most
probably present in the central nervous system. A and B
fractions were isolated, dialyzed and centrifuged in the
presence of TX100 or in the absence of detergent. We
demonstrated that the molecules contained in fraction A
sedimented as a well-defined peak (6.5S) in the presence
of 1% TX100; this component thus interacts with detergent:
it was called for this reason hydrophobic AChE. On the
contrary the component contained in fraction B sedimented
at 7.3S in the presence or in the absence of TX100: this
component does not interact with detergent and was called
detergent-insensitive, hydrophilic AChE (Arpagaus and
Toutant, 1985).

   The two components A and B are not resolved by
centrifugation analyses in the presence of TX100: this is
due to their close S values (6.5 and 7.3S). In order to
improve the separation of these components, we used
10-oleyl-ether (Brij 96, Sigma) a detergent which interacts
with hydrophobic proteins forming molecular associations
of lesser density than Triton X100. Figure 2 shows that
the two components of Pieris AChE are now clearly resolved
when a TS extract is centrifuged in the presence of Brij
96. The hydrophobic component sediments at 3S whereas the
sedimentation of hydrophilic AChE is unchanged (7.3S).
The use of Brij 96 in sedimentation analyses may be
therefore a useful method to visualize, in a single step,
the different components of Insect AChE.

   Components A and B were further analyzed in non-
denaturing electrophoresis (Figure 3). In the direct high-
salt detergent extract (TS) two electromorphs presented
strikingly different mobilities: they correspond
respectively to the hydrophobic component (fraction A) and
to the hydrophilic enzyme (fraction B). When deoxycholate
(a negatively-charged detergent) was included with TX100
(neutral detergent) in the electrophoresis gel and buffer,
the mobility of component B was not altered, whereas the
mobility of component A was increased from 25% to 75% of
the migration of component B (Arpagaus and Toutant, 1985).
This confirms the hydrophobic and hydrophilic characters
of components A and B.

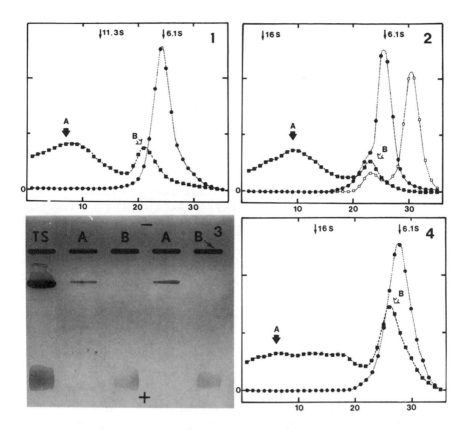

Figures 1, 2 and 4: Sedimentation analyses of AChE
activity present in TS extracts (1M NaCl, 1% Triton X100)
of Pieris heads (Figures 1 and 2) and Locusta brains
(Figure 4). Sucrose gradients (5-20%) contained 1M NaCl
and 1% Triton X100 (·●····●·), 0.5% Brij 96 (·O····O·,see
Figure 2) or no detergent (▬■▬ ▬ ▬■▬). Centrifugations were
run for 20h at 40,000 rpm at 4°C in SW41 rotors. The AChE
activity is plotted in arbitrary units (absorbance at 412
nm). E. coli β-galactosidase (16S) and alkaline phosphatase
(6.1S), and beef liver catalase (11.3S) were used as
internal sedimentation markers.

Figure 3: Components A (hydrophobic) and B
(hydrophilic) of Pieris AChE were analyzed in non-
denaturing PAGE (7.5% polyacrylamide gels containing 1%
Triton X100), in comparison with the total extract (TS).

In the hemolymph of Pieris, we observed the presence of only one detergent-insensitive component sedimenting at 7.3S, which probably corresponds to the hydrophilic component (7.3S) of brain AChE. The secretion of CNS enzyme may be thus the origin of hemolymph AChE.

Preliminary results obtained with dissected brains of Locusta migratoria (Orthoptera) and Carausius morosus (Phasmoptera) indicate that their AChE present also both hydrophobic and hydrophilic components (Figure 4). We suggest that this duality might be a general character of Insect AChE.

In conclusion, we suggest that the different components of Insect AChE, characterized by their direct interactions with detergents, correspond to distinct cellular locations in the nerve cells: membrane-bound for the hydrophobic component, cytoplasmic or secreted for the hydrophilic variant.

References

Arpagaus M. and Toutant J.-P.(1985) Polymorphism of acetylcholinesterase in adult Pieris brassicae heads. Evidence for detergent-insensitive and Triton X100-interacting forms. Neurochem. Int. 7,793-804.

Eldefrawi A.T. (1985) Acetylcholinesterases and anti cholinesterases. In Comprehensive Insect Physiology, Biochemistry and Pharmacology (Ed by Kerkut G. and Gilbert L.I.) vol. 12-1,115-130. Pergamon, Oxford.

Massoulié J. and Bon S. (1982) The molecular forms of cholinesterase and acetylcholinesterase in Vertebrates. Ann. Rev. Neurosci. 5,57-106.

BINDING OF SYNTHETIC INSECT PEPTIDE ANALOGS TO COMPONENTS OF INSECT MEMBRANE PREPARATIONS

D. K. Hayes, H. Jaffe, N. O. Morgan and
R. E. Redfern

Livestock Insects Laboratory, ARS, USDA,
Beltsville, MD 20705

Water-soluble neuropeptides serving as molecular messengers are believed to bind to specific receptor sites on the surfaces of their target cells (Alberts et al., 1983). Hormones in the adipokinetic hormone-red pigment concentrating hormone (AKH RPCH) family (Jaffe et al., 1986) of insect neuropeptides undoubtedly act at receptor sites on cells in the organs they affect. The present report summarizes studies to determine whether interactions exist between synthetic AKH-RPCH analogues and membrane preparations from the face fly, Musca autumnalis.

Face fly larvae (Beltsville colony) were reared on manure under LD 16:8, CW fluorescent illumination. Pupae were collected and held at LD 16:8, $25°C \pm 1°C$ until the abdomens of pharate pupae or of adults were dissected.

Abdomens from insects anesthetized with dry ice were placed immediately on dry ice until homogenized or until transferred to a SOLO® freezer at $-70°C$. Abdomens were homogenized in phosphate buffered saline (PBS) (0.2 g KCl, 0.2 g KH_2PO_4, 8.0 g NaCl and 2.16 g $Na_2HPO_4.7H_2O$ per l). Other additives are described in the legend for Table 1. When extracts were subjected to affinity

---

Mention of a commercial product in this paper does not constitute an endorsement by the USDA.

chromatography, 10 μl pepstatin (0.4 mg/ml) and 10 μl
leupeptin (1 mg/10 ml) were also added. For binding
studies the homogenate was centrifuged at about 10,000 x g
at 5°C in a microfuge and the pellet resuspended in the
homogenizing buffer and used as the crude membrane preparation. For affinity chromatography the solution used to
solubilize membranes contained 0.1% n-octyl-β-D-gluco-
pyranoside.

In what is a previously unreported technique for
isolating iodinated oligopeptide, the synthetic peptide,
Tyr-Leu-Asn-Phe-Thr-Pro-Asn-Trp-Gly-Thr-NH$_2$ was iodinated
by treating 20 μg of the peptide in 20 μl PBS with 30 μl
chloramine T (10 mg/ml in PBS) and 1 mci ^{125}I in 10
μl. After 5 min 190 μl of Na$_2$S$_2$O$_5$ (2.4 mg/ml PBS)
and 190 μl KI (10 mg/ml PBS) were added and the reaction
mixture was swirled and transferred to a C-18 Sep-Pak™ that
had been washed with three 3-ml portions of acetonitrile
and six 3-ml portions of distilled water. The loaded
Sep-Pak was washed with 2 ml PBS and five 1-ml portions of
distilled water and the iodinated peptide was eluted with
four 1-ml portions of acetonitrile and taken to dryness in
a Speed Vac™. The ^{125}I-labelled peptide was added to
the mixture, incubated with homogenates in an ice bath for
10 min and filtered through a Whatman 2.4 cm GFK glass
microfiber filter. Iodine-125 was counted on a crystal
scintillation counter.

Affinity gels were prepared using cyanogen bromide
activated Sepharose 4B described in Kohn and Wilchek
(1982). The activated columns were reacted with either 1%
albumin or a 1 mg/ml solution of the peptide pGlu-Leu-Ser-
Phe-Thr-Ser-Trp-Gly-Thr NH$_2$ in a ratio of 1 ml of
solution to 1 ml of activated Sepharose 4B. The conjugated
Sepharose was then treated with ethanolamine in a ratio of
one part beads to two parts ethanolamine, washed with PBS
and stored in PBS containing 0.1% sodium azide. The
membrane extract diluted 1:3 with PBS was applied and bound
material was eluted as in the legend for Fig. 1. The
eluates were concentrated using the Amicon Centricon™ -10
microconcentrator (10 K molecular weight cutoff) and
subjected to vertical sodium dodecyl sulfate polyacrylamide
gel electrophoresis (SDS-PAGE) followed by a silver stain.
In some cases protein was determined using a modified Lowry
technique (Lowry et al., 1951).

Table 1 shows binding of iodinated peptide to an insect homogenate. Because this conventional technique resulted in low binding, affinity columns were used to isolate proteins binding to non-iodinated peptide. Figure 1 illustrates the results obtained from such an experiment. From Fig. 1b-Lane 2, it is clear that the basic diethylamine has eluted proteins from the peptide affinity column of about 40 K, 65-70 K and 116 K daltons. These may have been bound selectively.

From these data we conclude only that materials obtained from the affinity column are proteins that bind and are released under appropriate conditions. However, the considerable concentration of the high molecular weight bands in Lanes 2 and 3 of Fig. 1b relative to their near absence in Lane 2 of Fig. 1a suggests that they are preferentially bound. The high molecular weight band that did not move in Lanes 2 and 3 of Fig. 1b may represent a membrane component of high molecular weight. Based on other work with myosin (205 K) we suggest that this material has a molecular weight around 200 K. That these bands represent one or more receptors for peptides of the AKH/RPCH family remains to be demonstrated.

TABLE 1. Binding of iodinated peptide.

Sample	^{125}I-bound % of added $\pm$ S.D.
No homogenate	1.0 ± 0.37
Boiled resuspended pellet	23.7 ± 1.5
Resuspended pellet	7.8 ± 0.85
Resuspended pellet + cold peptide (10 1)	6.3 ± 0.29
Supernatant from pellet	1.3 ± 0.61

Reaction mixture contained 100 µl of insect preparation [6 abdomens from refrigerated pharate face fly pupae into 1400 µl assay buffer prepared as follows: to each 1000 µl PBS, containing 45 mg BSA-Fract V, 5$\underline{M}$ in EDTA, were added 100 µl phenylthiourea (satd), 20 µl aprotinin (13 units/cc), and 10 µl phenylmethylsulfonylfluoride in DMSO (200 mM)] plus 200 µl distilled water, 200 µl of assay beffer above, and 10 µl of iodinated peptide (555,498 cpm). Reaction was stopped with 500 µl 50% saturated ammonium sulfate and the mixture filtered through GF/C glass microfiber filters soaked in 50% ammonium sulfate.

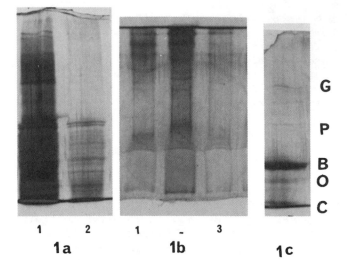

Fig. 1a - Initial samples - 1a1 - initial undiluted
homogenate; 1a2 - eluate albumin column. Fig. 1b - Peptide
column eluates. 1b1 - 2 M NaCl in PBS, 1b2 - 2 M
diethylamine, 1b3 - dilute PBS 1:3 (50 mM in NaCl) pH 3.0.
Fig. 1c - Chymotrypsin (C) (29 K), ovalbumin (O) (43 K),
BSA-V (B) (68 K), phosphorylase b (P) (93 K) and
β-galactosidase (G) (116 K) (faint).   .

REFERENCES

Alberts, B., Bray, D., Lewis, J., Raff, M., Roberts, K. and
        Watson, J. D. (1983) The Cell. Garland Publishing,
        Inc., New York.
Jaffe, H., Raina, A. K., Riley, C. T., Fraser, B. A.,
        Holman, G. M., Wagner, R. M., Ridgway, R. L. and
        Hayes, D. K. (1986) Isolation and primary structure
        of a peptide from the corpora cardiaca of Heliothis
        zea with adipokinetic activity. Biochem. and Biophys.
        Res. Commun. 135, 622-628.
Kohn, J. and Wilchek, M. (1982) A new approach
        (cyano-transfer) for cyanogen bromide activation of
        Sepharose at neutral pH, which yields active resins,
        free of interfering nitrogen derivatives. Biochem.
        and Biophys. Res. Commun. 107, 878-884.
Lowry, O. H., Rosebrough, A., Farr, A. L. and Randall, R.
        J. (1951) J. Biol. Chem. 193, 265-275.

SECOND MESSENGERS LINKED TO THE MUSCARINIC ACETYLCHOLINE

RECEPTOR IN LOCUST (Schistocerca gregaria) GANGLIA

Michael J. Duggan and George G. Lunt

Department of Biochemistry, University of Bath,

Bath,   BA2 7AY,   U.K.

Insects have both nicotinic and muscarinic acetylcholine receptors (mAChR) in the central nervous system (CNS) (Sattelle, 1985). In vertebrates mAChRs are linked to several different effector mechanisms, including a phosphatidylinositol (PtdIns) turnover and adenylate cyclase (Hirschowitz et al., 1984).

We have previously demonstrated the presence of mAChR in the supra-oesophageal ganglion of the locust (Schistocerca gregaria) (Aguilar and Lunt, 1984) and we now describe the linkage of mAChRs to both PtdIns turnover and adenylate cyclase.

PHOSPHATIDYLINOSITOL TURNOVER

Agonist-dependent phosphoinositide metabolism is known to be important in the second messenger production of many neurotransmitters and hormones (Berridge, 1984). Trimmer and Berridge (1985) have shown that the locust CNS shows incorporation of [^{3}H]inositol into the inositol phosphates thought to be the second messengers. However they were unable to show elevated levels of these compounds with several neurotransmitters, though the muscarinic antagonist atropine reduced their formation.

Supra-oesophageal ganglia were removed from the freshly decapitated heads of adult locusts (Schistocerca gregaria). The optic lobes and any connective tissue were

251

Table 1. PtdIns turnover

Ligand	Specific incorporation of [^{3}H]inositol (% of control $\pm$ SD)	n
None	100 $\pm$ 16	8
Carbachol (10^{-3}M)	164 $\pm$ 22	6
Carbachol (10^{-3}M) & Atropine (5 x 10^{-5}M)	91 $\pm$ 21	4

Experiments were done in duplicate, and results calculated as cpm/µg of phosphorus.

removed and the ganglia cut in half before being placed into 5ml gassed (O$_2$: CO$_2$; 95:5) phosphate buffered saline (pH6.8) containing: NaCl, 150mM; KCl, 10mM; CaCl$_2$, 2mM; MgCl$_2$, 5mM; NaHCO$_3$, 8mM; KH$_2$PO$_4$, 6mM; glucose, 10mM. The ganglia (10 per sample) were allowed to equilibrate for 2h at room temperature. Then the saline was changed to 2ml fresh gassed saline and 2µCi of myo [2-^{3}H]inositol (10–20Ci mmol^{-1}, Amersham International plc) were added in the prescence or absence of carbachol and atropine. The incubation lasted for 1h and was stopped by centrifuging the tubes, pouring off the supernatant and homogenising (30s, Ultraturrax) the ganglia in 10ml chloroform:methanol (2:1). The lipids were extracted by a further homogenisation, in a glass/glass handheld homogeniser and washed (Folch et al., 1957). The samples were divided, 4/5 taken for liquid scintillation counting and the remainder assayed for phosphate concentration (Bartlett, 1959).

The results of the [^{3}H]-inositol incorporation experiments show that label is incorporated to a significantly greater degree in the presence of the cholinergic agonist carbachol (Table 1). This effect is blocked by the muscarinic antagonist atropine.

In preliminary experiments in which the 2h equilibration period was not used there was no apparent effect of carbachol on [^{3}H] inositol incorporation. We speculate that the successful demonstration of agonist stimulated PtdIns turnover is a consequence of this 2h .

Table 2.   Adenylate cyclase

Ligand	cAMP production (% of control $\pm$ SD)	n
None	100 ± 10	4
Carbachol (10^{-3} M)	51 ± 8	4
Carbachol (10^{-3} M) & Atropine (10^{-5} M)	75 ± 5	4

Experiments were done in duplicate.

Table equilibration period. These studies provide the first direct confirmation of the prescence of a mAChR linked PtdIns turnover in the CNS of the locust.

ADENYLATE CYCLASE

Cyclic AMP (cAMP) is an important second messenger in nervous systems (Drummond, 1983). In mammals a sub-population of mAChR is linked to adenylate cyclase, and cholinergic agonists inhibit cAMP production (Onali et al., 1983). The presence of adenylate cyclase in locust nervous tissue has been shown (Morton, 1984).

Freshly dissected ganglia were immediately frozen with liquid nitrogen before storing at $-80°C$. Adenylate cyclase was assayed in homogenates of these ganglia. Ten ganglia were homogenized in 0.5ml of 50mM Tris buffer (pH 7.5) containing $MgCl_2$, 6mM; NaCl, 90mM; EDTA, 0.8mM; theophylline, 8mM. The homogenate was sieved through nylon mesh and the volume made up to 2ml. The assay was carried out in the same buffer, with the addition of ATP (1mM), GTP (0.1mM) and cholinergic ligands. A total volume of 0.5ml was used including 0.05ml of homogenate. The reaction was started with addition of the ATP and GTP and stopped by immersion in boiling water for 5 min. Protein was spun down, and the cAMP assayed by competition with [^{3}H]cAMP for a cAMP binding protein (BDH Ltd.) (Brown et al., 1971).

The results (Table 2) show that the cAMP accumulation of these homogenates is attenuated by carbachol. This can be prevented by incubation with the muscarinic antagonist .

atropine. This indicates the presence of a mAChR linked
adenylate cyclase.

The function of the mAChR in the insect CNS is poorly
understood. What evidence that there is has been taken to
indicate a presynaptic function (Sattelle, 1985). These
findings suggest that there is more than one type of mAChR
in the locust CNS.

REFERENCES

Aguilar J.O. and Lunt G.G. (1984) Cholinergic binding
    sites with muscarinic properties on membranes from the
    supraoesophageal ganglion of the locust (Schistocerca
    gregaria). Neurochem. Int. 6, 501–507.
Barlett G.R. (1959) Phosphorus assay in column
    chromatography. J. Biol. Chem. 234, 466–468.
Berridge M.J. (1984) Inositol trisphosphate and
    diacylglycerol as second messengers. Biochem. J. 220,
    345–360.
Brown B.L., Albano J.D.M., Ekins R.P. and Sgherzi A.M.
    (1971) A simple and sensitive saturation assay method
    for the measurement of adenosine 3':5'–cyclic
    monophosphate. Biochem. J. 121, 561–562.
Drummond G.I. (1983) Cyclic nucleotides in the nervous
    system. Adv. Cyclic Nucleotide Res. 15, 373–494.
Folch J., Lees M. and Sloane Stanley G.H. (1957) A simple
    method for the isolation and purification of total
    lipids from animal tissues. J. Biol. Chem. 226, 497–509.
Hirschowitz B.I., Hammer R., Giachetti A., Kierns J.J. and
    Levine R.R. (Editors) (1984) Subtypes of muscarinic
    receptors. Trends Pharmacol. Sci. (suppl.) 1–103.
Morton D.B. (1984) Pharmacology of the octopamine–stimulated
    adenylate cycladse of the locust and tick CNS. Comp.
    Biochem. Physiol. 78C, 153–158.
Onali P., Olianas M.C., Schwartz J.P. and Costa E. (1983)
    Involvement of a high–affinity GTPase in the inhibitory
    coupling of striatal muscarinic receptors to adenylate
    cyclase. Mol. Pharmacol. 24, 380–386.
Sattelle D.B. (1985) Acetylcholine receptors. In Kerkut,
    G.A. and Gilbert, L.I. (editors), Comprehensive Insect
    Physiol. Biochem. and Pharmacol. 2, 395–434.
Trimmer B.A. and Berridge M.J. (1985) Inositol Phosphates in
    the insect nervous system. Insect Biochem. 15 811–815.

NEUROTRANSMITTERS AND NEUROPEPTIDES IN THE OLFACTORY

PATHWAY OF THE SPHINX MOTH, *MANDUCA SEXTA*

John G. Hildebrand, Uwe Homberg, Timothy G.
Kingan, Thomas A. Christensen and Brian R.
Waldrop

University of Arizona

Tucson, AZ 85721

Like other animals, insects have many and diverse chemical "messengers" in their nervous systems. A growing list of synaptic neurotransmitters, neuromodulators, neuropeptides, and neurohormones -- collectively "transmitters" -- prompts efforts to seek physiological roles and mechanisms of action for these substances. An improved understanding of these chemical messengers in the insect nervous system promises to reveal key regulatory mechanisms, novel and accessible targets for pharmacological agents, and phyletic differences that can be exploited in new approaches to the manipulation of insect behavior and the selective destruction of harmful populations of insect pests and disease vectors. Toward such goals, we study the biochemistry, cellular distribution, and physiological actions of transmitter candidates in an experimentally favorable insect "model", the sphinx moth *Manduca sexta*. In contrast with significant advances that have been made in many laboratories investigating peripheral neural and neuromuscular systems, relatively much less is known about chemical signalling in the insect central nervous system (CNS). With this in mind, and building upon substantial previous and on-going studies of the anatomy, physiology, and development of the insect CNS in many laboratories including our own, we focus on the cellular neurochemistry of the CNS in *Manduca*. In particular we are exploring the olfactory pathway in the brain, for which we have accumulated

255

much information about the types of neurons and their functional organization and development (e.g. see recent reviews: Hildebrand, 1985; Hildebrand and Montague, 1986;
Christensen and Hildebrand, 1987).

ORGANIZATION OF THE OLFACTORY PATHWAY

Primary olfactory receptor cells in the antenna (e.g.
ca. 250,000 in the male antennal flagellum) send their axons
via the antennal nerve to the primary olfactory center in
the brain, the antennal lobe (AL) of the deutocerebrum. A
second population of olfactory receptors (ca. 1800) resides
in the pit organ of each labial palp. Like antennal olfactory receptors, these cells also send axons to the ALs.

Each AL consists of a central region of coarse neuropil
(largely principal neurites of AL neurons) surrounded by an
orderly array of spheroidal glomeruli (50–100 μm in diameter
and comprising terminals of primary olfactory axons, arborizations of CNS neurons, all of the recognized synapses in
the AL, and a glial investment) and 3 groups of somata of AL
neurons (large lateral, smaller medial, and still smaller
anterior groups).

Intracellular staining studies have revealed much about
the organization of the AL. Nearly all of the olfactory axons
from an antenna project to the glomeruli in the ipsilateral
AL, while the labial olfactory fibers project bilaterally to
a single glomerulus in each AL. Many of the ca. 1000 AL neurons are local, amacrine interneurons, which engage in extensive dendro-dendritic synaptic interactions (in the glomeruli) with other AL neurons. The rest of the AL neurons are
projection neurons that send their axons out of the AL to
other parts of the CNS -- usually to the calyces of the ipsilateral mushroom bodies and/or the ipsilateral lateral
protocerebrum, but sometimes to the contralateral AL or other
targets. All local interneurons (LNs) have multiglomerular
dendritic arborizations in most or all AL glomeruli, whereas
projection neurons (PNs) have either uniglomerular dendritic
arborizations or ramifications in a few glomeruli. The primary olfactory axons project to single glomeruli.

Ultrastructural observations have shown that the chemical synapses in the AL, which are confined to the glomeruli,

fall into at least 3 morphological classes characterized by
different types of synaptic vesicles in the presynaptic elements (Tolbert and Hildebrand, 1981). The most prominent
type, with round, electron-lucent vesicles, represents the
primary-afferent terminals and possibly some AL-neuron presynaptic elements as well. All of the synapses reconstructed
to date have been multiplads in which 2-7 presynaptic elements synapse upon a single postsynaptic element.

TRANSMITTERS IN THE ANTENNAL LOBES

Neurotransmitter "screening" experiments have revealed
that the principal transmitter candidates in the ALs are
acetylcholine (ACh), γ-aminobutyric acid (GABA), serotonin
(5-hydroxytryptamine, 5HT), histamine (HA), tyramine (TA),
and a number of apparent neuropeptides including substances
immunochemically similar to substance P (SP), locust adipokinetic hormone (AKH), FMRFamide, and corticotrophin-releas-
ing factor (CRF). A variety of evidence indicates that the
primary sensory cells of the antenna are cholinergic and
make nicotinic synapses upon their target neurons in the AL.
A few AL neurons may also be cholinergic. About 230 neurons
in each AL exhibit GABA-like immunoreactivity; most of these
cells are LNs, but a few are PNs (Hoskins et al., 1986).
5HT-like immunoreactivity is confined to 1 neuron in each AL
(Kent, 1985). Putative neuropeptides, revealed by immunocytochemistry, are distributed as follows: SP (>20 neurons);
FMRFamide (≧1 neuron, possibly a LN but not one containing
GABA); CRF (>80 neurons, probably LNs that do not contain
GABA); and AKH (>10 LNs). From these findings it is apparent
that the LNs, although morphologically largely similar to
each other, fall into several different neurochemical types
and therefore probably exert different physiological effects
upon other AL neurons.

Inhibitory Synaptic Transmission

Intracellular recording has revealed that inhibitory
synaptic transmission plays a prominent role in information
processing in the AL. Thus, in response to electrical stimulation of the ipsilateral antennal nerve, many PNs exhibit
complex postsynaptic responses typically including a relatively early, compound IPSP (graded with different stimulus
intensities) followed by a compound EPSP and burst of action

potentials. The inhibitory input to the PN appears to be indirect, via at least one inhibitory interneuron in the AL, which is consistent with the fact that we have no evidence for primary-afferent inhibition in the AL. The IPSP is due to chemical synaptic transmission and exhibits a reversal potential negative to the AL cell's apparent resting potential. Preliminary results of Cl^--substitution experiments suggest that the IPSP is mediated by Cl^-. Treatment of the desheathed AL with picrotoxin or bicuculline reversibly blocks the IPSP. All of these observations suggest that GABA, acting on a synaptic receptor that gates a Cl^- channel, mediates the inhibitory transmission. Injection of GABA into the AL neuropil can hyperpolarize PNs in a way that mimics the neurally evoked inhibitory influences.

REFERENCES

Christensen T. A. and Hildebrand J. G. (1987) Olfactory information processing in insects -- Functions, organization, and physiology of the olfactory pathways in the lepidopteran brain. In *Arthropod Brain: Its Evolution, Development, Structure and Functions* (edited by A. P. Gupta). John Wiley, New York, in press.

Hildebrand J. G. (1985) Metamorphosis of the insect nervous system: Influences of the periphery on the postembryonic development of the antennal sensory pathway in the brain of *Manduca sexta*. In *Model Neural Networks and Behavior* (edited by A. I. Selverston), pp. 129–148. Plenum, New York.

Hildebrand J. G. and Montague R. A. (1986) Functional organization of olfactory pathways in the central nervous system of *Manduca sexta*. In *Mechanisms in Insect Olfaction* (edited by T. L. Payne, M. C. Birch and J. S. Kennedy), pp. 279–285. Oxford Univ. Press, UK.

Hoskins S. G., Homberg U., Kingan T. G., Christensen T. A. and Hildebrand J. G. (1986) Immunocytochemistry of GABA in the antennal lobes of the sphinx moth *Manduca sexta*. *Cell Tiss. Res.* 244, 243–252.

Kent K. S. (1985) Metamorphosis of the antennal center and the influence of sensory innervation on the formation of glomeruli in the hawk moth *Manduca sexta*. Ph.D. Dissertation, Harvard University, Cambridge, MA.

Tolbert L. P. and Hildebrand J. G. (1981) Organization and synaptic ultrastructure of glomeruli in the antennal lobes of the moth *Manduca sexta*: A study using thin sections and freeze-fracture. *Proc. Roy. Soc. (Lond.)* B213, 279–301.

OCTOPAMINERGIC REGULATION OF AN INSECT VISCERAL MUSCLE

Ian Orchard and Angela B. Lange

Department of Zoology, University of Toronto,

Toronto, Ontario, Canada, M5S 1A1

Octopamine is now well established as a naturally occurring biogenic amine in the nervous system of invertebrates and vertebrates. While its physiological role within the vertebrate nervous system is still only speculative, its importance as a chemical messenger in the invertebrates is well established (see Orchard, 1982; Evans, 1985). However, despite the overwhelming body of knowledge covering its role as a neurotransmitter, neurohormone and neuromodulator, identification of octopamine containing neurons has only been made in a very limited number of cases. In insects, identified octopaminergic neurons have been shown to belong to a group of specialised cells located on the dorsal mid-line of the thoracic and abdominal ganglia (see Evans, 1985). These cells have bifurcating axons which project symmetrically into left and right peripheral nerve roots of the ganglion and consequently have been called DUM neurons. Examples of such neurons include DUMETi, in the metathoracic ganglion of locusts (Evans and O'Shea, 1978), and four DUM cells in the VIIIth abdominal ganglion of the firefly (Christensen et al., 1983). DUMETi modulates neuromuscular transmission and myogenicity of the extensor tibiae muscle in the hind leg, while the Photuris DUM cells regulate larval luminescence. While DUM neurons appear to be distributed extensively throughout the ventral nerve cord little is known of their function except from studies of the identified neurons mentioned above.

In the present paper we will describe the target site

259

of two identifiable DUM neurons located in the VIIth
abdominal ganglion of locusts. These neurons project axons
to the oviducts of the locust and are termed DUMOV. We will
present evidence for the octopaminergic nature of these
neurons, and illustrate their physiological role and
probable mode of action on the oviducts of Locusta
migratoria.

The oviducts of Locusta migratoria receive innervation
from a pair of oviducal nerve arising from the VIIth
abdominal ganglion. Cobalt backfilling of the branches of
the oviducal nerve which project to the oviducts revealed
two dorsal unpaired median (DUM) neurons lying at the
posterior end of the ganglion (Lange and Orchard, 1984).
Electrophysiological mapping confirmed that the DUM neurons
project to the oviducts.

The octopaminergic nature of these neurons was revealed
by radioenzymatic assay (see Orchard and Lange, 1985).
Octopamine is present in the isolated DUMOV neurons, the
oviducal nerve and innervated region of the oviducts. Areas
of oviduct which we believe receive little or no innervation
(the upper lateral oviducts) contain 22 fold less octopamine
(when comparing nmol per gram wet weight) than the
innervated (lower lateral and upper common) oviduct (Orchard
and Lange, 1985).

If octopamine plays a regulatory role in the oviducts,
than clearly it should be released from DUMOV and should
also have some physiological effect. Electrical stimulation
of the oviducal nerves resulted in the release of octopamine
into the perfusate (unpublished observation). Furthermore,
intrasomatic stimulation of one DUMOV neuron, sufficient to
elevate its firing frequency to 12 Hz, also resulted in an
increase in release of octopamine. The release of
octopamine was calcium-dependent, since calcium-free saline
was capable of reducing the high-potassium induced release
of octopamine.

The physiological effects of octopamine were revealed
following exogenous application of octopamine to an oviduct,
whilst it was attached to a force transducer (Orchard and
Lange, 1985). Electrical stimulation of the oviducal nerve
at 30 Hz for 2 sec resulted in a sustained contraction of
the oviducts. Application of octopamine had three effects
upon the oviducts. Firstly, there was an inhibition of

myogenic contractions; secondly, a relaxation of basal
tonus; and thirdly a reduction in the amplitude of
neurally-evoked contractions. The effect of octopamine upon
neurally-evoked contractions was studied in more detail and
shown to be dose-dependent with half maximal inhibition at 5
x 10^{-7} M and threshold lying between 5 x 10^{-10} M and 7 x 10^{-9}
M.

Evidence has accumulated over recent years for the
presence of octopamine-sensitive adenylate cyclase in
invertebrate tissues. The elevated levels of cyclic AMP
induced by octopamine are believed to act as the cellular
mediators for the physiological actions of octopamine. With
the discovery of identified octopaminergic neurons in
insects it is now possible to directly activate modulatory
octopaminergic neurons and examine for effects upon cyclic
AMP levels. Recently Evans (1984) demonstrated the ability
of DUMETi to increase the levels of cyclic AMP in skeletal
muscle. Furthermore it was shown that elevation of cyclic
AMP levels mimicked the physiological effects of octopamine
and DUMETi. To examine for a similar phenomenon in insect
visceral muscle we have performed a series of experiments to
test the hypothesis of an involvement of cyclic AMP in the
actions of DUMOV on locust oviduct muscle (Lange and
Orchard, 1986). The content of cyclic AMP in the lateral
oviducts was elevated by the phosphodiesterase inhibitor,
IBMX, and by octopamine. The action of octopamine in the
presence of IBMX was dose-dependent with the threshold at
about 10^{-8} M and half-maximal stimulation at about 7 x 10^{7} M
octopamine. In addition forskolin, a specific activator of
adenylate cyclase, was also capable of elevating cyclic AMP
content. If, as the results indicate, octopamine induces an
elevation in cyclic AMP in the lateral oviducts, then
release of octopamine from DUMOV should do the same.
Electrical stimulation of DUMOV either intrasomatically or
antidromically resulted in an elevation of cyclic AMP in the
lateral oviducts. Artificially elevating cyclic AMP levels
with the use of either IBMX, forskolin, or dibutyryl cyclic
AMP mimicked the major physiological effects of octopamine.
It seems reasonable to conclude from the above experiments
that the physiological effects of octopamine are indeed
mediated by cyclic AMP (Lange and Orchard, 1986).

In conclusion therefore we have described a new
octopaminergic system in insects. This system involves two
DUMOV neurons regulating neuromuscular transmission and

myogenicity of locust oviduct visceral muscle. We have
demonstrated the distribution of octopamine in neurons
projecting to the oviducts and have demonstrated
physiological release of octopamine. Furthermore octopamine
has specific physiological effects upon this muscle. At
least some of these effects appear to be mediated by cyclic
AMP. Locust oviducts therefore provide an ideal model for
examining an identified octopaminergic system.

References

Christensen T.A., Sherman T.G., McCaman R.E. and Carlson
A.D. (1983) Presence of octopamine in firefly photomotor
neurons. Neuroscience 9, 183-189.

Evans P.D. (1984) A modulatory octopaminergic neurone
increases cyclic nucleotide levels in locust skeletal
muscle. J. Physiol. 348, 307-324.

Evans P.D. (1985) Octopamine. In "Comprehensive Insect
Physiology and Pharmacology" (Ed. by Kerkut G.A. and Gilbert
L.I.), Vol. 11, pp. 499-530, Pergamon Press, Oxford

Evans P.D. and O'Shea M. (1978) The identification of an
octopaminergic neurone and the modulation of a myogenic
rhythm in the locust. J. Exp. Biol. 73, 235-260.

Lange A.B. and Orchard I. (1984) Dorsal unpaired median
neurons, and ventral bilaterally paired neurons, project to
a visceral muscle in an insect. J. Neurobiol. 15, 441-453.

Lange A.B. and Orchard I. (1986) Identified octopaminergic
neurons modulate contractions of locust visceral muscle via
adenosine 3',5'-monophosphate (cyclic AMP). Brain Res. 363,
340-349.

Orchard I. (1982) Octopamine in insects: neurotransmitter,
neurohormone, and neuromodulator. Can. J. Zool. 60,
659-669.

Orchard I. and Lange A.B. (1985) Evidence for
octopaminergic modulation of an insect visceral muscle. J.
Neurobiol. 16, 171-181.

NEUROCHEMICAL REGULATION OF LIGHT EMISSION FROM PHOTOCYTES

James A. Nathanson

Department of Neurology
Harvard Medical School

Massachusetts General Hospital
Boston, Massachusetts 02114

Physiological studies indicate that the initiation of a firefly's flash is neurogenically controlled through a pacemaker, present in the insect's brain, which sends periodic volleys of nerve impulses to the light organ (or lantern) located under the ventral cuticle of the distal abdominal segments. The onset of light production during a flash is associated with depolarization of the light emitting photocytes in the lantern. The terminal neurons in this activation pathway appear to be a group of octop-amine-containing dorsal unpaired median (DUM) cells, located in the distal two abdominal ganglia (Christensen and Carlson, 1981; Christensen et al, 1983). Putative octopaminergic nerve endings terminate near the redundant rosettes of photogenic cells which make up the light-emit-ting portion of the lantern. Octopamine has been shown to be present in significant amounts in the light organ (Robertson and Carlson, 1976), and octopamine (or its N-methyl derivative, synephrine) applied directly to light organs causes a bright glow both in normal and denervated lanterns (Carlson, 1968; 1972). These studies suggest that octopamine may be acting postsynaptically as the pri-mary neurotransmitter at the nerve-light organ junction. A number of biochemical experiments that we have carried out over the past few years support this hypothesis and suggest, further, that cyclic AMP may mediate octopamine's effect on light emission.

Prior evidence had indicated that, in many tissues, octopamine's receptor actions are mediated through the activation of adenylate cyclase and the synthesis of

263

cyclic AMP (Nathanson and Greengard, 1973, 1974; Nathanson, 1976). In 1979, we found that the light organ of the firefly contains an octopamine-activated adenylate cyclase of extraordinarily high specific activity. This enzyme is enriched in the photogenic layer of the lantern, and stimulation by octopamine results in as much as a 100-fold increase in enzyme activity (Nathanson, 1979; Nathanson & Hunnicutt, 1979a,b). Relevant to the possible involvement of cyclic AMP in stimulating light initiation has been our observation that the rank-order potency of various amines for activating adenylate cyclase (and cyclic AMP production) in lantern membrane fractions is nearly identical to the rank order potency of these same amines in eliciting light emission when injected into isolated lanterns. In particular: N-methyl-octopamine > octopamine > epinephrine > norepinephrine > tyramine > phenylethanolamine > phenylethylamine > dopamine.

More recently, we have found that there is also a correspondence between the ability of a variety of synthetic octopamine agonists to activate light organ adenylate cyclase and the ability of these compounds to elicit light emission (Nathanson and Hunnicutt, 1981; Nathanson, 1985a, b). For example, NC-5, a phenyliminoimidazolidine with about 20 times the potency of octopamine in stimulating light organ adenylate cyclase, is also about 15-20 times more potent than octopamine in producing light when injected into isolated lanterns. Also, demethylchlordimeform or NC-7, both partial agonists of the adenylate cyclase, are, similarly, partial agonists of the light response. Interestingly, at high concentrations, both compounds substantially inhibit enzyme activity and, likewise, cause inhibition of the light response.

If cyclic AMP mediates the response of octopamine in the light organ, one would predict that inhibition of cyclic AMP breakdown should enhance the potency of agonists acting at the octopamine receptor. This is what we have observed when using inhibitors of cyclic nucleotide phosphodiesterase (PDE), such as IBMX (Nathanson, 1984a,b; 1985a,b). Thus, when 0.1mM IBMX is injected together with octopamine, we find about a 5-fold increase in the potency of octopamine to elicit glowing. (Because methylxanthines are known, also, to be able to mobilize calcium ion, these experiments are carried out with saline containing no calcium and a high concentration of manganese ion.)

Additional evidence that cyclic AMP, itself, is sufficient to elicit light emission has come from a series of
experiments using cholera toxin which, in vertebrates, is
known to cause a prolonged and irreversible activation of
adenylate cyclase. We have found that cholera toxin
causes similar changes in light organ adenylate cyclase,
and, when injected into living fireflies, causes a prolonged and permanent activation of light emission (Nathanson, 1985c). More specifically, cholera toxin works by
slowly penetrating cells and causing ADP-ribosylation of
the GTP-binding protein (G-protein or regulatory protein
of adenylate cyclase). This covalent modification markedly slows the ability of the G-protein to hydrolyze GTP
and, in the presence of GTP, the G-protein causes a continuous activation of adenylate cyclase. The onset of the
irreversible activation by cholera toxin can be hastened
by applying hormone, since it is hormone-receptor binding
which allows the G-protein to initially take on GTP. Interestingly, these biochemical events are exactly mimicked
by the effects of cholera toxin on light emission. Thus,
a few hours after the toxin is injected into fireflies, we
observe a gradual, spontaneous increase in light output
from the lantern, resulting, in an hour or two, in a bril-
liant glow which lasts for many hours. This is associated
with a marked increase in lantern cyclic AMP levels and
basal adenylate cyclase activity. Furthermore, if the
firefly flashes several times (thereby releasing endogenous octopamine and pushing more G-proteins into an activated state) at a time when the toxin is first taking
affect, then the rate of irreversible light output is
markedly accelerated. Indeed, with each flash, there is a
rachet-like effect, with a normal onset but a prolonged
and incomplete inactivation of light output.

Further evidence indicating an involvement of cyclic
AMP in light emission has come from experiments in which
we have used the diterpene, forskolin. As in vertebrates,
we have found that this compound causes a direct activation of the catalytic subunit of adenylate cyclase. When
similar concentrations are injected into isolated lanterns, light emission is elicited. We have also found
that light emission results from the injection of lipid-
soluble, PDE-resistant cyclic AMP analogs, such as p-chlo-
rophenylthio-cyclic AMP. In other studies, we have
measured cyclic AMP levels in lanterns at various times
following injection of octopamine and have found that

there is an excellent correlation between light output and cyclic nucleotide level. Recently, in experiments carried out with Albert Carlson, we have found that cyclic AMP levels are also correlated with light emission evoked by nerve stimulation of the lantern nerve. Taken together, this evidence, as well as that described above, strongly supports an involvement of octopamine and cyclic AMP in the mediation of light initiation in the firefly light organ.

ACKNOWLEDGEMENTS

This work was supported, in part, by USDA Grant 8600090, the JLN-Daniels Research Fund, and by a grant from the McKnight Foundation.

REFERENCES

Carlson, A.D. (1968) J. exp. Biol. 49, 195-199.
Carlson, A.D. (1972) ibid. 57, 737-743.
Christensen, T. & Carlson, A.D. (1981) ibid. 93, 133-147.
Christensen, T., Sherman, T.G., McCaman, R.E. and Carlson, A.D. (1983) Neuroscience 9, 183-189.
Nathanson, J. (1976) In Trace Amines and the Brain, (Ed. by Usdin and Sandler) pp. 161-190, Marcel Dekker, N.Y.
Nathanson, J. (1979) Science 203, 65-68.
Nathanson, J. (1984a) Science 226, 184-187.
Nathanson, J. (1984b) Soc. Neurosci. Abstr. 10, 899.
Nathanson, J. (1985a) Mol. Pharmacol. 28, 254-268.
Nathanson, J. (1985b) Proc. Natl. Acad. Sci. USA 82, 599-603.
Nathanson, J. (1985c) J. Cyclic Nucl. Prot. Phosphor. Res. 10, 157-166.
Nathanson, J. and Greengard, P. (1973) Science 180, 308-310.
Nathanson, J. and Greengard, P. (1974) Proc. Natl. Acad. Sci. USA 71, 797-801.
Nathanson, J. and Hunnicutt, E. (1979a) J. Exp. Zool. 208, 255-62.
Nathanson, J. and Hunnicutt, E. (1979b) Soc. Neuroscience Abstr. 5, 346.
Nathanson, J. and Hunnicutt, E. (1981) Mol. Pharmacol. 20, 68-75.
Robertson, H. and Carlson, A.D. (1976) J. Exp. Zool. 195, 159-164.

PROTEIN PHOSPHORYLATION ASSOCIATED WITH SYNAPTOSOMAL MEMBRANES OF THE PUPAL BRAIN OF MANDUCA SEXTA

Wendell L. Combest, Mark J. Birnbaum,
Timothy J. Bloom, Stephen T. Bishoff and
Lawrence I. Gilbert

Dept. of Biology, Univ. of North Carolina
Chapel Hill, NC 27514

INTRODUCTION

Protein phosphorylation/dephosphorylation reactions are central to the action of many, if not all, second messengers involved in the regulation of cellular metabolism. Several protein kinases have been identified that are regulated by second messengers such as Ca^{2+}, cyclic nucleotides, phospholipids, and polyamines. The intracellular levels of these messengers are in turn controlled specifically by various peptide and amine hormones as well as other external stimuli e.g. light and membrane depolarization. Indeed, studies on Aplysia neurons have implicated protein phosphorylation in the regulation of ion channels (Lemos et al., 1985), and causal relationships have been established between protein phosphorylation and neurotransmitter synthesis and release in the mammalian nervous system (Nestler and Greengard, 1984). On the other hand, research on the role of protein phosphorylation in the insect nervous system is almost nonexistent due in large measure to a lack of basic data on the protein kinases and the phosphorylated endogenous substrates in the insect CNS. We report here properties of several protein kinases and their endogenous protein substrates in synaptosomal enriched membranes from the day 1 pupal brain of Manduca.

Cyclic AMP-dependent protein kinase activity (cAMP-PK), assayed with histone H_2b as exogenous substrate, was shown to be associated with the synaptosomal enriched membrane fraction, and represented 12% of the total brain cAMP-PK. Cyclic AMP-PK could be completely solubilized with 1% Triton X-100 in the presence of 0.5 M NaCl. DEAE-cellulose chromatography of solubilized membrane fractions revealed a single peak of cAMP-PK activity eluting at 0.15-0.3 M NaCl. The phosphorylation of histone was enhanced by cAMP (Ka $\sim$ $1x10^{-7}M$) and inhibited > 95% by the protein kinase inhibitor (PKI) from bovine heart, which is specific for the catalytic subunit of cAMP-PK. Photoaffinity labeling of synaptosomal membranes with 8 - azido [^{32}P] - cAMP resulted in the labeling of a single 55 kDa peptide we presume to be the regulatory subunit of cAMP-PK. These data indicate that the membrane associated cAMP-PK of the Manduca brain resembles the well characterized vertebrate type II isozyme of cAMP-PK from mammalian brain (Rubin et al., 1983).

Synaptosomal membranes incubated under phosphorylating conditions in the presence of [$\gamma^{32}P$] ATP resulted in the labeling of four prominent peptides (280, 34, 21, and 19 kDa) whose phosphorylation was enhanced by cAMP or by the addition of the purified catalytic subunit of cAMP-PK, and blocked by PKI. The three low molecular weight cAMP-PK substrates were also present in membrane fractions from thoracic and abdominal ganglia, connectives and peripheral nerves as well as in membranes of non-neuronal tissues such as the prothoracic gland, wing epidermis, gut, and fat body. These phosphoproteins were also found in the post-20,000 x g microsomal membrane fraction but were not present in brain cytosol. Of interest was the observation that the phosphorylation of the cAMP-PK substrates was markedly inhibited by physiological concentrations of the polyamines spermine and spermidine. Thus, polyamine inhibition of cAMP-PK promoted phosphorylation may be an important mechanism by which cAMP signals are attenuated in the insect brain.

We reported previously the presence of a prominent cyclic nucleotide and Ca^{2+}-independent 62 kDa phosphoprotein in brain homogenates of day 5 fifth instar larvae (Combest and Gilbert, 1986). A 62 kDa phosphoprotein is also present in synaptosomal as well as microsomal membrane preparations from day 1 pupal brains.

The phosphorylation of the 62 kDa peptide in EDTA and EGTA washed membranes is dependent upon Ca^{2+} ($EC_{50} \sim 50\ \mu M$) and exogenous calmodulin and is blocked by the calmodulin antagonist trifluoperazine. The 62 kDa phosphoprotein is also present in ganglia and connectives. This protein and its associated protein kinase are resistant to solubilization with 0.5% Triton X-100 but can be completely extracted as an active complex with 1% Triton X-100 and 0.25 M NaCl. Several 60 kDa Ca^{2+}/calmodulin dependent phosphoproteins have been described in synaptosomal membranes from rat brain and have recently been identified as the autophosphorylated 60 kDa subunit of Ca^{2+}/calmodulin protein kinase II (Kelly et al., 1984) and the 63 kDa subunit of Ca^{2+}/calmodulin phosphodiesterase (Sharma and Wang, 1986).

Two peaks of Ca^{2+} and cAMP-independent protein kinase activity were separated from solubilized membranes by DEAE-cellulose chromatography and displayed elution patterns similar to those of mammalian casein kinase I and II. The phosphorylation of exogenous casein by the second peak of activity (90% of total membrane casein kinase activity) was enhanced 2-3 fold in the presence of spermine, to a lesser extent by spermidine, and was inhibited by low concentrations of heparin. Both effects are characteristic of vertebrate casein kinase II (Hathaway and Traugh, 1982). The first peak of casein kinase activity was unresponsive to both polyamines and heparin, characteristic of vertebrate casein kinase I. Four synaptosomal membrane peptides (330, 320, 270 and 260 kDa) showed enhanced phosphorylation in the presence of micromolar concentrations of spermine or spermidine, presumably via endogenous casein kinase II. These high molecular weight phosphoproteins and associated casein kinase were solubilized as an active complex by the detergent n-octyl-β-D-thioglucopyranoside.

These data have provided the foundation for our current studies on the developmental and endocrinological roles of protein phosphorylation in the insect brain.

ACKNOWLEDGEMENTS

This work was supported by grants from the National Institutes of Health (AM-30118), the National Science

Foundation (DCB-8502194), and the Monsanto Company.

REFERENCES

Combest W.L. and Gilbert L.I. (1986) Phosphorylation of endogenous substrates by the protein kinases of the larval brain of Manduca sexta. Insect Biochem. 16, 607-616.

Hathaway G.M. and Traugh J.A. (1982) Casein kinases: Multipotential protein kinases. Curr. Top. Cell. Reg. 21, 101-107.

Kelly P.T., McGuinness T.L., and Greengard P. (1984) Evidence that the major postsynaptic density protein is a component of a Ca^{2+}/calmodulin-dependent protein kinase. Proc. natn. Acad. Sci. U.S.A 81, 945-949.

Lemos J.R., Novak-Hofer I. and Levitan I.B. (1985) Phosphoproteins associated with the regulation of a specific potassium channel in the identified Aplysia neuron R15. J. biol. Chem. 260, 3207-3214.

Nestler E.J. and Greengard P. (1984) Protein Phosphorylation in the Nervous System. Wiley and Sons, New York.

Rubin C.S., Sarkar D., Stein J.C. and Erlichman J. (1983) Characterization of neural-specific cAMP-dependent protein kinase. In Posttranslational Covalent Modification of Proteins. (Ed. by Johnson, B.C.), pp. 81-104. Academic Press, New York.

Sharma R.K. and Wang J.H. (1986) Calmodulin and Ca^{2+} dependent phophorylation and dephosphorylation of 63-kDa subunit - containing bovine brain calmodulin-stimulated cyclic nucleotide phosphodiesterase isozyme. J. biol. Chem. 261, 1322-1328.

THE CONTROL OF CASEIN KINASE II ACTIVITY AND PROTEIN

PHOSPHORYLATION IN THE BRAIN OF MANDUCA SEXTA BY POLYAMINES

Mark J. Birnbaum, Wendell L. Combest,
Timothy J. Bloom and Lawrence I. Gilbert

Dept. of Biology, Univ. of North Carolina
Chapel Hill, NC   27514

INTRODUCTION

The cationic, aliphatic polyamines are essential for normal cell function. Intracellular levels of the polyamines putrescine, spermidine and spermine become elevated during rapid cell growth and division and have been shown to influence macromolecular syntheses. In non-neural tissues, they appear to govern the phosphorylation of proteins catalyzed by casein kinase II, a cAMP and Ca^{2+} independent-protein kinase which prefers acidic protein substrates (Hathaway and Traugh, 1982). Within the vertebrate nervous system, polyamine metabolism, distribution, and roles in neurotransmitter release have been examined. However, no information exists regarding the polyamines in insect neural tissue. Further, the putative polyamine regulation of protein phosphorylation in the nervous system of any animal species has been virtually unexplored. The study of the control of neural protein phosphorylation is regarded as critical to an understanding of the mechanisms of neuronal function (Nestler and Greengard, 1984). Therefore, we have studied the effects which the polyamines exert upon protein phosphorylation in cytosolic fractions from the day 1 pupal brain of Manduca. We have characterized both the endogenous protein substrates, as well as a protein kinase which support polyamine directed protein phosphorylation in the brain of Manduca.

271

TABLE 1

Characteristics of Casein Kinase II from Manduca Brain,
Drosophila, and Calf Thymus

	Calf Thymus[a]	Manduca brain cytosol	Drosophila[b]
Native molecular weight	138 kDa	130 kDa[c]	130 kDa
Sedimentation coefficient	7.8 S	6.5 S[d]	6.4 S
Stokes radius	ND[e]	49 $\overset{\circ}{A}$[f]	48 $\overset{\circ}{A}$
Km: ATP	14 μM	34 μM[g]	17 μM
GTP	30 μM	55 μM[g]	66 μM
Heparin IC$_{50}$	unpublished	30 ng/ml	21 ng/ml

a   From Dahmus, 1981.
b   From Glover et al., 1983; enzyme isolated from nuclear
    and cytosolic fractions of whole embryos and Kc cells.
c   Calculated relative to standards using data from both d
    and f independently.
d   Measured relative to standards by 2.5-15% linear
    sucrose density centrifugation.
e   Not determined.
f   Measured relative to standards by Sephadex G-200
    exclusion chromatography.
g   Calculated from Lineweaver-Burke analysis between the
    concentrations of 1.0 to 100 μM nucleotide.

        Casein kinase II has been purified and characterized
in only one other insect species, Drosophila melanogaster
(Glover et al., 1983).   The striking physiochemical
similarities between the enzymes found in the cytosol of
the Manduca brain, whole embryos or Kc cells of Drosophila,
and calf thymus (Dahmus, 1981) are outlined in Table 1.
The Stokes radii, sedimentation coefficients and native

molecular weights of the three enzymes are virtually identical.   All three utilize either ATP or GTP as phosphoryl donors, and the Km for GTP is consistently higher than that observed for ATP. Also, all three enzymes are inhibited in the presence of very low concentrations of heparin.   In cytosolic fractions of the Manduca brain, we find that polyamines increase the phosphorylation of exogenous casein 2-3 fold.   This response is sensitive to ionic strength, being reduced at higher salt concentrations.

We have also studied the endogenous protein substrates present in the brain cytosol which appear to be regulated by polyamines via casein kinase II.   Low concentrations of spermine (100 µM) strongly enhance the phosphorylation of three high molecular proteins (305 kDa, 340 kDa, and 360 kDa).   These phosphoproteins are largely restricted to cytosolic fractions from neural tissues.

It has been postulated that the hormones which govern insect metamorphosis may influence intracellular polyamine concentrations.   In vertebrate systems, polyamine levels have been shown to be under endocrine control (Combest and Russell, 1983).   Qualitative and quantitative changes in polyamine content during the postembryonic development of a small number of insect species have been reported (e.g. Hamana et al., 1984) and there is some evidence to suggest that the molting hormone, 20-hydroxyecdysone, may regulate polyamine biosynthesis (Wyatt et al., 1973).   In addition, it has been hypothesized that juvenile hormone might also be involved in this process (Willis, 1981).   Therefore, information concerning the intracellular effects of the polyamines, such as modulation of protein phosphorylation, should provide insights into the mechanisms by which these hormones act.

ACKNOWLEDGEMENTS

This work was supported by grants from the National Institutes of Health (AM-30118), the National Science Foundation (DCB-8502194), and the Monsanto Company.

REFERENCES

Combest W.L. and Russell D.H. (1983) Alteration in cyclic AMP-dependent protein kinases and polyamine biosynthetic enzymes during hypertrophy and hyperplasia of the thyroid in the rat. Molec. Pharmacol. 23, 641-647.

Dahmus M.E. (1981) Purification and properties of calf thymus casein kinases I and II. J. Biol. Chem. 256, 3319-3325.

Glover C.V.C., Shelton E.R. and Brutlag D.L. (1983) Purification and characterization of a type II casein kinase from Drosophila melanogaster. J. Biol. Chem. 258, 3258-3265.

Hamana K., Matsuzaki S. and Inoue K. (1984) Changes in polyamine levels in various organs of Bombyx mori during its life cycle. J. Biochem. 95, 1803-1809.

Hathaway G.M. and Traugh J.A. (1982) Casein kinases: Multipotential protein kinases. Curr. Top. Cell. Regul. 21, 101-127.

Nestler E.J. and Greengard P. (1984) Protein phosphorylation in the nervous system. Wiley and Sons, New York.

Willis J.H. (1981) Juvenile hormone: The status of "status quo". Amer. Zool. 21, 763-773.

Wyatt G. R. Rothaus K., Lawler K. and Herbst E.J. (1973) Ornithine decarboxylase and polyamines in silkmoth pupal tissues: Effects of ecdysone and injury. Biochim. Biophys. Acta 304, 482-494.

Neurophysiology

OVERVIEW: PROGRESS IN INSECT NEUROPHYSIOLOGY

Michael E. Adams, Division of Toxicology and
Physiology, Department of Entomology,
University of California, Riverside, CA 92521

In recent years, neurohormonal mechanisms governing
insect development, metabolism and behavior have been
brought into sharper focus by key conceptual and technological advances. These include the introduction of
antibodies, development of more sensitive and specific
bioassays and improved liquid chromatography and peptide
sequencing technologies. If a dominant theme existed at
ICINN '86, it was a continued focus on the structure elucidation of endogenous hormones and transmitters. Two
strategies underlie the sequencing of novel peptides. One
begins with the knowledge that a process such as eclosion,
primary urine secretion or ecdysone release is controlled
by a "factor" or hormone. The hormone is isolated using
its action on the physiological target as a biological
assay. Structure determination follows prior knowledge of
the neuropeptide's physiological action. The second, more
general strategy is to choose a convenient bioassay such
as hindgut contraction or heartbeat and to purify excitatory or inhibitory substances from tissue extracts. This
strategy was taken for proctolin, Periplanetin CC1/CC2 and
for the leucokinins reported at this conference (see
below), all of which have myotropic actions on muscle. In
this case, the novel peptides often are "in search of a
mission." Although they surely do affect muscle, the
question of the bona fide physiological action remains to
be answered.

MYOTROPIC PEPTIDES. Five new myotropic peptides (leucokinins) were announced by Cook and Holman. Although these peptides were isolated from the heads of cockroaches, they may be distributed in nerve endings in other organ systems. Their functions have yet to be established through distribution, localization and phy- siological studies in model nerve-muscle preparations.

Proctolin cotransmitter actions were summarized by O'Shea, who described direct and modulatory actions on muscle targets by modulating intracellular calcium through novel second messenger mechanisms. Cotransmission at synapses may be a general phenomenon involving numerous other peptides and classical transmitters. For example, AKH-like peptides have been localized in what appear to be effector neurons in the gut. The leucokinins also may prove to be neuromuscular or central transmitters. But questions remain regarding the multiplicity of peptides. Are so many needed, or is there a certain level excess baggage resulting from gene duplication? Are they all released as chemical signals? Are their actions important to the organism's survival? Are there joint actions involving synergism or antagonism between released neuro- peptides? In this era of "sequence mania", these questions deserve increasing attention if sense is to be made of the plethora of peptides that are emerging.

METABOLIC PEPTIDES. Diuretic hormones, two from Locusta migratoria and one from Heliothis zea were reported at ICINN '86. Proux and co-workers described the sequencing of an arginine vasopressin-like diuretic hormone from subesophageal and thoracic ganglia. Rafaeli et al., demonstrated the presence of an ACTH-like diuretic hormone from the corpus cardiacum of the same species. A new type of diuretic hormone from lepidoptera was reported by Bushman et al., on a posteclosion diuretic factor in Heliothis zea. The Proux and Rafaeli hormones illustrate fascinating instances of parallel evolution with ver- tebrate peptides, since both vasopressin and ACTH are involved in water balance functions in the vertebrates.

Functions for diuretic hormones in insects having substantial water loads are straightforward, but for insects in arid environments, this is less well under- stood. In a cogent and fascinating lecture, Maddrell reviewed evidence that flying insects actually generate a significant water load through metabolic processes, and that urination can be measured during flight.

DEVELOPMENTAL PEPTIDES.   Several papers dealt with brain peptides regulating production of ecdysone (PTTH; Suzuki et al., Masler et al., Kelly et al.), juvenile hormones (allatotropins/statins; Rankin and Stay, Granger et al., Gadot and Applebaum) and pheromones (PBAN; Raina et al). The allatotropins and statins appear to act either humorally or by direct delivery to the corpora allata via nervous innervation, and are in initial stages of characterization.   Pheromone biosynthesis activating neuropeptides (PBAN) from the brain of the corn earworm, H. zea (Raina et al.) are in advanced stages of purification and sequencing. Sequences for two classes of prothoracicotropes, 4K and 22K PTTH were reported by Suzuki and coworkers.   Alpha and beta forms of 4K PTTH occur in multiple forms having heterogeneity in N-terminal sequences. The 4K PTTH hormones bear substantial sequence homology to insulin. Immunohistochemical localization studies have been used to pinpoint PTTH in specific brain neurons of B. mori (Suzuki et al.) and Manduca sexta (O'Brien et al.).

At the last ICINN meeting, the neuropeptides proctolin and adipokinetic hormone were the only insect neuropeptides of known amino acid sequence. In the intervening time period the number has gone to eleven. This number will probably again double within the next 1-2 years.

RECEPTORS AND ION CHANNELS.   Reports at this conference indicated substantial progress in the characterization of insect central and peripheral ion channels using synaptosomes (Breer, Dwivedy) and cultures of central neurons and skeletal muscle (Beadle, Pichon et al.). Synaptosomes from the insect CNS are now in routine use by Breer and co-workers for the analysis of cholinergic receptor pharmacology. Breer described the first successful reconstitution of a functional insect cholinergic receptor into an artificial membrane, and translation of mRNA coding for that receptor into Xenopus oocytes.

The application of whole and patch clamp methods to insect central neurons has been greatly facilitated by the efforts of Beadle et al., who reported on the establishment of primary cultures of cockroach central neurons at

the ICINN '83. In this volume, Beadle, Pichon and co-
workers demonstrate the power of these approaches to
obtain elegant pharmacological data on ion channels in the
membranes of central neurons. Analyses of the properties
of cholinergic and GABAergic receptors and of voltage
sensitive channels show the potential for analyzing drug
effects on well-defined classes of individual channel proteins. To complement this central nervous pharmacology,
skeletal muscle cultures from cockroach embryos have been
used to measure glutamate-induced single channel currents
by Bermudez et al.

MOLECULAR BIOLOGY. Although many groups around the world
are working toward the identification of genes for neuropeptides, receptors and ion channels, there were no such
reports presented at this meeting. Nevertheless, it seems
certain that such genes in insects will be identified
within the next few years and that this will greatly accelerate our abilities to define the synthesis, processing
and regulation of chemical messenger systems in insects.
Some progress in the identification of neuropeptide precursors was reported by O'Shea, using in vitro amino acid
incorporation to identify precursors for adipokinetic hormone.   Recent results show the presence of at least two
precursor proteins which appear to be processed to AKH I
and AKH II.   Because of the enormous levels of AKH peptides present in the corpus cardiacum, it appears to be
well-suited for such incorporation studies as well as for
eventual cloning of neuropeptide genes.
     How might molecular biology be used to manipulate
pest insect populations? A futuristic strategy recently
was translated into reality by the Rohm and Haas Corp.
with their successful incorporation of the toxic protein
from Bacillus thuringiensis into tobacco plants, a manipulation which afforded dramatic protection from pest
insects. This breakthrough points to a possible genera-
lized strategy of incorporating peptides or proteins dele-
terious to insects into either plants or pathogens.
Whether these approaches can ever be translated into economic insect control methods will depend on the resolution
of many additional technological as well as social and
political issues. It nevertheless seems certain that the
scientific wherewithal for the manipulation of genetic
messages in agricultural insect control is fast upon us.

THE MATURATION OF ADULT LOCUST CORPORA ALLATA AND

DEVELOPMENT OF RESPONSIVENESS TO ALLATOTROPIN

Shalom W. Applebaum and Michal Gadot

Department of Entomology, The Hebrew University

Rehovot 76100, ISRAEL

Reproductive maturation in the adult female locust is associated with an increase in the volume and number of cells of the endocrine corpora allata (CA) and in concurrent ultrastructural changes. We propose the term "allatal maturation" to denote the exact events occurring in the developmental biochemistry of the CA, as related to their acquisition of competence to synthesize juvenile hormone III (JH-III). Within the biosynthetic pathway leading to JH-III, terminal epoxidation and methylation of farnesoic acid (FA) is regarded as non-rate-limiting in mature CA, but the increase in these activities is possibly a part of the maturation process occurring during the first few days after adult emergence. The state of allatal maturation does predict the maximal level of JH-III production, but even when full competence is attained, maximal capacity for JH-III synthesis is not necessarily expressed. It is supposed, but has never been directly demonstrated, that rate-limiting enzyme(s) involved in the synthesis of precursors of FA are responsible for the variability of JH-III production observed with individual CA of mature locust adults in vitro. It has been shown previously that a cerebral allatotropin, also present in the corpora cardiaca (CC), is able to rapidly induce enhanced production of JH-III and it is assumed that its action involves the activation of the presumptive rate-limiting enzyme(s).

281

In order to evaluate allatal competence, excised CA are subjected in vitro to stimulation by saturating levels of exogenous farnesoic acid (FA) or allatotropin, and the level of JH-III synthesis evoked is measured by the radiochemical assay. The parameters of CA response in vitro to either FA or allatotropin are identical, indicating that allatotropin acts on a rate-limiting stage preceeding FA biosynthesis. The in vivo activity of individual CA is influenced by multiple factors and does not necessarily express the allatal level of competence. In locusts these factors include:

The basal level of nonstimulated activity

To some degree, basal activity represents the state of non-rate limiting enzymes involved in terminal epoxidation and methyation of endogenous FA. Basal activity usually is less than the maximal level of JH-III synthesis which can be evoked by either FA or allatotropin. In all cases of significantly active CA, basal activity is highly correlated to the level of competence of the individual gland examined. This indicates that variability between the two CA of an individual locust is influenced by the inherent competence of each gland, and is not a consequence of a mere pulsatile on/off activation (Gadot and Applebaum, 1986a).

Allatotropic stimulation

This stimulation is superimposed on the basal level of activity, and cannot exceed the level of competence of the individual gland. Allatotropin has been described from locust brain and corpora cardiaca (CC) as a methanol-soluble, heat resistant peptide, degraded by various proteolytic enzymes (Ferenz and Diehl, 1983, Ferenz, 1984; Gadot and Applebaum, 1985 and unpublished).

Allatotropin and FA are equivalent in probing the developmental maturation of the locust CA. The validity of the routine partition assay has been established for allatotropin-stimulated JH-III synthesis. Regression analysis demonstrates that the linear regression curves of FA-stimulated and allatotropin-stimulated CA are identical and with a high coefficient of correlation ($r = 0.97$) (Gadot and Applebaum, 1986a). Addition of either FA or allatotropin, or both together, elicits identical levels of stimulated JH-III production (Fig. 1).

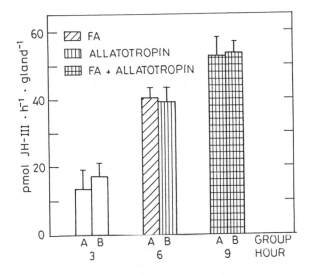

Fig. 1. Effect of FA (40μM) and allatotropin (0.1 brain
equivalents) on JH-III production by locust CA in vitro

We have found that reserpine treatment severely
retards reproductive maturation in locusts. A single dose
of 50 μg on day 0-1 after adult emergence effectively
inhibits the development of allatal competence to respond
to allatotropin. Basal CA activity levels recover within
two weeks, but the development of responsiveness to
allatotropin does not. This demonstrates the independence
of the different stages of control. Temporary starvation
(day 0-4) slightly delays somatic growth but does not
retard CA maturation (Gadot et al., 1986b).

The activation of CA by allatotropin is rapid, and
quickly declines when the CA are removed to
allatotropin-free media. Therefore, JH-III synthesis in
vitro, as measured by the RCA, does not reflect the
activity of the gland in vivo, and should be evaluated by
direct measurement of hemolymph JH-III titers and tissue
pools.

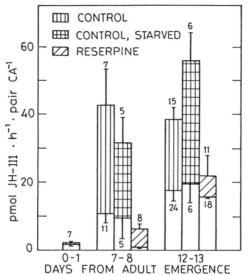

Fig. 2. Effect of reserpine and starvation on basal and allatotropin-stimulated synthesis of JH-III by locust CA

REFERENCES

Ferenz, H.J. (1984) Isolation of an allatotropic factor in Locusta migratoria and its effect on corpus allatum activity in vitro. In: Biosynthesis, Metabolism and Mode of Action of Invetebrate Hormones. Hoffmann J.A. and Porchet M., eds.) Springer-Verlag, Berlin. pp. 92-96.

Ferenz, H.J. and Diehl, I. (1983) Stimulation of juvenile hormone biosynthesis in vitro by locust allatotropin. Z. Naturforsch. 38C, 856-858.

Gadot, M. and Applebaum, S.W. (1985). Rapid in vitro activation of corpora allata by locust brain allatotropic factor. Arch. Insect Biochem. Physiol. 2, 117-129.

Gadot, M. and S.W. Applebaum (1986a) Farnesoic acid and allatotropin stimulation in relation to locust allatal maturation. Mol. Cell. Endocr. (in press).

Gadot, M., Faktor, O. and Applebaum, S.W. (1986b) Maturation of locust corpora allata during the reproductive cycle: Effect of reserpine on allatotropic activity, juvenile hormone-III biosynthesis, and oocyte development. Arch Insect Biochem. Physiol. (in press).

DIFFERENTIAL SYNTHESIS AND ALLATOTROPIC STIMULATION OF

JH-III AND JH-III DIOL IN LOCUST CORPORA ALLATA

M. Gadot[1], A. Goldman[1], M. Cojocaru[2] and
S.W. Applebaum[1]

[1]Departments of Entomology and Biochemistry
The Hebrew University, Rehovot 76100, ISRAEL

[2]Department of Chemistry, Bar-Ilan University
Ramat Gan, ISRAEL

Locust allatotropin, isolated from brains or corpora
cardiaca of adult females, is assayed by measuring the
amount of juvenile hormone-III (JH-III) produced in vitro
by excised corpora allata (Ferenz and Diehl, 1983; Gadot
and Applebaum, 1985). JH-III is the only form of juvenile
hormone produced by locust CA, but other compounds
exhibiting similar partition characteristics and mass
spectra have been observed in hexane extracts of Locusta
migratoria CA (Mauchamp et al., 1985) and are terminally
methylated in the radiochemical assay (RCA) (Gadot and
Applebaum, 1985). They are more polar than JH-III when
subjected to reversed phase HPLC on an RP-18 column. This
report deals with one of these, the major polar compound
accounting for most of the non-JH-III radiolabel produced
in the RCA, which we have identified as JH-III diol. We
present evidence that locust allatotropin preferentially
stimulates the production of JH-III, but only slightly -
that of JH-III diol.

Using a double-labelling technique with CA of day
3-5 females, we found that [2-^{14}C]acetate and
[methyl-^{3}H]methionine were incorporated in similar ratios
into both JH-III and into the polar compound, and that the
molar ratios were as predicted for structurally related
acetate-derived and terminally-methylated compounds (Fig.
1 and Table 1).

Table 1. Acetate/[Methyl]methionine ratio of in vitro incorporation into JH-III and polar compound by locust CA

INCUBATION PERIOD	JH-III	POLAR COMPOUND
0-3 hr (control)	11.7	9.4
3-6 h (Allatotropin-stimulated)	9.1	11.1

----

12 CA were incubated for 3 hrs in the presence of radio-labelled precursors, and transferred for a second incubation, again with radiolabelled precursors but supplemented with allatotropin (4 brain equivalents). Media were hexane-extracted and processed (see Fig. 1).

Chromatographic analysis of terminally methylated radiolabelled products released by locust CA in vitro indicates that almost all the radiolabel is incorporated into JH-III and into the major polar compound (Fig. 1). The ratio of these compounds is related to the degree of allatal activation, and inactive CA of adult females (wherein the level of synthesis is below 10 $pmol.h^{-1}.gland^{-1}$) produce 3-fold more of the polar compound, whereas active glands from adult females (wherein the level of synthesis exceeds 15-20 $pmol.h^{-1}.gland^{-1}$) predominantly produce JH-III (Fig. 2).

The retention time of the more polar compound on the RP-18 column was found to coincide with that of synthetically-derived JH-III diol, and the chromatographic properties of acetylated synthetic diol and acetylated radiolabelled polar compound were identical as well.

Mass spectra were obtained by direct probe inlet in a Finnigan 4021 quadrupole instrument equipped with a data system. Chemical ionization conditions were: ammonia or methane as reagent gas and ionizing potential of 70 eV. Analysis of derivatives and fragmentation products of JH-III, synthetic diol and acetylated derivative, and the native compounds produced by the CA and their acetylated derivative - identify the polar compound as JH-III diol.

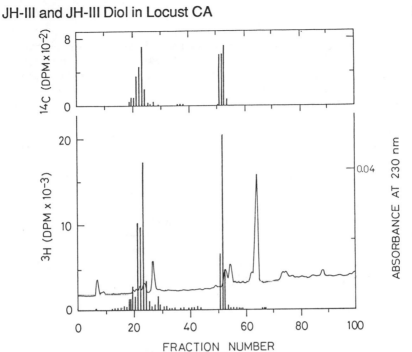

Fig. 1. HPLC profiles and distribution of radiolabel incorporated and released by CA _in vitro_. Radiolabelled incubation media were hexane-extracted, dried; reconstituted in water and separated on an RP-18 column with a 60-80% acetonitrile gradient. Fractions were counted in a scintillation counter and molar ratios of incorporation calculated.

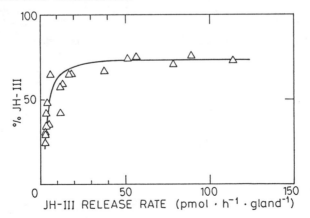

Fig. 2. Percentage of authentic JH-III released, of the total radiolabelled JH-III calculated from incorporation studies with individual adult female locust CA

JH-III diol is generally regarded as a degradation product of JH-III, but is not known to be produced intrinsically in CA. We have previously demonstrated that radiolabelled JH-III is not degraded when incubated with locust CA (Gadot and Applebaum, 1985), precluding the possibility that the appearance of JH-III diol is a consequence of exogenous degradation of JH-III secreted into the medium.

Table 2. Differential stimulation of JH-III and diol synthesis in inactive locust CA by allatotropin and FA

TREATMENT	$pmol.h^{-1}.gland^{-1}$			
	JH-III		JH-III DIOL	
	BASAL	STIMULATED	BASAL	STIMULATED
Allatotropin	3.7	13.2	5.0	6.8
FA	0.3	15.1	1.3	8.8

Allatotropin (0.2 brain equivalents) and farnesoic acid (FA) (6-200µM) preferentially stimulate the synthesis of JH-III (Table 2). These observations corroborate the specificity of allatotropin, identify its action as prior to the synthesis of FA, and suggest that the endogenous conversion of JH-III to JH-III diol is rate limited. This conversion may be an additional system for the regulation of JH-III levels in the CA. This does not preclude the possibility that JH-III diol performs some as yet undescribed physiological function in the locust.

REFERENCES

Ferenz, H.J. and Diehl, I. (1983) Stimulation of juvenile hormone biosynthesis in vitro by locust allatotropin. Z. Naturforsch. 38C, 856-858.

Gadot, M. and Applebaum, S.W. (1985) Rapid in vitro activation of corpora allata by locust brain allatotropic factor. Arch. Insect Biochem. Physiol. 2, 117-129.

Mauchamp, B., Couillard, F. and Malosse, C. (1985) Gas chromatography - mass spectroscopy analysis of juvenile hormone released by insect corpora allata. Anal. Chem. 145, 251-256.

OCCURRENCE OF ALLATOSTATIN IN THE COCKROACH,

DIPLOPTERA PUNCTATA, AND ITS EFFECT IN VITRO ON

CORPORA ALLATA FROM FEMALES IN DIFFERENT

REPRODUCTIVE STAGES.

Susan M. Rankin and Barbara Stay

Department of Biology
University of Iowa
Iowa City, Iowa 52242

In most insects, factors from the brain inhibit juvenile hormone synthesis by corpora allata toward the end of larval development and between gonadotrophic cycles (see Tobe and Stay, 1985, for review). In adult females of the viviparous cockroach, Diploptera punctata, juvenile hormone synthesis is inhibited at the end of oocyte development and remains low during the 60 day gestation period which follows (Rankin and Stay, 1985). During pregnancy the corpora allata are inhibited by intact nerves from the brain and by factors from the brain effective through the hemolymph (Rankin and Stay, 1985). Extracts of protocerebrum of the brain were found to inhibit the corpora allata in vitro (Rankin et al., 1986). Since the extracted allatostatin was inactivated by trypsin, it is presumed to be a neuropeptide (Rankin et al., 1986). In those studies the source of the allatostatin was non-reproductive (virgin) females and it was tested on glands of low activity from similar animals.

Now the presence of allatostatin in protocerebra of females at the height of vitellogenesis and the ability of allatostatin to regulate corpora allata of near-maximal activity is investigated. Rates of juvenile

289

hormone synthesis were measured in pmol h^{-1} per pair
using the in vitro radiochemical method of Tobe and
Clarke (1985). The activity of glands was established
after a 3 h incubation in normal medium, then after
subsequent incubation in medium containing extract of
brain. Saline extracts were prepared from protocerebra
of the brain and administered as described by Rankin et
al. (1986).

We asked first whether brains from reproductive
females contain the allatostatin. Extract of
protocerebra from 4-day mated females, at maximal
vitellogenesis, was tested at a concentration of 0.3
protocerebral equivalents/sample on glands of low
activity from 2-day virgins. For comparison extract
from non-reproductive females (previtellogenic 7-day
virgins) was tested in similar glands at the same
concentration. The results are shown in Table 1.
Extract from reproductive females was as effective as
control extract from non-reproductive females in
inhibiting juvenile hormone synthesis by these relatively
inactive test glands. At the end of the treatment
period, juvenile hormone synthesis by corpora allata
incubated with extract from either source was only about
half that of glands in untreated medium (Table 1).
Juvenile hormone synthesis by corpora allata incubated
in untreated medium for the test period showed no decline
from the pretreatment rate (data not shown).

Since protocerebra from reproductive females appear
to contain substantial quantities of allatostatin
(Table 1), one might hypothesize that allatostatin is
continually released but that glands of high activity
are refractory to it. The relative sensitivity of
glands of high and low activity to allatostatin was
tested by incubating glands from 4-day mated females and
glands from 2-day virgins in normal medium for 3 h, then
for 1 h and a subsequent 2 h in medium containing 0.5
protocerebral equivalents/sample (from 2-day virgins).
The results are shown in Table 2. After 1 h in treated
medium, glands of low activity had declined over 60% and
remained inhibited to about that extent in the succeeding
2 h test period (total of 3 h in treated medium). In
contrast, glands of high activity did not respond

Table 1. Effectiveness of protocerebral extract from females of different reproductive stages in inhibiting juvenile hormone synthesis by corpora allata of low activity.

Protocerebral donor		Corpus allatum response*		
Age & Stage	Juvenile hormone synthesis+	Juvenile hormone synthesis		
		n	3h Untreated	3h Treated
7-day virgin non-vitellogenic	15-20	5	28.2±6.9	10.6±1.1
4-day mated, vitellogenic	80-100	5	28.1±2.8	11.5±3.1

+ pmol h^{-1} per pair
* Corpora allata from 2-day virgins were incubated for 3 h in untreated medium, then for 3 h in medium containing extract of 0.3 protocerebra. Results are expressed as mean ± standard error of the mean.

Table 2. Sensitivity of corpora allata of low and high activity to extracts of protocerebra.

Corpus allatum			% Inhibition ± SEM* treatment time	
activity	source	n	1 h	3 h
low	2-day virgin	6	62.5±7.4	56.3±4.6
high	4-day mated	5	4.6±3.9	28.5±8.2

* Corpora allata were incubated for 3 h in untreated medium, then in medium containing extract of 0.5 protocerebra from 2-day virgin females. These results are expressed as % Inhibition = [1-(treated rate/untreated rate)] x 100 ± standard error of the mean of the ratios.

significantly to the allatostatin in 1 h, and declined in juvenile hormone synthesis only about 30% by the end of the treatment period. Glands of both low and high activity were able to recover from treatment with protocerebral extract within 1 h after being placed in normal medium (data not shown).

These experiments show that extracts of protocerebra from females undergoing rapid vitellogenesis in which corpora allata are near-maximally active contain as much allatostatin as those of females with non-vitellogenic oocytes. However, highly active corpora allata are less sensitive to the allatostatin in vitro than are glands of low activity in that they respond more slowly and to a lesser degree. Further study is needed to determine whether allatostatin is released from protocerebra of reproductive females in which corpora allata are relatively unresponsive to it, or whether allatostatin is stored for subsequent release at the end of the gonadotrophic cycle and during the extended period of pregnancy.

REFERENCES

Rankin S.M. and Stay B. (1985) Regulation of juvenile hormone synthesis during pregnancy in the cockroach, Diploptera punctata. J. Insect Physiol. 31, 145-157.
Rankin S.M., Stay B., Aucoin R.R. and Tobe S.S. (1986) In vitro inhibition of juvenile hormone synthesis by corpora allata of the viviparous cockroach, Diploptera punctata. J. Insect Physiol. 32, 151-156.
Tobe S.S. and Clarke N. (1985) The effect of L-methionine concentration on juvenile hormone biosynthesis by corpora allata of the cockroach, Diploptera punctata. Insect Biochem. 15, 175-179.
Tobe S.S. and Stay B. (1985) Structure and regulation of the corpus allatum. Adv. Insect Physiol. 18, 305-432.

INHIBITION OF JH I SYNTHESIS IN VITRO BY A CEREBRAL

ALLATOSTATIC NEUROPEPTIDE IN MANDUCA SEXTA

Noelle A. Granger and William P. Janzen

Department of Anatomy
University of North Carolina at Chapel Hill

Chapel Hill, North Carolina   27514

Regulation of juvenile hormone (JH) synthesis by the corpora allata (CA) is a complex process integrating both stimulatory and inhibitory mechanisms to achieve developmentally precise changes in gland activity. These changes in turn contribute to the fluctuations in the JH hemolymph titer which govern post-embryonic development. Recent evidence indicates that stimulation of the CA is effected in some systems by neuropeptides termed allatotropins, and inhibition of the CA by allatostatic factors has also been suggested (Tobe and Stay, 1985). However, comparatively little is known of the nature of these moieties.

The mechanism(s) by which the CA of the tobacco hornworm, Manduca sexta, are inactivated during the fifth (last) larval instar has recently assumed critical importance because of the discovery of additional possible roles for JH during larval-pupal development. It is generally accepted that a drop in the JH titer early in the fifth instar is permissive for PTTH release (Rountree and Bollenbacher, 1986), and that the resulting small ecdysteroid peak in the absence of detectable JH results in the reprogramming of the larval epidermis for pupal syntheses (Riddiford, 1978). It is now believed that the drop in the JH titer also causes 1) a decrease in the titer of a hemolymph protein which stimulates ecdysone synthesis (Watson et al., 1986), and 2) the acquisition of competence by the prothoracic glands to respond to both PTTH and the hemolymph stimulatory protein (Watson

293

and Bollenbacher, unpublished). Thus the mechanism(s)
inhibiting CA activity during this period represents the
most upstream event in the cascade resulting in larval-
pupal metamorphosis.

Recent information on the regulation of the CA has
been generated using in vitro experimental systems. One
of these, which uses radioimmunoassays for JH I and JH
III to quantify the synthesis of these homologs or their
acids in vitro (Granger et al, 1981), has been used to
titer CA activity during larval-pupal development in
Manduca (Granger et al., 1982) and to characterize an
allatotropic neuropeptide which specifically stimulates
JH III synthesis (Granger et al., 1984). In the former
study, it was observed that the synthesis of JH I by CA
incubated as a complex with the brain-corpora cardiaca
was generally lower than that by CA incubated alone,
suggesting an inhibition of the CA by the brain (Granger
et al., 1982). Subsequent investigations have indicated
that a cerebral allatostatic neuropeptide which specifically affects JH I synthesis elicits this inhibition.
The existence of this factor was first demonstrated by
co-incubating brain-corpora cardiaca from larvae on day 4
of the fifth instar (V/4), when JH synthesis is low, with
CA from day 0 of the instar (V/0), when JH synthesis is
high. JH I synthesis by the V/0 CA was reduced by 50%,
while JH III synthesis was unaffected.

If this factor were indeed a neuropeptide acting in
a hormonal fashion, then its inhibition of the CA would
be dose-dependent. To demonstrate a dose response of inhibition, V/4 brains were homogenized in Grace's medium
and the homogenate centrifuged to obtain a post-micro-
somal supernatant, which served as a crude brain extract.
Inhibition of the CA by this extract was demonstrated
with an in vitro assay. For this assay, a pair of left
CA was incubated with the brain extract (experimental),
while the contralateral right glands from the same two
larvae were incubated in Grace's medium only (control).
This paired gland approach was possible because right and
left glands from the same larva synthesize comparable
amounts of JH I. Moreover, it minimized the inherent
variability in basal gland activity between animals which
could obscure an inhibitory effect. The effect of the
brain extract was expressed as an inhibition ratio (I_r),
which was calculated by dividing experimental gland

synthesis by control gland synthesis. Determination of
the I_rs for various dosages of brain extract revealed a
dose dependent inhibition of JH I synthesis by V/0 CA; JH
III synthesis was unaffected. A maximum I_r of 0.45 was
obtained with 0.5 brain equivalents (BE) of V/4 brain
extract, with a dynamic concentration range of 0.1-0.5BE.
A time course study of inhibition revealed that maximum
inhibition was achieved within 4 hr in vitro, and this
incubation time was used in subsequent studies.

Although homolog-specific, the observed inhibition
could have been due to a non-specific toxic effect of the
extract on the glands, rather than to a JH I allatostatic
factor (ASF). Thus it was necessary to demonstrate that
the inhibition was reversable. This was accomplished by
preincubation of CA with the ASF for 1 min, a time sufficient to elicit maximal inhibition of the glands. The
time course of JH I synthesis by inhibited CA postincubated in medium without ASF was then compared to that
of CA both pre- and postincubated with the ASF. Results
showed that reversal of inhibition occurred within 4-6 hr
of postincubation without ASF. By this time, the rate of
JH I synthesis by the previously inhibited glands
equalled the basal rate of JH I synthesis by CA never
exposed to the ASF. The specificity of inhibition by the
apparent JH I ASF was further supported by the fact that
crude extracts of either muscle or thoracic and abdominal
ganglia did not inhibit JH I synthesis in the effective
range of concentrations of brain extract. Thus the JH I
ASF appears to be localized specifically to the brain.

Various chemical properties of the JH I ASF were
assessed to further define the nature of the moiety. The
ASF was found to be heat labile: treatment at 100°C for
less than 2 min destroyed essentially all activity. However, storage for up to 2 weeks at -70°C did not affect
activity, unless the extract were thawed more than once.
The proteinaceous nature of the ASF was indicated by its
sensitivity to proteolytic digestion by pronase, which
completely destroyed all activity under defined assay
conditions. Ultrafiltration and gel filtration of the
crude extract indicated two apparent molecular weights
for the JH I ASF of 6.8 and 13 kD.

Investigations are currently in progress to elucidate more fully the developmental significance and the

mechanism of action of the JH I ASF. Preliminary data
from a titer of ASF activity in the brain during the last
larval instar show that activity can first be detected by
late day 2, that it peaks on day 4 and that it is unde-
tectable by late on day 5. In summary, there is com-
pelling evidence to suggest that a neuropeptide which
specifically inhibits JH I synthesis exists in the Manduca
larval brain at a time when the activity of the CA must be
reduced to permit the drop in the JH titer requisite for
larval-pupal metamorphosis.

ACKNOWLEDGEMENTS

    This work was supported by grant NS-14816 from the
National Institutes of Health to N.A. Granger.

REFERENCES

GRANGER N.A., BOLLENBACHER W.E., and GILBERT L.I. (1981)
    An in vitro approach for investigating the regulation
    of the corpora allata during larval-pupal metamorpho-
    sis. In Insect Endocrinology and Nutrition (Ed. by
    BHASKARAN G., FRIEDMAN S., and RODRIGUEZ J.G.), pp.
    83-105, Plenum Press, New York.
GRANGER N.A., NIEMIEC S.M., GILBERT L.I., and BOLLEN-
    BACHER W.E. (1982) Juvenile hormone synthesis in vit-
    ro by larval and pupal corpora allata of Manduca
    sexta. Mol. Cell. Endocrinol. 28, 587-604.
GRANGER N.A., MITCHELL, L.G., JANZEN, W.P. and BOLLEN-
    BACHER, W.E. (1984) Activation of Manduca sexta
    corpora allata in vitro by a cerebral neuropeptide.
    Molec. Cell. Endocr. 37, 349-358.
RIDDIFORD L.M. (1978) Ecdysone-induced change in cellu-
    lar commitment of the epidermis of the tobacco horn-
    worm, Manduca sexta, at the initiation of metamorpho-
    sis. Gen. Comp. Endocrinol. 34:438-446.
ROUNTREE D.B. and BOLLENBACHER W.E. (1986) Release of
    the prothoracicotropic hormone in the tobacco horn-
    worm, Manduca sexta, is controlled intrinsically by
    juvenile hormone. J. exp. Biol. 120, 41-58.
WATSON R.D., WHISENTON L.B., BOLLENBACHER W.E. and
    GRANGER N.A. (1986) Interendocrine regulation of the
    corpora allata and prothoracic glands of Manduca
    sexta. Insect Biochem. 16,149-155.
TOBE S.S. and STAY B. (1985) Structure and regulation
    of the corpus allatum. Adv. Insect Physiol. 18,305-431.

PARASITE REDIRECTION OF NEUROHORMONALLY DRIVEN DEVELOPMENTAL PATHWAYS THAT ARE ASSOCIATED WITH SIZE THRESHOLDS

Davy Jones

Department of Entomology, University

of Kentucky, Lexington, KY 40546

The metamorphosis of insects provides an interesting and useful model with which to study how neural and neurohormonal systems regulate the transition from the immature to the adult stage. One level of regulation involves or is closely correlated with size thresholds. Classical and recent work with hemipterans has shown that the stimulus for molting is derived from stretch receptors which are sensitive to gut distension (Nijhout 1981). Research on lepidopteran growth and development has identified several size thresholds which regulate expression of certain developmental programs. The cycle of larval-larval molts continues until the insect reaches a body or exoskeletal size which signals attainment of the final larval instar. The size of the sclerotized head capsule is an index of this critical size, and a critical head capsule width has been determined for several species. During the final feeding phase, size thresholds have been correlated with commitment of certain tissues toward metamorphosis. The feeding stage ends when a final size threshold is reached, which is a function of the size of the insect at ecdysis to the final instar (Nijhout 1981; Jones et al. 1981).

Several hormones interact with growth thresholds for regulation of metamorphosis. Attainment of the critical size threshold for ending the final stadium feeding phase

triggers a neurohormonal inhibition of juvenile hormone (JH)
synthesis. The JH titer declines, permitting release
of prothoracicotropic hormone (PTTH) by the brain and ecdysone production. The larvae are then committed toward the
metamorphic developmental program, and wandering behavior is
initiated. The PTTH/ecdysteroid system becomes highly
active again during the prepupal stage to promote prepupal
development toward the pupal molt (Roundtree and
Bollenbacher 1986 ; Safranek et al. 1980).

Parasites in the genus Chelonus cause precocious
initiation of metamorphosis in host Lepidoptera, through a
mechanism resembling a reduction in the size threshold
associated with attainment of the final instar. Actual
pupation is blocked as the larvae cease PTTH/ecdysteroid
driven development early in the prepupal stage (Jones et
al. 1986). Reported here are results for experiments on the
nature of the Chelonus effect on the host template, and on
use of these hosts as experimental systems for the study of
expression of developmental programs in normal larvae.

Eggs of the cabbage looper, Trichoplusia ni (Heubner)
were stung by female Chelonus sp. near curvimaculatus.
'Pseudoparasitized' larvae, which show symptoms of
redirected development in the absence of an internal
parasite, were used for study (Jones et al. 1986).

Normally growing larvae of T. ni attain the final
instar when they reach a head capsule width of 1.66 mm or
greater (Jones et al. 1981). The most accurate predictor
of the size at which their feeding stage ends is the size
ratio (weight, mg)$^{1/3}$/(head capsule width, mm). In normal
larvae this ratio is approximately 3.29 (Table 1), and is
achieved after 3-4 days. When the JH titer in feeding larvae
is maintained by topical JH treatment, the delay in ecdysteroid production delays wandering behavior or head capsule
slippage and lengthens the feeding stage. The size ratio
at eventual ecdysteroid release is also higher (Table 1).

When normally growing 4th instar larvae are similarly
treated with JH, there is no effect on either the body size
necessary to trigger the next molt or the timing of the
molt. However, when 4th instar pseudoparasitized larvae

are treated with JH, there is a significant delay in the
time of the next ecdysteroid release (as indexed by the
time of wandering behavior or head capsule slippage) (Table
1). This lengthening of the feeding stage causes the size
ratio to be increased (Table 1).

Some of the JH treated larvae molted to viable 'supernumerary' instar larvae, instead of attempting pupation during the 4th stadium. These larvae were then used to answer
the question "does the feeding-stage program for sensitivity of the PTTH/ecdysteroid system to JH stay on, once it
has been activated?" Supernumerary pseudoparasitized
larvae were sensitive to application of a JH analog
(fenoxycarb) during the feeding stage, irrespective of
whether or not their head capsule width was above or below
the width for normal attainment of the final instar (Table
1). The JH analog treated larvae showed a delay in
ecdysteroid release, and reached much higher size ratios
than the control larvae.

Several experiments shed light on the nature of the
developmental lesions suffered by pseudoparasitized larvae.
All supernumerary pseudoparasitized larvae which (despite
an extra larval molt) still possessed a head capsule width
below the 1.66 mm threshold for metamorphosis again attempted metamorphosis. Further, all supernumerary pseudoparasitized larvae which reached the wandering stage ceased prepupal development, irrespective of their head capsule width.

The results permit the following conclusions:
1. The reduction in the size threshold for attainment
of the pupation instar is permanent. It is not erased by
passage through an extra larval molt.
2. Pseudoparasitized larvae express (feeding stage)
sensitivity of the ecdysteroid production system to JH .
They do so during their 4th and all supernumerary stadia.
3. The latent lesion for suppressed ecdysteroid
production appears specific to prepupae, and is not erased
by passage through an extra larval molt.

Supported, in part, by NIH GM 33995 & USDA 58-43YK-5-0034.
Approved by Director, Ky. Agr. Exp. Stn. (No. 86-7-160).

Table 1. Size ratios correlated with PTTH/ecdysteroid driven developmental events during the feeding stage of normal and pseudoparasitized larvae.

Treatment	Pupation Instar	$\bar{x}\pm SE$ (Weight $^{1/3}$)/ Head Capsule Width	Days to Wandering or Next Larval Molt
Normal Larvae			
EtOH	5	3.29+0.02	3-4
JH II	5	3.88+0.17	5-7
Pseudoparasitized			
EtOH	4	3.45+0.03	3-4
JH II	4	3.89+0.05	5-7
*EtOH	5	3.00+0.05	3-4
*Fenoxycarb	5	3.27+0.06	4-6
**EtOH	5	2.89+0.09	3-4
**Fenoxycarb	5	3.56+0.10	5-7

All JH/ fenoxycarb treatments (10 nmoles twice daily) different from immediately preceeding EtOH treatment, t test, $p < 0.05$. Asterisks indicate 'supernumerary' 5th instar with head capsule width < 1.66mm (*) or > 1.66 mm (**). The supernumerary larvae were generated by treatment of 4th instar pseudoparasitized larvae with JH.

REFERENCES

Jones, D., Jones, G., Rudnicka, M., Click, A., Reck-Malleczewen, V. and Iwaya, M. (1986) Pseudoparasitism of host Trichoplusia ni as a new model system for parasite regulation of host physiology. J. Insect Physiol. 32, 315-328.

Jones, D., Jones, G., and Hammock, B. D. (1981) Growth parameters associated with endocrine events in larval Trichoplusia ni and timing of these events with developmental markers. J. Insect Physiol. 27, 779-788.

Nijhout, H. F. (1981) Physiological control of molting in insects. Amer. Zool. 21, 631-640.

Roundtree, D. B. and Bollenbacher, W. E. (1986) The release of the prothoracicotropic hormone in the tobacco hornworm Manduca sexta is controlled intrinsically by juvenile hormone. J. Exp. Biol. 120, 41-58.

Safranek, L, Cymborowski, B. and Williams, C. M. (1980) Effects of juvenile hormone on ecdysone-dependent development in the tobacco hornworm, Manduca sexta. Biol Bull. 158, 248-256.

ROLE OF JUVENILE HORMONE AND JUVENILE HORMONE ESTERASE IN RELATION TO CUTICLE TANNING IN SPODOPTERA LITURA

M. Aruchami, R. Jeyaraj, C.A. Vasuki and
T. Thangaraj
Department of Zoology

Kongunade Arts & Science College
Coimbatore 641 029, INDIA.

The larval-pupal transformation in the lepidopteran forms is triggered by the decline in the juvenile hormone titre during the last larval stage (Riddiford and Truman, 1978; Granger and Bollenbacher, 1981) which allows the release of larval steroid hormone, ecdysone (Sparks and Hammock, 1980). The combination of higher titre of ecdysone and low level of juvenile hormone prevents the larval moult and favours the transition from larval to pupal stage. The declined juvenile hormone titre remains low until the early prepupal phase of development and a brief burst of juvenile hormone has been found to appear in the haemolymph of many lepidopterans (Varjas et al., 1976; Yagi and Kuramochi,1976; Hsiao and Hsiao, 1977; Riddiford and Truman, 1978; Jones and Hammock, 1983, 1984). The appearance and degradation of the prepupal juvenile hormone has been implicated for the proper development of pupal features and the timing of ecdysis to pupa (Kiguchi and Riddiford, 1978; Cymborowski and Stolarz, 1979; Hiruma, 1980; Safranek et al., 1980; Sieber and Benz, 1977).

Among various events linked with the appearance and removal of juvenile hormone by the juvenile hormone esterase, synthesis and mobilization of tyrosine glucoside are related to the juvenile hormone titre (Ahmed et al., 1985). Declining titre of juvenile hormone during prepupal stage by juvenile hormone esterase forms an important step in the removal of juvenile hormone during prepupal stage which results in the regulation of the synthesis of precursor material to sclerotin formation. The present investigation provides evidences for the relationship of the declining titre of juvenile hormone by the juvenile hormone esterase and the regulation of glucoside synthesis during larval-pupal transformation.

Tyrosine glucoside level was found to be higher during the early prepupal period and the glucoside level declines as the pupation advances and low level of tyrosine glucoside was noticed as the cuticle tanned (brown cuticle stage). The juvenile hormone esterase activity was observed to be higher during the prepupal stage, indicating the appearance of higher juvenile hormone in the haemolymph and its subsequent degradation by juvenile hormone esterase (Table 1). The appearance of juvenile hormone in the haemolymph at the prepupal stage has been suggested to induce the juvenile hormone esterase activity directly because, induction of juvenile hormone esterase by juvenile hormone via the brain would involve a lag period (Jones and Hammock, 1984). This has been observed in the present study when the juvenile hormone was applied as the external source. Thus the inducted juvenile hormone

Table 1. Tyrosine, Tyrosine glucoside levels
in the haemolymph (mg/ml) and juvenile
hormone esterase and phenoloxidase
activities (%) during larval pupal
transformations of Spodoptera litura

Stages	Tyrosine (mg/ml)	Tyrosine glucoside (mg/ml)	Juvenile hormone esterase (%)	Phenol oxidase (%)
Feeding	0.852	0.684	33	40
Wandering	0.872	1.236	24	86
Prepupal	0.340	2.272	100	100
Pupal	1.614	0.992	20	41

esterase acts upon the juvenile hormone and thereby regularising the titre of juvenile hormone. This modulation of the enzyme which degrades the juvenile hormone appears to play a key role in the initiation of different biochemical events leading to the larval-pupal transformation.

The present study supports the contention of Ahmed et al. (1985) that tyrosine glucoside level is determined by the juvenile hormone titre in the hameolymph. When juvenile hormone was applied as an external source after ecdysis to maintain a higher titre of hormone level, tyrosine glucoside level was found to be low, indicating that juvenile hormone level is acting as a negative factor during the synthesis of tyrosine glucoside (Table 2). It has been suggested that tyrosine glucoside synthetase enzyme may be inhibited (Ahmed et al., 1985). Moreover, declining titre of juvenile hormone by the degradative action of juvenile hormone esterase favours the release of the hormone, ecdysone, which in turn initiates the activity of the enzyme, tyrosine glucoside hydrolase to convert tyrosine glucoside to tyrosine. Now the available tyrosine is used to synthesis the precursor material (i.e. NADA) for the sclerotin formation during the pupal cuticle synthesis.

Table 2. Tyrosine glucoside level in
the haemolymph of sixth instar larvae
of Spodoptera litura after treatment
with juvenile hormone-II

Time (Hrs.)	Tyrosine glucoside (Normal)	Tyrosine Glucoside (After treatment JH-II)
24	0.684	0.323
48	1.236	0.154
72	2.272	0.574
96	0.992	0.397

Acknowledgement: Thanks are due to Mr K. Logankumar,
Department of Zoology, Kongu Nadu Arts & Science College,
for technical assistance.

REFERENCES

Ahmed R.F., Hopkins T.L. and Kramer K.J.(1985) Role of
   juvenile hormone in regulating tyrosine glucoside synthesis
   for pupal tanning in Manduca sexta (L.) J. Insect Physiol.
   31: 341-345

Cymborowski B. and Stolarz G. (1979) The role of juvenile
   hormone larval-pupal transformation of Spodoptera littoralis:
   Switchover in the sensitivity of the prothorasic gland to
   juvenile hormone. J. Insect Physiol. 25, 939-942.

Granger N.A. and Bollenbacher W.E. (1981) Hormonal control of
   insect metamorphosis. In Metamorphosis A problem in
   DevelopmentalBiology(Ed. by Gilbert L.I. and Frieden E.)
   pp. 105-137. Plenum Press, New York.

Hiruma K. (1980) Possible roles of juvenile hormone in the
   prepupal state of Mamestra brassicae. Gen Comp. Endocr.
   41, 329-399.

Hsiao T.H. and Hsiao C. (1977) Simultaneous determination
   of molting and juvenile hormone titers of the greater wax
   moth. J. Insect Physiol. 23, 89-93.

Jones G. and Hammock B.D. (1983) Prepupal regulation of
   juvenile hormone esterase through direct induction by
   juvenile hormone. J. Insect Physiol. 29, 471-475.

Jones G. and Hammock, B.D. (1984) Regulation of juvenile
   hormone esterase activity in larvae of the cabbage looper
   Trichoplusia ni In Insect Neuro-chemistry and Neurophysiology
   (Ed. by Borkovec A.B. and Kelly T.J.) Plenum Press,
   New York, London.

Kiguchi K. and Riddiford L.M.(1978) A role of juvenile hormone
   in pupal development of the tobacco hornworm, Manduca sexta.
   J. Insect Physiol. 24, 673-680.

Riddiford L.M. and Truman J.W. (1978) Biochemistry of
    insect hormones and insect growth regulators. In
    Biochemistry of Insects (Ed. by Rockstein), pp. 307-357.
    Academic Press, New York.

Safranek L., Cymborowski B. and Williams C.M. (1980) Effects
    of juvenile hormone on ecdysonedependent development in
    the tobacco hornworm. Manduca sexta. Biol. Bull. mar.
    biol. Lab. Woods Hole 158, 248-256.

Sieber R. and Benz G. (1977) Juvenile hormone in larval
    diapause of the codling moth, Laspeyresia pomonella L
    (Lepidoptera: Tortricidae). Experientia 33, 1598-1599.

Sparks T.C. and Hammock B.D. (1980) Comparative inhibition
    of the juvenile hormone esterases from Trichoplusia ni,
    Musca domestica and Tenebrio molitor. Pestic. Biochem.
    Physiol. 14, 290-302.

Varjas L., Paguia P. and de Wilde. J. (1976) Juvenile hormone
    titers in penultimate and last instar larvae of Pieris
    brassicae and Barathra brassicae, in relation to the
    effect of juvenoid application. Experientia 32, 249-251.

Yagi S. and Kuramochi K. (1976) The role of juvenile
    hormone in larval duration and spermiogenesis in relation
    to phase variation in the tobacco cutworm, Spodoptera
    litura (Lepidoptera : Noctuidae) Appl. ent. Zool. 11,
    133-138.

EFFECT OF PRECOCENE-II ON ENDOCRINES, FEEDING AND

DIGESTION IN THE SEMILOOPER CATERPILLAR, ACHOEA JANATA

D. Muraleedharan, Abraham Varghese, George Abraham and Reema A. Mathews

Department of Zoology, University of Kerala, Kariavattom, **India**

Ageratochromenes of plant origin and their synthetic analogues (precocenes) received almost equal attention by entomological researchers both doing fundamental and applied work due to their twin advantage; using as a physiological probe in the former avoiding surgical allatectomy and as an effective tool in devising 'fourth generation insecticides' in the latter. Recent literature on the topic suggests a selective sensitivity to these chromene compounds by different insect groups and even among the sensitive groups, intensity of sensitivity varies (Feyereisen et al., 1981; Pratt, 1983). According to Bowers (1980) precocenes are generally unable to induce precocius metamorphosis in lepidopterans. However, Kiguchi (1982) demonstrated precocius development in the larvae of Bombyx mori by the treatment of precocene. In addition to the conventional proallatocidal changes, several other physiological alterations have also been attributed to precocene treatment in several insect species. Many of the hemipterans showed diminished neurosecretory activity, ovicidal and sterilisation effects (Unnithan et al., 1977; Fagoonee and Umrit, 1981; In certain coleopterans and homopterans precocene influenced migratory behaviour sex and castemorph determination (Mackauer et al., 1979; Rankin and Rankin, 1980). Diapause induction was noticed in precocene-treated coleopterans (Bowers, 1976) while the same treatment prevented diapause in lepidopterans (Sieber and Benz, 1980). Precocenes interfere with the feeding activity in some species (Kelly and Fuchs, 1978; Hodkova and Socha, 1982). Slama (1978) attaches an antifeedant effect to these compounds. In the present studies we have tried to analyse

307

precocene-induced light microscopic and ultra-structural
changes on the endocrines (brain NS cells, SOG and CA)
and also the influence imparted by this compound on feeding
and digestive capacity in the 5th instar caterpillars of castor
semilooper, Achoea janata.

For all the experiments, 5th instar caterpillars of A.
janata isolated from the stock colony maintained under controlled conditions in our laboratory on castor (Ricinus communis)
leaves were used. Precocene-II(6,7-dimethoxy-2,2-dimethyl
chromene) were tried by single topical application (50µg/
animal) and controls were applied with only acetone. Experi
mental and control animals were sacrificed and 12 hr later
the endocrine organs (Brain, CA and SOG) recovered were
either processed for light microscopic studies using the routine
neurosecretory staining techniques or for EM studies. Neurosecretory index from stained sections were calculated as
done by Jalaja and Prabhu (1977). Light microscopic measurements were done using a calibrated ocular micrometer.
For EM studies tissues were fixed in gluteraldehyde (2.5%),
post-fixed in OsO4 (1%) and embedded in Spurr's resin as
done by Feyereisen et al. (1981) with minor modifications.
Sections were taken on KL B- Broma ultratome, doubly stained
with uranyl acetate and lead citrate and examined using
a Carl Zeiss 109S microscope at an accelerating voltage
of 80kv.

Amount of food consumed were quantified by the area
of leaf fed traced on graph sheets as described earlier (Muraleedharan and Prabhu, 1981) at different time intervals for
the entire 5th instar period. For allatectomy the method
of Schneiderman (1967) was followed and controls were sham-
operated. Methoprene (ZR515) was the JHa used for evaluating
the effect of JH on precocene-treated larvae and single
topical application of 10µg/larvae in acetone was performed.
Midgut protease activity was biochemically estimated using
partially purified enzyme extracted from gut contents and
gut tissues separately and the method followed was as described earlier (Muraleedharan and Prabhu, 1979).

Lightmicroscopic and ultra-structural changes were studied in the A-type pars intercerobralis NS cells. Significant
reduction in neurosecretory index and in the diameters of
NS cells and their nuclei were noticed (Table.1). EM studies
revealed that the elementary secretory granules present

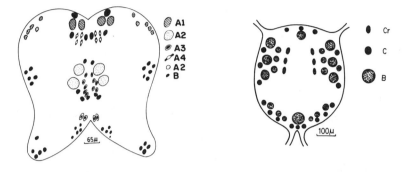

FIG. 1   FIG. 2

Schematic drawing of larval brain and suboesophageal ganglion of A.janata. Different types of neurosecretory cells are denoted by the corresponding letter and number.

Table.1. Effect of Precocene-II on the endocrines of 5th instar larvae of A.janata after 12 hr.

	NS index (arb.Units)	Cell Diam. (in uµ)	Nu. Diam (in µ)	Diam of CA (in µ)
A-type	PINSC			
EXP	32	@ 22 ± 0.6	@ 8 ± 0.2	-
CON	43	25 ± 0.8	10 ± 0.3	-
CA				
EXP	-	- -	-	@ 118±3
CON	-	-	-	186±7
SOG				
EXP	38	26 ± 0.9	9 ± 0.4	-
CON	35	24 ± 0.8	9 ± 0.2	-

@ Each value is mean from 8 separate individuals Mean values of adjacent pairs significantly different ($P < 0.05$)

FIG.3

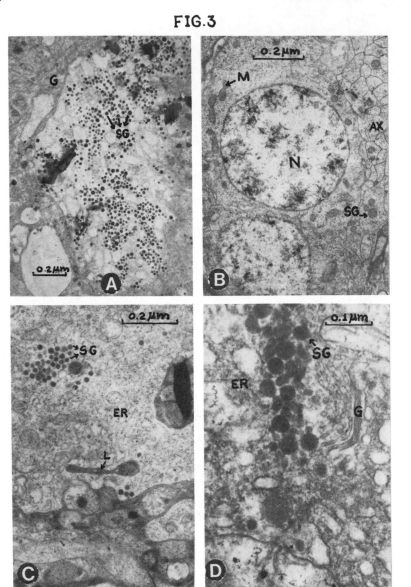

Fig. 3.   Electronmicrographs showing Precocene effect on
          endocrines. 3A, A-type NS cell axonal region and
          nucleus (control). 3C, A-type cell 12 hr treatment
          of Precocene 3D, Copus allatum of Control

FIG.4

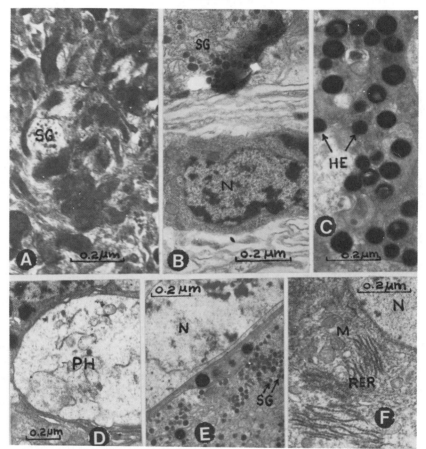

Fig. 4   showing precocen treatment on corpus allatum and
          suboesophageal ganglion. 4A to D, CA of treated
          larva showing extreme degeneration. E,F showing
          unaffected SOG of precocene treated larva.

at the axonal region in control larvae, indicative of rapid
release of NS material was absent in treated larvae. An
abundance of mitochondria with well-defined cristae noticed
in control insects were also lacking in treated animals. The
golgi complex in control was well equipped with stacks of
smooth membrane and vesicles indicating their involvement
in the formation of secretory granules were less evident
in treated larvae. Frequent presence of phagosomes and
the ill-defined nature of granular ER in treated cells suggested
rapid degradation and lower rate of neurosecretory synthesis
(Figs. 3A to 3C).CA size was smaller in the treated insects
than that in controls (Table.1). Ultrastructural details also
indicated an extensive damage to the CA cytoarchitecture
(Fig.3D and 4A to D). Compact distribution of nerve fibers
and the glial tissue was apparently due to a decrease in
CA volume consequent to degeneration. Matrix surrounding
the central mass containing remnance of degenerating pare-
nchyama cells with picnotic nuclei. Condensation of cytoplasm
resulting in more electron density due to closer packing
of ribosomes and other cytoplasmic constituents. Nuclei
were small and irregular in shape chromatin coagulated and
became dense. A-type cells were absent in the SOG and
only B,C and Cr type NS cells were noticed in this. No significant difference were noticed in the NS index in the B-cells
between those present in treated and control larvae. EM
studies were restricted to anterio-lateral B-cells. Both treated
and control cells showed almost the same activity level
as evidenced by the abundance of granulated ER and secretory
granules and mitochondria (Figs. 4E and F).

Precocene treatment reduced the rate of feeding (Fig.5)
and the differences between the values in the experimental
and control animals were significant (P $<$ 0.05). The total
amount of food consumed by a treated larva during the entire
5th instar life were also significantly lower than that in
control (P$<$ 0.5). A significant reduction was noticed in the
food consumption of allatectomised larvae 8, 16, 24 and
32 hr after operation (P $<$ 0.01). Midgut protease activity
in treated larvae showed a significant reduction both in gut
tissue and in the gut conetnts after precocene supply (Fig.6).
However, when JHa was supplied to precocene-treated larvae,
enzyme activity both in gut contents and in the tissues showed
a substantial elevation.

The above findings clearly demonstrate that in the 5th
instar caterpillars of A. janata single topical application

of precocene-II induces degenartive changes and retards the activity of both A-type NS cells in the brain and the activity of CA while it does not affect the activity of the B-cells of SOG. Many of the earlier reports suggested that lepidopetarns are generally insensitive to precocenes (Bowers,

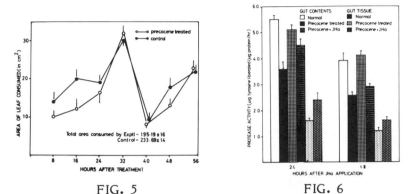

FIG. 5   FIG. 6

FIG. 5. Graph showing feeding by Precocene treated larvae of A.janata. Each point denotes mean of 8 separate determinations and the bars are ± SEM. Fig.6. Graph representing the effect of Precocene on midgut protease activity in Larval A.Janata. Each value denotes mean of 8 separate determinations and the bars ± SEM.

1980). However Kiguchi (1982) demonstrated that in Bombyx mori precocene induced precocius moulting. It became also evident from the present studies that precocene treatment in A. janata larvae reduced both food consumption and midgut pretease activity. Socha and Hodkova (1982) demonstrated that in D.cingulatus precocene treatment reduced food consumption and according to Slama (1978) precocenes act as an atifeedant in certain insect species. Median neurosecretory cells were found to stimulate food consumption and the amount of food consumed stimulated midgut protease and invertase activity in D.cingulatus through a secretogogue mechanism (Muraleedharan and Prabhu, 1979). So the present findings suggest that in the caterpillers of A.janata, precocene-II retards the activity of A-type NS cells of the brain and the activity of CA while it is insensitive to the B-type NS cells of SOG. As in some other insects viz: D.cingulatus, Hyblaea puera, brain NS cells and CA regulate food consumption in A.janata also, the reduced level of feeding noticed in precocene treated A.janata can be well explained. A reduction in food consumption effected through these endocrines must have lowered the midgut protease activity as well.

Acknowledgements: Authors are thankful to Dr. J.J. Solomon, Scientist, C.P.C.R.I., Kayamkulam for the help rendered in EM work. Research grants received from the University Grants Commission, India also is gratefully acknowledged.

REFERENCES

Bowers W.S (1976) Discovery of insect antiallatotropins. In The Juvenile Hormones (Ed. by Gilbert L.I.)pp. 349-408. Plenum Press, New York.

Bowers W.S.(1980) Chemistry of Plant/Insect Interactions. In Insect Biology in the Future (Ed. by Locke M. and Smith D.S.) pp.613-633. Academic Press, New York.

Fagoonee I and Umrit G. (1981) Antigonadotropic hormones from the goatweed, ageratum conyzoides. Insect. Sci.Appl.1, 373-376.

Feyereisen R., Johnson G., Koener J., Stay B. and Tobe S.S. (1981) Precocenes as pro-allatiocidins in adult female Deploptera punctata; A functional and ultrastructural study. J. Insect Physiol.27, 855-868.

Kelley T.J. and Fuchs M.S. (1978) Precocene is not a specific anti-gonadotropic agent in adult female Aedes aegypti Physiol. Ent.3, 297-301.

Kiguchi K. (1982) Effects of an juvenile hormones on the development of the silkworm. The 5th Interant. Congr. Pesticide Chem, Kyoto, Japan.

Mackauer M., Nair K.K. and Unnithan G.C. (1979) Effect of precocene-II on alate production in the pea aphid, acyrthsiphon pisum. Cand.J.Zool,57, 856-859.

Muraleedharan D. and Prabhu V.K.K. (1979) Role of median neurosecretory cells in the secretion of protease and invertase in the red cotton bug, Dysdercus cingulatus. J.Insect Physiol. 25, 237-240.

Muraleedharan, D and Prabhu V.K.K.(1981) Hormonal influence on feeding and digestion in a plant bug, Dysdercus cingulatus and a caterpillar, Hyblaea puera. Physiol Ent.6, 183-189.

Pratt G.E.(1983) The mode of action of Pro-allatocidins. In Natural Products for Innovative Pest Management.

STUDIES ON THE REGULATION, FUNCTION AND PROPERTIES

OF JUVENILE HORMONE ESTERASES OF <u>TRICHOPLUSIA NI</u>

Grace Jones

Department of Entomology, University of

Kentucky, Lexington, Ky 40546

The demonstration of juvenile hormone esterase (JHE) activity in insect hemolymph during the early 1970's stimulated interest in degradation as a potentially important level of regulation of JH titers. Early studies laid the foundation for the initial model of JHE (Gilbert et al. 1980). The functions, properties and mechanisms of regulation of JHE postulated by that model prompted subsequent research on these topics. Unresolved aspects included: experimental demonstration that JHE is necessary for metamorphosis; whether JHE activity is due to a single protein; the degree of 'specificity'of JHE; whether endogenous JH or neurohormonal factors regulate JHE; the biochemical properties of JHE, etc. These questions have been recently addressed, primarily with lepidopteran systems, and in particular with <u>Trichoplusia ni</u>. In the present review, our recently published work on this model system will be examined, as well as research currently being conducted in our laboratory.

<u>Necessity of JHE activity for metamorphosis</u>. JHE hydrolysis of the JH ester to acid led to the proposition that the enzyme functions to reduce endogenous JH. There is an important difference between the ability to hydrolyze JH and the actual necessity of this enzymatic activity for metamorphosis. Recent data show that application of nonpharmacological levels of a JHE inhibitor to <u>T</u>. <u>ni</u> causes a supernumerary larval molt (Jones 1985), through accumulation of

315

endogenous JH (unp. data). Blockage of JH degradation also
delays wandering behavior, a behavior symptomatic of metamorphosis. After metamorphic commitment, removal of the
corpora allata (CA) blocks prepupal development in T. ni,
but JH application to these prepupae restores development
(Jones and Hammock 1983). Careful examination of ecdysteroid titers in neck-ligated prepupae shows that JH application accelerates the prepupal ecdysteroid increase. Through
its effect on JH, JHE has been demonstrated to influence the
timing of the prepupal ecdysteroid peak (Jones et al. 1986c).

     Properties of JH esterase of T. ni. The purification
of JHE from T. ni remained elusive for several years.
Actually, high resolution isoelectric focusing has identified at least two major electrophoretic forms (pI 5.5, 5.3)
and two minor forms (pI 5.6, 5.1) (Jones et al. 1986a;
Jones and Click 1986). Purification procedures have
been developed for a preparation homogeneous for the
electrophoretic forms (Rudnicka and Jones 1986) and for
individual isolation of the two major forms (unp. data).
The latter material is of high yield (ca. 40%) and has very
high specific activity (ca. 2 umoles/min/mg). All of
these electrophoretic forms have a molecular weight near
66,000, and tests with antibodies to T. ni JHE demonstrate
that they are immunologically related (unp. data).
Initial studies on T. ni JHE by various laboratories have
provided disparate values for the apparent K_m. It is now
known that in crude hemolymph preparations JHE functions
kinetically as two active sites, with apparent K_ms in the
10^{-8} M and 10^{-7} M regions (Jones et al. 1986c). Partially or
highly purified JHE from T. ni does not exhibit the lower K_m
activity. Each of the two major electrophoretic forms are
kinetically very similar (Jones et al. 1986c), and also
accept alpha-naphthyl acetate (αNA) as a substrate
(Rudnicka and Jones 1986). This result does not mean that
JHE is a major contributor to hemolymph αNA esterase
activity. Proposed transition state mimic inhibitors will
be useful tools, although it is unclear whether their
effectiveness is due 1) only to a strong positive polarity
on the carbonyl carbon interacting with the active site
serine hydroxyl group (not a true transition state mimic),
2) to a conformation attained at the catalytic site after
interaction with the initial binding site (Hammock et al.
1982), or 3) to a hydrated conformation attained prior to
the initial binding interaction (Abdel-aal et al. 1984).

Each of these three alternatives could require different strategies in inhibitor design.

Regulation of T. ni JHE. During the final larval stadium of Lepidoptera, JHE occurs as a large peak during the late feeding phase and as a second prepupal peak (Jones et al. 1982). Induction of JHE activity by exogenous JH in post-feeding stage Lepidoptera suggested that, in vivo, JH induces its own degrading enzyme (Whitmore et al. 1974; Sparks and Hammock 1980). This proposal was confirmed in T. ni (Jones and Hammock 1983) and then in other insects by the effects of surgical removal and implantation of the CA. The fat body is the major source of JHE and its preparation as a source of in vitro enzyme activity was demonstrated using Manduca sexta (L.) (Whitmore et al. 1974). We have found that both the JH induction of JHE activity in a T. ni in vitro system and its inhibition by actinomycin D occur in a dose-dependent manner (Jones et al., unp. data). All four electrophoretic forms are produced. During the feeding stage, neural tissue (primarily the brain) appears directly responsible for regulation of JHE activity (Jones et al. 1981). Using the in vitro system, we have demonstrated a direct action of the neurohormonal material on the fat body. This system should be a useful assay procedure for the purification of the trypsin and heat sensitive factor.

Supported by NIH grant GM 33995. Published with the approval of the Director, Ky. Agr. Exp. Stn (86-7-158).

REFERENCES

Abdel-aal, Y. A. I, Roe, R. M. and Hammock, B. D. (1984) Kinetic properties of the inhibition of juvenile hormone esterase by two trifluoromethylketones and O-ethyl, S-phenyl phosporamidothiolate. Pest. Bio. Phys. 21, 232-241.
Gilbert, L. I., Bollenbacher, W. E., Goodman, W., Smith, S. L., Agui, N., Granger, N. and Sedlak, B. J. (1980) In Recent Progress in Hormone Research, Acad. Press, Inc.
Hammock, B. D., Wing, K. D., McLaughlin, J., Lovell, V. M. and Sparks, T. C. (1982) Trifluoromethyketones as possible transition state analog inhibitors of juvenile hormone esterase. Pest. Biochem. Physiol. 17, 76-88.
Jones, D., Jones, G., Rudnicka, M., Click, A. J. and Sreekrishna, S. C. (1986a) High resolution isoelectric focus-

ing of juvenile hormone esterase activity from the haemolymph of Trichoplusia ni (Heubner). Experientia 42, 45-47.

Jones, D., Jones, G., Click, A., Rudnicka, M. and
Sreekrishna, S. (1986b) Multiple forms of JHE active sites
in the hemolymph of larvae of Trichoplusia ni. Comp.
Biochem. Physiol. (in press).

Jones, D., Jones, G., Wing, K. D., Rudnicka, M. and Hammock,
B. D. (1982) Juvenile hormone esterases of Lepidoptera I.
Activity in the hemolymph during the last instar of 11
species. J. Comp. Physiol. 148, 1-10.

Jones, G. (1985) The role of juvenile hormone esterase in
terminating larval feeding and initiating metamorphic development in Trichoplusia ni. Entomol. Exp. Appl. 39,
170-176.

Jones, G. and Click, A. (1986) Developmental regulation of
juvenile hormone esterase in Trichoplusia ni: its multiple
electrophoretic forms occur during each larval ecdysis. J.
Insect Physiol. (in press).

Jones, G., Click, A., Reck-Malleczewen, V., Loeb, M. J.,
Brandt, E. P. and Woods, C. W. (1986c) Haemolymph ecdysteroid titre in larvae of the cabbage looper Trichoplusia
ni. J. Insect Physiol. 32, 561-566.

Jones, G. and Hammock, B. D. (1983). Prepupal regulation
of juvenile hormone esterase through direct induction by
juvenile hormone. J. Insect Physiol. 29: 471-475.

Jones, G., Wing, K. D., Jones, D. and Hammock, B. D. (1981)
The source and action of head factors regulating juvenile
hormone esterase in larvae of the cabbage looper
Trichoplusia ni. J. Insect Physiol. 27, 85-91.

Rudnicka, M. and Jones, D. (1986) Characterization of
homogeneous juvenile hormone esterase from larvae of
Trichoplusia ni. Insect Biochem. (in press).

Sparks, T. C. and Hammock, B. D. (1980) Comparative
inhibition of the juvenile hormone esterases from
Trichoplusia ni, Musca domestica and Tenebrio molitor.
Pestic. Biochem. Physiol. 14, 290-302.

Whitmore, D., Gilbert, L. I. and Ittycheriah, P. I. (1974)
The origin of hemolymph carboxylesterase induced by the
insect juvenile hormone. Molec. Cell. Endoc. 1, 37-54.

DISSOCIATION OF THE PROTHORACIC GLANDS OF MANDUCA SEXTA INTO HORMONE-RESPONSIVE CELLS

Wendy A. Smith

Dept. of Biology, Northeastern University

Boston, MA 02115

Dorothy B. Rountree, Walter E. Bollenbacher, and Lawrence I. Gilbert

Dept. of Biology, University of North Carolina

Chapel Hill, NC 27514

INTRODUCTION

Prothoracicotropic hormone (PTTH) is a cerebral neuropeptide which stimulates the synthesis of the steroid prohormone, ecdysone, by the insect prothoracic glands. In the tobacco hornworm, Manduca sexta, both calcium and cAMP have been found to play important roles in the steroidogenic action of PTTH (see Smith and Gilbert, 1986). These second messengers appear to act sequentially, with PTTH stimulating a calcium-dependent increase in cAMP formation, and the cyclic nucleotide in turn activating cAMP-dependent protein kinase and protein phosphorylation in a calcium-independent manner.

Experiments to date have employed an in vitro assay of intact prothoracic glands, a protocol originally described by Bollenbacher et al. (1983). The intact-gland protocol was designed primarily for the bioassay of active fractions of PTTH during purification and for studies of developmental changes in ecdysone synthesis. Such assays are best conducted under conditions that are as close as possible to

319

physiological conditions, i.e., with minimal manipulation of
tissue between the time of animal sacrifice and the time of
challenge with the peptide fractions of interest. The
utility of this assay has been demonstrated repeatedly
and has led to the isolation and characterization of 29 kD
and 7 kD peptides with stage-specific steroidogenic effects in
Manduca (Bollenbacher et al., 1984). However, for detailed
studies regarding the mode of action of PTTH, we required an
assay preparation with the following characteristics: 1) reduced
variability among replicate samples, 2) enhanced accessibility
of cells to hormones and pharmacological test agents, and 3)
homogeneity and manipulability, required for procedures that
involve rapid handling of large number of cells (for example,
studies of calcium uptake). For these reasons, a technique
was developed whereby the intact prothoracic glands of Manduca
sexta could be dispersed into viable, hormone-responsive cells.

RESULTS

The following procedure was found to provide single
cells and groups of 2-5 cells, >95% of which remained viable
for at least 4 hr as judged by trypan blue exclusion (all
incubations carried out at 25° C): 1) prothoracic glands from
day 0 pupal Manduca sexta were incubated for 20 min in
Grace's medium containing 0.4% trypsin/chymotrypsin/elastase
(Sigma #T5266); 2) cells were dissociated by drawing the
tissue in and out of a 27 gauge needle attached to a 1 ml
plastic syringe 20 times; 3) cells were rinsed 3 times by
centrifuging through Grace's medium (100 x g, 3 min), and
resuspended in Grace's medium at a concentration of 2-4 gland
equivalents per 0.05 ml. Aliquots of the cell suspension
(0.05 ml) were placed in polystyrene microfuge tubes and
challenged for 90 min in control Grace's medium or medium
containing PTTH or pharmacological agents. Following
centrifugation at 10,000 x g, ecdysone synthesis was measured
by radioimmunoassay of the supernatant (see Bollenbacher et
al., 1983). It should be noted that enzyme preparations
consisting primarily of collagenase (used for dissociating
locust fat body, Asher et al., 1983), or elastase (used for
dissociating insect Malpighian tubules and imaginal discs,
Levinson and Bradley, 1984) were not effective in dissociating
prothoracic glands of Manduca. Bovine serum albumin (Fraction
V, 0.3%) was found to enhance both basal, and PTTH-
stimulated, ecdysone synthesis. However, significant,
reproducible responses (2 to 3-fold stimulation of pupal glands

in response to HPLC-purified big PTTH) were obtained in the absence of albumin, thus this protein was not added in subsequent experiments.

Sensitivity to big PTTH was statistically equivalent in dispersed-cell and intact-gland preparations (ED50 for dispersed cells = 2 Units big PTTH/ml; ED50 for intact glands = 2.5 Units PTTH/ml). Further, as has previously been shown for intact prothoracic glands (see Smith and Gilbert, 1986), the dispersed cells exhibited a significant (2 to 3-fold) increase in ecdysone synthesis in response to the calcium ionophore, A23187 (0.01 mM), and dibutyryl cAMP (5 mM). In all cases, the coefficient of variance of 5-10 replicate samples was less than 10%.

The effects of several additional agents on prothoracic gland activity were tested. The tested agents were: 1) phorbol myristate acetate (PMA), a potent activator of protein kinase C; 2) BAY K8644 and nitrendipine, which act as calcium-channel agonists and antagonists, respectively, in mammalian cells; and 3) cycloheximide, a protein synthesis inhibitor. As shown in Table 1A, PMA did not stimulate basal ecdysone synthesis, nor did this compound potentiate or inhibit the steroidogenic action of PTTH. As indicated in Table 1B, exposure to the calcium channel agonist BAY K8644 led to a slight, but significant stimulation of ecdysone synthesis ($p < 0.02$). Exposure to the calcium-channel antagonist nitrendipine led to a slight, but significant inhibition of PTTH-stimulated ecdysone synthesis ($p < 0.01$), and cycloheximide strongly inhibited PTTH-stimulated ecdysone synthesis ($p < 0.001$).

Table 1. Effects of pharmacological agents on basal and PTTH-stimulated ecdysone synthesis by dispersed cells.

A. Treatment	Dose	Ecdysone synthesis (ng/90 min) (Mean $\pm$ S.E.M, n = 5-10/treatment)
Control	----	2.76 ± 0.07
PMA	0.05 mM	2.78 ± 0.12
PTTH	2 U/ml	4.21 ± 0.33
PTTH + PMA		4.36 ± 0.22

B. Treatment	Dose	Ecdysone synthesis (ng/90 min)
Control	----	2.85 + 0.30
BAY K8644	0.02 mM	4.05 + 0.12
PTTH	4 U/ml	8.40 + 0.48
PTTH + Nitrendipine	0.02 mM	7.03 + 0.45
PTTH + Cycloheximide	0.10 mM	4.48 + 0.48

The following conclusions can be drawn from these results: 1) prothoracic glands can be dispersed into viable cells which remain responsive to hormones and pharmacological agents; 2) variability among replicates is reduced in dispersed cell preparations, permitting the detection of subtle changes in cellular response; 3) the results of pharmacological studies suggest that protein kinase C is not involved in the acute stimulation of ecdysone synthesis; but favor roles for both calcium and protein synthesis in neuropeptide-stimulated steroidogenesis.

REFERENCES

Asher C., Moshitzky P., Ramachandran J. and Applebaum S. W. (1984) The effects of synthetic locust adipokinetic hormone on dispersed locust fat body cell preparations. Gen. Comp. Endocrinol. 55, 167-173.

Bollenbacher W. E., O´Brien M. A., Katahira E. J. and Gilbert L. I. (1983) A kinetic analysis of the action of the insect prothoracicotropic hormone. Molec. Cell. Endocrinol. 32, 27-46.

Bollenbacher W. E., Katahira E. J., O´Brien M. A., Gilbert L. I., Thomas M. K., Agui N. and Baumhover A. H. (1984) Insect prothoracicotropic hormone: evidence for two molecular forms. Science 224, 1234-1245.

Levinson G. and Bradley T. J. (1984) Removal of insect basal laminae using elastase. Tissue and Cell 16, 367-375.

Smith W. A. and Gilbert L. I. (1986) Cellular regulation of ecdysone synthesis by the prothoracic glands of Manduca sexta. Insect Biochem. 16, 143-147.

ACKNOWLEDGMENTS: BAY K6844 and nitrendipine were generously provided by Dr. A. Scriabine (Miles Laboratories, New Haven, CT). This work was supported by NIH grants AM-37435, NS-18791, and AM-30118; and by an RSDF Award from Northeastern University.

GYPSY MOTH PROTHORACIC GLAND: RESPONSE TO A SYNTHETIC PEPTIDE ANALOG

Chih-Ming Yin, Shu-Xia Yi and John H. Nordin

University of Massachusetts

Amherst, MA   01003,   U.S.A.

Insect prothoracicotropic hormones (PTTHs) are brain neuropeptides which act on prothoracic glands (PGs) to regulate secretion of ecdysone. Partial sequencing (19 amino acids) of three 4K-PTTH in silkworms (Nagasawa et al., 1984), and complete sequencing and identification of A- and B- chains of 4K-PTTH (Nagasawa et al., 1986) have been accomplished. This paper reports the prothoracicotropic effect of a synthetic peptide containing the first 15 amino acids (from N-terminus) of the silkworm 4K-PTTH-II.

Gypsy moths and southwestern corn borers were reared using published methods (Hollander and Yin, 1985; Yin et al., 1985). Conditions for PG-PTTH incubations and the radioimmunoassay for ecdysone were described (Yin et al., 1985). The pentadecapeptide containing 3 cysteines was synthesized at Children's Hospital, Boston. To encourage -S-S-bond formation between cysteines at A6 (position 6, A-chain) and A11, thus imitating that of insulin, the -SH at A7 was blocked with an acetamidomethyl group. The choice was made prior to the discovery (Nagasawa et al., 1986) that 4K-PTTH-II has A- and B-chains. The synthetic peptide contains 83.1% peptide by weight and after hydrolysis, amino acid molar ratios are: Asp, 1.95; Ser, 0.99; Glu, 0.94; Pro, 1.86 (may contain Cys); Gly, 0.94; Cys, N.D.; Val, 2.81; Ile, 0.50; Leu, 1.08; Arg, 1.10.

We measured the basal in vitro ecdysteroid secretion
from PGs of 3, 5, 8 and 10 day-old female last instar gypsy
moths, and pharate and newly-formed pupae. PGs of pharate
and new pupae, and of day-3 and -5 larvae, secreted either
too much (more than 6.8 ng) or too little (less than 0.2
ng) ecdysone. We decided that PGs from day-8 larvae
produced an average of 2.7 ng of ecdysone per 6 hr incubation is suitable for PTTH testing. We found that left
and right PGs from day-8 last instar female larvae secreted
comparable amounts of 3.53 + 0.41 and 3.73 + 0.48 ng
ecdysteroid/gland/ 6 hr, respectively (t-test, df=10, t=
0.32, P>0.05), a prerequisite for the bioassay. Since
gypsy moths were used to test a PTTH analog modeled after
the silkworm hormone, gypsy moth PGs were monitored for
their response to PTTH from mature 6th instar corn borers.
When treated with two, one, and 0.25 brain equivalents of
PTTH extract from corn borers, gypsy moth PG secreted 5.00
+ 0.36, 2.31 + 0.56, and 1.28 + 0.07 fold more ecdysteroid
than controls.

Tests showed that 10 ng of this synthetic analog evoked
a 100% increase in ecdysteroid secretion (activation ratio=
2.05 + 0.25), while 1 ng (Ar=1.22 + 0.17) and 100 ng (Ar=
0.75 + 0.12) caused no such effect. This 100% increase is
significantly different from controls (t-test, t=2.83, df=
18, p<0.05). To evaluate the importance of the -S-S-bond
in the pentadecapeptide's activity, PGs were incubated with
(10 ng) or without it in 50 ul Grace's medium containing
3.1 ng of dithioerythritol (DTE) to reduce the -S-S-bonds.
At the level of 3.1 ng DTE, control and test glands secreted 3.05 + 0.41 and 2.43 + 0.28 ng ecdysteroid/gland/6 hr.
The difference is not significant (t-test, df=18, t=1.26,
p>0.05).

Our results agree with earlier reports that there are
cross-reactivities among PTTHs and PGs of different moths
(Bollenbacher and Granger, 1985). The activity (Ar=2.05)
of the pentadecapeptide is not very high, but considering
that it contains only 1/3 of the amino acids of 4K-PTTH-II,
expectation of high activity is unrealistic. We believe the
Ar of 2.05 can be increased if PG of lower basal activity

are used. It is known that PGs of different basal activities respond to PTTHs or brain-extract differently (Yin et al., 1985).

The prothoracicotropic effect of the pentadecapeptide indicates that it may bind to PTTH-receptors. Study of insulin receptor-binding suggested that amino acids at A1 (Gly), A5 (Gln), A19 (Tyr) and A21 (Asn) of insulin may be involved (Pullen et al., 1976). It is not known if A1 (Gly) and A5 (Glu) of the pentadecapeptide are also involved in PTTH receptor-binding. If they are, insulin and 4K-PTTH should also be compared for their affinity, and specificity at the receptor level. DTE eliminated the prothoracicotropic effect of pentadecapeptide, but did not appear to alter the basal ecdysone secretion of control PGs. Basal activities were 3.05 ± 0.41 ng/gland/6 hr (with) or 3.53 ± 0.41 ng/gland/6 hr (without) DTE. The difference is not significant (t-test, t=0.78, df=14, p >0.05) and suggests that disulfide bond mediated structural features of PGs may not be essential for the biosynthesis and secretion of ecdysone.

Acknowledgements -- This work is supported by an NIH grant (RR07048-20) and Hatch project 570 and is published as Massachusetts Agricultural Experiment Station journal No. 2793.

REFERENCES

Bollenbacher W.E. and Granger N.A. (1985) Endocrinology of the prothoracicotropic hormone. In Comprehensive Insect Physiology, Biochemistry, and Pharmacology, Vol. 7 (Ed. by Kerkut G.A. and Gilbert L.I.), pp. 109-151.

Hollander A.L. and C. -M.Yin (1985) Lack of humoral control in calling and pheromone release by brain, corpora cardiaca, corpora allata and ovaries of the female gypsy moth, Lymantria dispar (L.). J. Insect Physiol. 31, 159-163.

Nagasawa H., Kataoka H., Isogai A., Tamura S., Suzuki A.,

Ishizaki H., Mizoguchi A., Fujiwara Y. and Suzuki A. (1984) Amino-terminal amino acid sequence of the silkworm prothoracicotropic hormone: homology with insulin. Science 226, 1344-5.

Nagasawa H., Kataoka H., Isogai A.. Tamura S., Suzuki A., Misoguchi A., Fujiwara Y., Suzuki A., Takahashi S. and Ishizaki H. (1986) Amino acid sequence of a prothoracicotropic hormone of the silkworm Bombyx mori. PNAS USA 83, in press.

Pullen R.A., Lindsay D.G., Wood S.P., Tickle I.J., Blundell T.L., Wollmer A., Krail G., Brandendurg D., Zahn H., Gliemann J., and Gammeltoft S. (1976) Receptor-binding region of insulin. Nature 259, 369-373.

Yin C. -M., Wang Z. -S. and Chaw W. -D. (1985) Brain neurosecretory cell and ecdysiotropin activity of the non-diapausing, pre-diapausing and diapausing southwestern corn borer, Diatraea grandiosella. J. Insect Physiol. 31, 659-667.

PROTHORACICOTROPIC HORMONE STIMULATION OF ECDYSONE SYNTHESIS

BY THE PROTHORACIC GLANDS OF THE GYPSY MOTH, LYMANTRIA DISPAR

Thomas J. Kelly, Edward P. Masler, *Belgaum S. Thyagaraja, Robert A. Bell and Alexej B. Borkovec

Insect Reproduction Laboratory, ARS, USDA, *RSRS Central Silk Board and Department of Zoology, University of Maryland

Beltsville, MD, USA, *Kollegal, Karnataka, India and College Park, MD, USA

Kopec (1922), using Lymantria, first demonstrated that a factor(s) from the head was necessary for insect metamorphosis. This factor was later termed prothoracicotropic hormone (PTTH) (Kobayashi et al., 1965). PTTH stimulates ecdysone synthesis by the paired prothoracic glands (PGs) in lepidopterans, and ecdysone is subsequently converted by other tissues to the active hormone, 20-hydroxyecdysone (see Gilbert et al., 1980). In vitro bioassays for PTTH have been established in both Bombyx mori (Nagasawa et al., 1984; Okuda et al., 1985) and Manduca sexta (Bollenbacher et al., 1979) and have been utilized to quantitate PTTH activity in various tissues and fractions (see Ishizaki and Suzuki, 1984; Bollenbacher and Granger, 1985).

In Lymantria, both members of a pair of PGs produce nearly equal amounts of ecdysone in vitro and can respond to brain extracts containing PTTH activity (Fig. 1). Similar to Bombyx (see Fig. 9, Okuda et al., 1985), the PGs of last-instar Lymantria respond to PTTH at any day of the last instar. When the level of response is quantitated by comparing the

---

Mention of a commercial or proprietary product in this paper does not constitute an endorsement of this product by the USDA

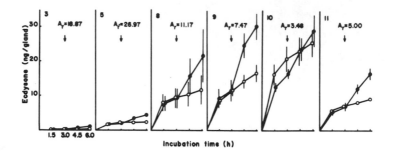

Fig. 1. Kinetics of ecdysteroid synthesis by 5th-instar
female Lymantria PGs. Glands were dissected (on the day indicated in the upper left) in insect Ringer's (Okuda et al.,
1985), transferred to Grace's medium (GIBCO, Grand Island, NY;
25ul per gland in polystyrene dishes) within 15-30 min of
dissection and incubated for the indicated times at 25°C,
constant light in a humidified chamber. The medium was exchanged for fresh medium or extract every 1.5h, and 1- or 10-
ul samples were analyzed by ecdysteroid radioimmune assay
according to Kelly et al. (1986) except that ecdysone (Sigma
Chemical Co., St. Louis, MO) was used as the unlabelled lig-
and. One gland (●) of each pair received brain extract (0.25
brain equiv./25ul - prepared from pooled brains of day 5-8
5th-instar females by gently homogenizing, placing in boiling water for 2min and centrifuging at 10,000xg for 1min) beginning at 3.0 h (arrow). Control gland (○). Each point
represents the mean + SE from at least 3 determinations. The
4th-5th larval ecdysis occurred on the night before day 1 and
pupation occurred on the night following day 11.

amount of ecdysone synthesis by extract-activated glands from
3.0 - 4.5 h (Fig. 1) with that of control glands (activation
ratio, A_r), the glands show a maximum response at day 5 with
an A_r of 26.97 + 3.36. This correlates with the time at which
maximum weight is attained and with the onset of wandering
behavior. Using A_r versus brain equivalent dose, we construc-

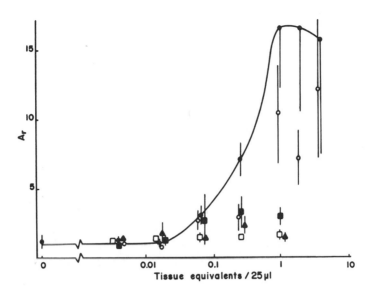

Fig. 2. Ecdysone synthesis dose-response by <u>Lymantria</u> PGs
(day 5-8) to various extracts. Gland dissection, incubation
and tissue preparation were as indicated in Fig. 1 except
that fresh medium or extract was only exchanged one time at
2.0 h and the 2-4-h medium was used to determine the A_r.
Each point represents the mean + SE from at least 3 determinations. Brain (●), retrocerebral complex (O), 3rd thoracic ganglion (■), abdominal leg muscle (□), fat body (△).

ted a dose-response (Fig. 2) that is saturable and thus indicative of specific hormonal stimulation (Stone and Mordue,
1980). The response is specific to extracts from the brain
and retrocerebral complex with extracts of the third thoracic
ganglion, abdominal leg muscle and fat body showing minimal
activity (Fig. 2). This <u>in vitro</u> bioassay, established for
<u>Lymantria</u> PTTH, will be useful in purification and compara-
tive physiological studies of this neurohormone.

REFERENCES

Bollenbacher W.E., Agui N., Granger N.A. and Gilbert L.I.
(1979) In vitro activation of insect prothoracic glands
by the prothoracicotropic hormone. Proc. Natl. Acad. Sci.
76, 5148–5152.

Bollenbacher W.E. and Granger N.A. (1985) Endocrinology of
the prothoracicotropic hormone. In Comprehensive Insect
Physiology, Biochemistry and Pharmacology, Vol.7,
Endocrinology I (Ed. by Kerkut G.A. and Gilbert L.I.), pp.
109–151, Pergamon Press, Oxford.

Gilbert L.I., Bollenbacher W.E. and Granger N.A. (1980)
Insect endocrinology: regulation of endocrine glands, hormone titer, and hormone metabolism. Ann. Rev. Physiol.
42, 493–510.

Ishizaki H. and Suzuki A. (1984) Prothoracicotropic hormone
of Bombyx mori. In Biosynthesis, Metabolism and Mode of
Action of Invertebrate Hormones (Ed. by Hoffmann J. and
Porchet M.) pp. 63–77, Springer–Verlag, Berlin.

Kelly T.J., Masler E.P., Schwartz M.B. and Haught S.B. (1986)
Inhibitory effects of oostatic hormone on ovarian maturation
and ecdysteroid production in Diptera. Insect Biochem.
16, 273–279.

Kobayashi M., Saito M. and Yamazaki M. (1965) Prothoracicotropic hormone in an insect, Bombyx mori. XXIII Int. Cong.
Physiol. Sci. Abstract papers, p. 246.

Kopec S. (1922) Studies on the necessity of the brain for
the inception of insect metamorphosis. Biol. Bull. Wood's
Hole 42, 322–342.

Nagasawa H., Kataoka H., Hori Y., Isogai A., Tamura S., Suzuki A., Guo F., Zhong X., Mizoguchi A., Fujishita M., Takahashi S., Ohnishi E. and Ishizaki H. (1984) Isolation and
some characterization of the prothoracicotropic hormone
from Bombyx mori. Gen. Comp. Endocr. 53, 143–152.

Okuda M., Sakurai S. and Ohtaki T. (1985) Activity of the prothoracic gland and its sensitivity to prothoracicotropic
hormone in the penultimate and last-larval instar of Bombyx
mori. J. Insect Physiol. 31, 455–461.

Stone J.V. and Mordue W. (1980) Adipokinetic hormone. In
Neurohormonal Techniques in Insects (Ed. by Miller T.), pp.
3–80, Springer–Verlag, New York.

DISCOVERY AND PARTIAL CHARACTERIZATION OF PROTHORACICOTROPIC HORMONES OF THE GYPSY MOTH, LYMANTRIA DISPAR

Edward P. Masler, Thomas J. Kelly, *Belgaum S. Thyagaraja, Charles W. Woods, Robert A. Bell and Alexej B. Borkovec

Insect Reproduction Laboratory, ARS, USDA, *RSRS, Central Silk Board and Department of Zoology and University of Maryland

Beltsville, MD, USA and *Kollegal, Karnataka, India and College Park, MD, USA.

The brain neuropeptide prothoracicotropic hormone (PTTH) is an ecdysiotropin that drives post-embryonic growth and development through the stimulation of ecdysteroid production by the prothoracic glands (PG) (Gilbert et al., 1980). Research to isolate PTTH has utilized primarily two species, Bombyx mori (Ishizaki and Suzuki, 1984) and Manduca sexta (Bollenbacher and Granger, 1985). Characterization and isolation studies have used both in vivo (Bombyx, Ishizaki et al, 1983, Nagasawa et al, 1984a; Manduca, Kingan, 1981) and in vitro (Bombyx, Nagasawa et al, 1984a; Okuda et al, 1985; Manduca, Bollenbacher et al, 1979) bioassays. Recent work on Bombyx and Manduca PTTH demonstrates that biological activity is associated with two molecular-weight ranges of ca. 20-28 kD and ca. 4-7 kD (Ishizaki et al., 1983, Bollenbacher et al., 1984). A family of 4 kD PTTH's has been isolated from Bombyx (Nagasawa et al., 1984b).

We used a two-phase in vitro PG bioassay (see Fig. 1) coupled with an ecdysteroid radioimmunoassay to detect PTTH activity in neural extracts of Lymantria. A dose response by day

---

Mention of a commercial or proprietary product in this paper does not constitute endorsement by the USDA.

331

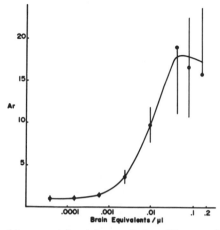

Figure 1. In vitro activation of L. dispar PG by L. dispar
brain extract. Five to 8-day, 5th instar female brains were
homogenized in Grace's medium (Gibco, Grand Island, NY)
heated (100°C, 2 m), centrifuged (1 m, 10,000xg) and diluted
in Grace's. Paired PG were dissected from 5-8-day, 5th instar females and incubated in 25 ul drops of Grace's at 26°C
in a humid chamber for 2 h (phase-1). Medium was removed
and replaced with fresh medium only (control gland), or brain
extract (test gland), and incubations continued for 2 h
(phase-2). Aliquots (10 ul) of phase-2 medium were tested
for ecdysteroid titer by radioimmunoassay (RIA) (Masler and
Adams, 1986). The ratio of the amount of ecdysteroid present in test medium to that present in control medium (activation ratio, Ar) was used as an index of PTTH activity.
Each point is the average of 4 to 9 determinations ± SE.

5-8, 5th instar female PG to increasing concentrations of
brain extract was observed (Fig. 1). There was an approximate 16-fold linear activation range with a maximum Ar of
16-19. Brain extracts fractionated by high performance size
exclusion chromatography (HP-SEC) with a physiological buffer
yielded a single peak of activity near 20 kD (Fig. 2a).
Brain extracts prepared and chromatographed in acidic organic
solvents (Fig. 2b) showed two activity peaks near 15-20 kD
(PTTH-I) and 4-6 kD (PTTH-II). Thus, Lymantria PTTH appears
to exist in multiple molecular sizes. The biochemistry of
these forms and the kinetics of PG activation by PTTH-I and
-II are being examined.

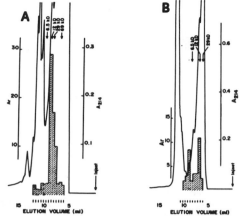

Figure 2. HP-SEC fractionations of L. dispar brain extract. Tissues and bioassay were as described for Figure 1. A. Brains were homogenized in 0.2M KH$_2$PO$_4$, pH 7.0, centrifuged (1 m, 10,000xg) and the supernatant fractionated on a Waters I-125 size-exclusion column (Waters Assoc., Milford, MA) at 1ml/min, with extraction buffer as mobile phase. One-half ml fractions were collected, dialyzed against HPLC-water (2000 molecular weight cut-off membrane), dried under vacuum and retentate dissolved in Grace's for bioassay. B. Brains were homogenized in acidic methanol (90% CH$_3$OH, 0.1% tri-fluoracetic acid, TFA, in water), centrifuged as above, and supernatant dried under vacuum. The residue was dissolved in 40% CH$_3$CN/0.1% TFA and fractionated (as for 2A) with 40% CH$_3$CN/0.1% TFA as the mobile phase. Dried fractions were dissolved in Grace's for bioassay. Molecular weight markers used were bovine serum albumin (69 kD), carbonic anhydrase (29 kD), cytochrome C (12 kD), aprotinin (6.5 kD). Shaded area represents PTTH activity (Ar).

REFERENCES

Bollenbacher W.E., Agui N., Granger N.A. and Gilbert L.I. (1979) In vitro activation of insect prothoracic glands by the prothoracicotropic hormone. Proc. Nat. Acad. Sci. USA 76, 5148-5152.

Bollenbacher W.E. and Granger N. (1985) Endocrinology of the prothoracicotropic hormone. In Comprehensive Insect Physiology, Biochemistry and Pharmacology vol. 7 (Ed. by Kerkut, G.A. and Gilbert, L.I.), pp. 109-151. Pergamon Press, Oxford.

Bollenbacher W.E., Katahira E.J., O'Brien M., Gilbert L.I., Thomas M.K., Agui N. and Baumhover A.H. (1984) Insect prothoracicotropic hormone: evidence for two molecular forms. Science 224, 1243-1245.

Gilbert L.I., Bollenbacher W.E., Goodman W., Smith S.L., Agui N., Granger N. and Sedlak B.J. (1980) Hormones controlling insect metamorphosis. Rec. Prog. Horm. Res. 36, 401-449.

Ishizaki H., Mizoguchi A., Fujishita M., Suzuki A., Moriya I., O'Oka H., Kataoka H., Isogai A., Nagasawa H., Tamura S. and Suzuki A. (1983) Species specificity of the insect prothoracicotropic hormone (PTTH): The brain of Bombyx mori. Develop. Growth Differ. 25, 593-600.

Ishizaki H. and Suzuki A. (1984) Prothoracicotropic hormone of Bombyx mori. In Biosynthesis, Metabolism and Mode of Action of Invertebrate Hormones (Ed. by Hoffmann, J. and Porchet, M.) pp 63-77. Springer-Verlag, Berlin.

Kingan T.G. (1981) Purification of the prothoracicotropic hormone from the tobacco hornworm Manduca sexta. Life Sci. 28, 2585-2594.

Masler, E.P. and Adams T.S. (1986) An Ecdysiotropin from Musca domestica. In Host Regulated Developmental Mechanisms in Vector Arthropods (Ed. by Borovsky D. and Speilman A.) pp 60-65. Univ. of Florida, Vero Beach.

Nagasawa H., Kataoka H., Hori Y., Isogai A., Tamura S., Suzuki A., Guo F., Zhong X., Mizoguchi A., Fujishita M., Takahashi S., Ohnishi E. and Ishizaki H. (1984a) Isolation and some characterization of the prothoracicotropic hormone from Bombyx mori. Gen. Comp. Endocrinol. 53, 143-152.

Nagasawa H., Kataoka H., Isogai A., Tamura S., Suzuki A., Ishizaki H., Mizoguchi A., Fujiwara Y. and Suzuki A. (1984b) Amino-terminal amino acid sequence of the silkworm prothoracicotropic hormone: homology with insulin. Science 226, 1344-1345.

Okuda M., Sakurai S. and Ohtaki T. (1985) Activity of the prothoracic gland and its sensitivity to prothoracicotropic hormone in the penultimate and last-larval instar of Bombyx mori. J. Insect Physiol. 31, 455-461.

EFFECT OF BOMBYX 4K-PTTH AND BOVINE INSULIN ON TESTIS DEVELOPMENT OF INDIAN SILKWORM, PHILOSAMIA CYNTHIA RICINI

Zong-Shun Wang, Weng-Hui Zheng and Fu Guo

Institute of Zoology, Academia Sinica

Beijing, China

Recent studies have shown that Bombyx adult brain contains two types of PTTHs with different molecular weights, 4K-PTTH and 22K-PTTH. The 4K-PTTH is composed of three heterogeneous molecular species, 4K-PTTH-I, -II and -III and their NH_2-terminal amino acid sequences have been determined (Nagasawa et al., 1984). It is of great interest that the sequences of the 4K-PTTHs contain regions homologous with that of insulin A chain, showing that Bombyx 4K-PTTH and vertebrate insulin are homologous in structure. What are the functions of 4K-PTTH and vertebrate insulin ? To determine if vertebrate insulin have a 4K-PTTH activity, we used Indian silkworm, Philosamia cynthia ricini, debrained pupae as experimental animal to examine its activity.

In Philosamia no diapause occurs at whichever developmental stage under any environmental condition. When the brain is removed within 24 h after pupation, adult development, including the reproductive organ, is blocked, and such pupae are most appropriate for testing the activity of 4K-PTTH. We have found that the development of testes in debrained Philosamia pupae is accompanied with the adult development after treatment with bovine insulin and Bombyx 4K-PTTH.

335

Adult development

It is shown in table 1 that the bovine insulin at a
dosage of 0.10 or 0.05 μg is available for initiating the
debrained pupae of Philosamia to develop, but at higher
dosage (above 1 μg), it turns out to be no significant
effect. As compared with Bombyx 4K-PTTH, the action of
bovine insulin on the debrained pupae is less efficient
than that of Bombyx 4K-PTTH; even so its action leads to
adult development in the brainless pupae. It is reasonable
to explain the response as a result of activation of PGs
by the insulin. Nagasawa and his colleagues (1984) reported
that porcine insulin failed to show PTTH activity at a dose
of 1 μg, which is similar to our results. However, we succeeded in using lower dosage of insulin to provoke the development of debrained Philosamia pupae.

Table 1. The response of Philosamia brainless pupae
        to serially diluted bovine insulin

Amount of insulin injected (μg)	No. positive/ No. negative	Response[*]
100.00	3/4	--
10.00	3/4	--
1.00	3/4	--
0.10	5/2	++
0.05	5/2	++

[*]The assay result is regarded as positive when more
than 4 pupae display adult characters.

Testis growth

In insects the task of the testis is to manufacture
sperm. Several studies have implicated JH and ecdysone in
the control of spermatogenesis in diapausing larvae and
pupae (Dumser, 1980). In our laboratory it has been demonstrated that Bombyx 4K-PTTH plays an important role in the
management of ovary development, oogenesis and chorion
formation in Indian silkworm (Chou et al., 1982; Xia and
Guo, 1985). It is very interesting that the action of bovine
insulin on testis development in debrained pupae of Philo-

samia is similar to that of Bombyx 4K-PTTH (Table 2), although insulin shows lower efficiency than Bombyx 4K-PTTH's.
Both Bombyx 4K-PTTH and bovine insulin have the capability
of initiating spermatogenesis in the debrained pupae of
Philosamia.

Table 2. Effect of Bombyx 4K-PTTH and bovine insulin
   on testicular growth of debrained Philosamia pupae
   in two strains (10 days after treatment)

	Testis size (V. mm^3)*	
Treatment	Guang-Baihuang	Cheng-Huangbai 681
Crude PTTH	229.02+0.14	265.91+0.08
Bovine insulin (0.10 μg)	159.28+0.20	271.43+0.14
Control debrained pupae	19.00+0.08	109.48+0.27

*Formula: $V = \frac{4}{3} ab^2 \pi$; a-major axis, b-minor axis.

Spermiogenesis

   Morphological changes in the follicles were observed
with phase microscopy. Insects have particularly cystic
spermatogenesis. The germ cells develop by synchronous division within a cyst. So it is most convenient to examine the
extensive morphological alteration of the spermatid termed
spermiogenesis. After treatment with bovine insulin and
Bombyx 4K-PTTH for 10 days, about 30% and 50% elongated
cysts appeared respectively in a follicle, whereas no elongated cyst was found in dauer pupae (Fig. 1), suggesting
that both the insulin and Bombyx 4K-PTTH have an accelerated
effect on spermiogenesis in the follicles of the debrained
pupae.

   Our results demonstrated that Bombyx 4K-PTTH and bovine
insulin can stimulate the testis development and spermiogenesis process of debrained pupae of Philosamia cynthia
ricini. It shows that insect PTTH and vertebrate insulin
not only are structurally homologous, but also are functionally iso-hormones.

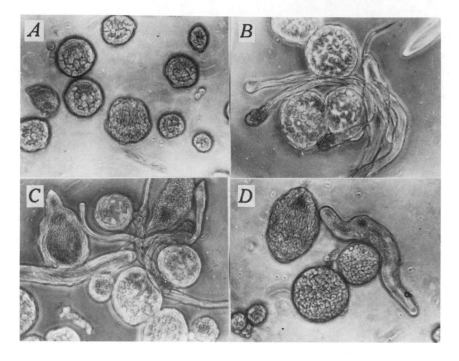

Fig.1. The morphological alteration of cysts in the
testis of debrained Philosamia pupae after treatment
for 10 days. (A) control. (B) Bombyx 4K-PTTH. (C) 0.10
μg bovine insulin. (D) 1.0 μg bovine insulin. A-D
micrographs x214.

REFERENCES

Chou Xu-Jia.,Wei Ding-Yi and Guo Fu. (1982) Effects of bra-
    in hormone of the silkworm, Bombyx mori,on metamorphosis
    and ovary development of eri-silkworms. Sinozoologia 2,
    183-190.
Dumser J. B. (1980) The regulation of spermatogenesis in
    insects. Annu. Rev. Entomol. 25, 341-369.
Nagasawa H., et al. (1984) Amino-terminal amino acid sequen-
    ce of silkworm prothoracicotrophic hormone: homology with
    insulin. Science 226, 1344-1345.
Xia Bang-Ying and Guo Fu. (1985) Chorion formation during
    pupal ovary development of Philosamia cynthia ricini and
    its regulation by brain hormone. Acta Biochim. Biophys.
    Sinica. 17, 208-214.

PRESENCE OF GONADOTROPIC AND PROTHORACICOTROPIC FACTORS
IN PUPAL AND ADULT HEADS OF MOSQUITOES

LaVern R. Whisenton and Walter E. Bollenbacher

Department of Biology, University of North
Carolina, Chapel Hill, NC 27514

During insect postembyronic development, ecdysteroids
stimulate the molecular events in target tissues that
culminate in molting and metamorphosis (Bollenbacher and
Granger, 1985). The cerebral peptide prothoracicotropic
hormone (PTTH) regulates the hemolymph titer of
ecdysteroids through its activation of the prothoracic
glands to synthesize ecdysone. In recent years, insect
reproduction has been shown to be controlled by
ecdysteroids as well. This is a surprising finding
considering that in adult insects the prothoracic glands
are absent (see Hagedorn, 1985). The role of ecdysteroids
in insect reproduction has been most extensively
investigated in the mosquito, Aedes aegypti, where the
cerebral peptide egg development neurosecretory hormone
(EDNH) stimulates the ovaries to synthesize ecdysone
(Hagedorn et al., 1979). This results in an increased
hemolymph ecdysteroid titer which evokes vitellogenesis
and ovary maturation.

Although PTTH and EDNH function as neurohormones at
different stages of the insect's life, their immediate
functions are the same, i.e., they stimulate ecdysone
biosynthesis. Given the steroidogenic function of these
peptides, they could be similar or the same molecules,
which during evolution have acquired different biological
functions. Extracts of brains from different insect
species at different stages of development and from within
the same insect (Aedes aegypti) at different stages of

339

its life cycle appear to possess EDNH-like activity
(Hagedorn et al., 1979), an observation suggesting PTTH
and EDNH may be related structurally. This report reviews
research from our laboratory that suggests structural and
functional similarities exist between steroidogenic
cerebral peptides from pupal (PTTH) and adult (EDNH)
brains of the mosquitoes, Aedes aegypti and Toxorhynchites
brevipalpis.

EDNH-like peptides in pupal and adult brains of Aedes
were identified using two bioassays, an in vitro ovary
assay which monitors the stimulation of ecdysone synthesis
(Hagedorn et al., 1979; Whisenton et al.,1986 a, b) and an
in vivo assay which monitors the induction of ovarian
maturation (Kelly et al., 1986). Extracts of pupal and
adult Aedes heads were active in both bioassays. These
activities have apparent molecular weights (M_r) of ~11 kD
(the predominant activity) and ~24 kD. These moieties are
proteins and the M_rs and functional properties of
corresponding EDNH activities between pupal and adult
heads are strikingly similar. Therefore, EDNH appears to
exist in at least two forms (also alluded to by: Masler
et al., 1983; Wheelock et al., 1985), each present in
pupal and adult brains of Aedes. That EDNH may exist in
several molecular forms is not surprising since this is
the case for other insect neuropeptides (see Bollenbacher
and Granger, 1985) and for vertebrate neuropeptides in
general.

While EDNHs are apparently present in pupal and adult
brains of a mosquito, it is not known if this is the case
for PTTH(s). To determine this, a bioassay for the
mosquito PTTH has been developed using the tree hole
mosquito, Toxorhynchites brevipalpis. This insect was
used instead of Aedes because of its large size and
greater experimental tractability. In this assay, late
fourth instar larval prothoracic glands (3 pair), which
exist in complex with the corpora allata, are incubated
for 5 h at 25°C in a 0.015 ml standing drop of Grace's
culture medium alone or in medium containing a sample of
brain extract. After incubation, the medium is assayed
for ecdysone synthesized/secreted by micro-
radioimmunoassay (Bollenbacher et al., 1983). The
presence of PTTH in a sample is denoted by increased
ecdysone synthesis by sample exposed glands over the
synthesis by unexposed, control glands. A heat treated

extract of Toxorhynchites pupal brains activates the prothoracic glands in a dose-dependent manner. The dynamic range of gland response was from 0.06 to 0.5 pupal brain equivalents. Thus it appears a bioassay exists with which the nature of the PTTH in mosquito brains can be investigated. As a means of determining preliminarily if the Toxorhynchites pupal brains possess EDNH activity, an aliquot of the PTTH brain extract was assayed for EDNH activity with the in vitro Aedes ovary assay. Steroidogenesis was stimulated by the extract and the effect was dose dependent; the response range was from 0.5 to 2.0 brain equivalents. Therefore, the pupal brains of Toxorhynchites possess both PTTH- and EDNH-like molecules. Similarly, extracts of Aedes pupal or adult heads activated Toxorhynchites prothoracic glands, indicating these bioassays could be used to characterize the PTTH activity present in the pupal and adult brains of Aedes.

In summary, the above data indicate that in vitro assays have been developed with which the PTTHs and EDNHs in mosquitoes can be characterized chemically and functionally. The properties of these two families of steroidogenic neurohormones can then be compared directly, enabling an assessment of the possible structural and functional similarities between them. If these different hormones are ultimately shown to be the same peptides, then during the evolution of an insect a given family of peptide neurohormones would have evolved to fulfill different physiological functions through their activation of ecdysone synthesis by different endocrine glands at different stages of the insect's life cycle.

Acknowledgments—The authors thank Ms. Kathy Luchok for preparing this manuscript. This research was supported by research grants NS-18791 and AM-31624 from the National Institutes of Health to W.E.B.

REFERENCES

Bollenbacher W.E., O'Brien M.A., Katahira E.J. and Gilbert L.I. (1983) A kinetic analysis of the action of the insect prothoracicotropic hormone. Mol. cell. Endocrinol. 32, 27-46.

Bollenbacher W.E. and Granger N.A. (1985) Endocrinology
of the prothoracicotropic hormone. In Comprehensive
Insect Physiology, Biochemistry and Pharmacology, Vol.
7, Endocrinology I (Ed. by G.A. Kerkut and L.I.
Gilbert), pp. 109-152, Pergamon Press, Oxford.

Hagedorn H.H. (1985) The role of ecdysteroids in
reproduction. In Comprehensive Insect Physiology,
Biochemistry and Pharmacology, Vol. 7, Endocrinology I
(Ed. by G.A. Kerkut and L.I. Gilbert), pp. 205-262,
Pergamon Press, Oxford.

Hagedorn H.H., Shapiro J.K. and Hanaoka K. (1979)
Ovarian ecdysone secretion is controlled by a brain
hormone in an adult mosquito. Nature (London) 282, 92-
94.

Kelly T.J., Whisenton L.R., Katahira E.J., Fuchs M.S.,
Borkovec A.B. and Bollenbacher W.E. (1986) Inter-species
cross-reactivity of the prothoracicotropic hormone of
Manduca sexta and egg development neurosecretory hormone
of Aedes aegypti. J. Insect Physiol., in press.

Masler E.P., Hagedorn H.H., Pretzel D.H. and Borkovec
A.B. (1983) Partial purification of egg development
neurosecretory hormone with reverse-phase liquid
chromatographic techniques. Life Sciences 33, 1925-
1931.

Wheelock G.D. and Hagedorn H.H. (1985) Egg maturation
and ecdysiotropic activity in extracts of mosquito
(Aedes aegypti) heads. Gen. Comp. Endocrinol. 60,
196-203.

Whisenton L.R., Kelly T.J. and Bollenbacher W.E. (1986)
Multiple molecular forms of steroidogenic peptides in
pupal and adult brains of mosquitoes. In Host Regulated
Developmental Mechanisms in Vector Arthropods (Ed. by D.
Borovsky and A. Speilman), pp. 73-77, University of
Florida-IFAS, Vero Beach, Florida.

Whisenton L.R., Kelly T.J. and Bollenbacher W.E. (1986)
Multiple forms of cerebral peptides with steroidogenic
functions in pupal and adult brains of the yellow fever
mosquito, Aedes aegypti. Mol. cell. Endocrinol., in
press.

IN VITRO ACTIVATION OF PROTHORACIC GLANDS FROM DIAPAUSE AND NON-DIAPAUSE DESTINED SARCOPHAGA ARGYROSTOMA AND CALLIPHORA VICINA

David S. Richard and David S. Saunders.
Dept of Zoology, Edinburgh University,
West Mains Rd., Edinburgh EH9 3JT, U.K.

The in vitro activation of isolated prothoracic glands from Sarcophaga argyrostoma (Robineau-Desvoidy) to produce ecdysone was accomplished in a dose related manner (Figure 1) with extracts of prepupal brains, presumably containing prothoracicotropic hormone (PTTH) (Roberts et al.,1984). Ecdysone was measured according to Richard and Saunders (1986). The steroidogenic action of PTTH was mimicked by dibutyryl cAMP, and the phosphodiesterase inhibitor MIX (Figure 2), suggesting the involvement of a cyclic nucleotide mediated secondary messenger system in PTTH action (Smith et al. 1984).

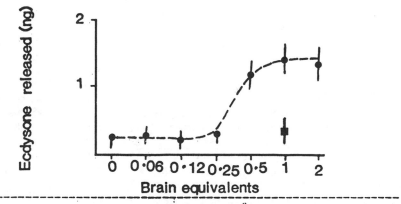

Figure 1: Dose response curve for the in vitro activation of Sarcophaga post-feeding larval (●---●) ring glands incubated with PTTH extracts from 4h prepupal brains in Grace's medium plus 10mM $CaCl_2$ for 4h at 25°C. An extract of ventral ganglion was used as a control (■). Each point represents the mean ± SEM of 5 to 10 separate

activation analyses. Results expressed as ecdysone
equivalents.

The removal of calcium ions by EDTA reduced PTTH action, as
noted by Smith et al (1985) in Manduca sexta. Cyclic
GMP and its derivatives had no effect on ecdysone release.
PTTH extracted from Sarcophaga and dibutyryl cAMP were
found to be effective at stimulating ecdysone release by
isolated ring glands from Calliphora vicina (Meigen)
larvae. Roberts et al (1986) showed that PTTH from
Sarcophaga and Manduca could stimulate ecdysone release
interspecifically, and suggested a great degree of
uniformity in PTTH structure and function in insects.

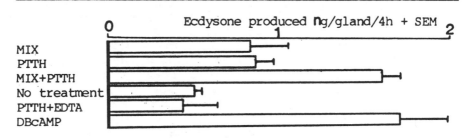

Figure 2: The effects of 0.1mM 1-methyl-3-isobutylxanthine
(MIX) and 10mM EDTA on the action of PTTH (0.5 brain
equivalents) and the effect of 10mM dibutyryl cAMP on
Sarcophaga post-feeding larval ring glands in vitro.
Each figure is the mean ± SEM of 6 to 10 separate activation
analyses. Results expressed as ecdysone eqivalents.

    Sarcophaga argyrostoma enters pupal diapause in
winter (Fraenkel and Hsiao,1968), after the formation of the
phanerocephalic pupa. Diapause in this species is induced by
short day photoperiods experienced by the developing embryo
and larva (Denlinger, 1971). Calliphora vicina however,
enters a larval diapause induced principally by short days
experienced maternally (Vinogradova and Zinovjeva, 1972). In
the diapause state, both species display a very low
haemolymph ecdysteroid titre (Richard et al,1986;
Richard and Saunders,1986), presumably as a result of
neuroendocrine regulation, which interrupts the release of
PTTH from the brain/retrocerebral complex (Bowen et
al.,1984). As the Sarcophaga population entered pupal
diapause at 96h post-pupariation at 18°C, the
prothoracic glands became refractory to PTTH, taking 24h to
lose their competency to respond (Figure 3A). The levels of

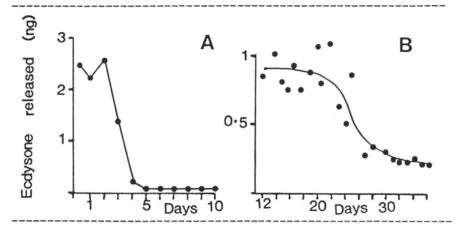

Figure 3: The competency of ring glands taken from post-pupariation diapause destined Sarcophaga (maintained under LD12:12 at 18°C) (A), and ring glands taken from post-wandering diapause destined Calliphora larvae (maintained at 11°C) (B) to synthesise ecdysone in response to incubation with 0.5 brain equivalents of 4h Sarcophaga prepupal brain extract.

ecdysone released in diapause were below the limits of detection of the radioimmunoassay used. The glands could not be stimulated by the cyclic nucleotides indicating that the block to PTTH action occurred beyond the proposed stage of cyclic nucleotide mediation. The Calliphora population, which has a larval diapause, took six days to reach this state (Figure 3B). Unlike Sarcophaga, their prothoracic glands continued to release ecdysone at a basal rate refractory to stimulation by PTTH or cyclic nucleotides. Following a temperature increase from 11°C to 25°C, all diapausing Calliphora had regained prothoracic gland competency within 24h and had pupariated within 36h (Figure 4). When isolated brain/ring gland complexes from diapausing Calliphora were incubated at 25°C, no recovery of ring gland competency was noted, indicating the requirement for in vivo reactivation after diapause. This suggested a different mechanism for diapause termination than for diapause induction. The levels of PTTH in pre-diapause and day 35 post-pupariation diapause brains of Sarcophaga were shown to be similar to those of non-diapause destined prepupal brains.

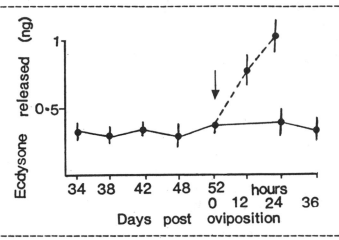

Figure 4: The competency of diapausing <u>Calliphora</u> larval ring glands to respond to PTTH when maintained at 11°C (●——●). At day 52/0h (arrow) a group of larvae were transferred to 25°C and tested with PTTH (●---●).

REFERENCES

Bowen,M.F., Bollenbacher,W.E., and Gilbert,L.I. (1984) J. Exp. Biol. <u>108</u>, 9-24.
Denlinger,D.L. (1971) J. Insect Physiol.<u>17</u>,1815-1822.
Fraenkel,G. and Hsiao,C. (1968) J. Insect Physiol. <u>14</u>, 689-705
Richard,D.S. and Saunders, (1986) Submitted to J. Insect Physiol.
Richard,D.S. Warren,J.T.,Saunders,D.S. and Gilbert,L.I.(1986) J. Insect Physiol. In Press.
Roberts,B., Gilbert,L.I. and Bollenbacher,W.E. (1984) Gen. and Comp. Endocrin. <u>54</u>, 469-477.
Roberts,B., Gilbert,L.I. and Bollenbacher,W.E. (1986). Experientia.In press.
Smith.W.A. Gilbert,L.I. and Bollenbacher,W.E. (1984) Mol. Cell. Endocrinol. <u>37</u>,285- 294.
Smith,W.A. Gilbert,L.I. and Bollenbacher,W.E. (1985) Mol. Cell. Endocrinol. <u>39</u>, 71-78.
Vinogradova,E.B. and Zinovjeva,K.B. (1972) J.Insect Physiol. <u>18</u>, 2401- 2409.

HIGH TEMPERATURE INDUCED TERMINATION OF DIAPAUSE

IN LEPIDOPTEROUS INSECTS

Robert A. Bell

Insect Reproduction Laboratory
ARS, USDA
Beltsville, MD   20705 USA

Pupal diapause in the tobacco hornworm, Manduca
sexta, is induced by the cumulative physiological effects of
daily photoperiods (13.5 hrs or less) experienced during the
late embryonic and subsequent larval feeding stages.
However, photoperiodic induction of the ultimate diapause
response occurs most effectively when the eggs and larvae
(photosensitive stages) are exposed to temperatures in the
range of 20 to 30°C which is considered optimal for this
experimental population. Higher temperatures (33°C)
strongly inhibited the diapause inducing effects of short
photoperiods (Bell et al., 1975). These results then
prompted further experiments to examine the effects of
temperature on the maintenance and termination of diapause
in this insect.

Tobacco hornworms used in these studies originated from
North Carolina field populations and were reared in the
laboratory on an artificial diet in accordance with techniques described previously (Bell and Joachim 1976). Diapausing pupae were obtained by exposing eggs and larvae to
daily photoperiods of 12 hours at 25-26°C; developing pupae
were produced by exposing the larvae to a photoperiod of 15
hours. Oxygen uptake was measured with a recording differential respirometer (Gilson Medical Electronics). Termination of pupal diapause was indicated by various criteria including rate of oxygen consumption, weight loss and the
usual morphological manifestations of development such as
the time of appearance of pigment in the developing

347

Table 1.   Influence of temperature on respiratory activity
            in developing and diapausing pupae of <u>Manduca</u>

Temperature	MM³ O₂/g live wt/hr	
(°C)	Diapausing	Developing
5	---	8.1 (2.1)
10	---	59.1 (7.7)
17	4.8 (1.3)	86.7 (15.3)
27	7.9 (1.2)	168.4 (33.3)
35	30.0 (8.2)	251.1 (42.0)

Oxygen consumption rates are expressed as treatment means
(N = 10 pupae/treatment) with SD in parentheses.

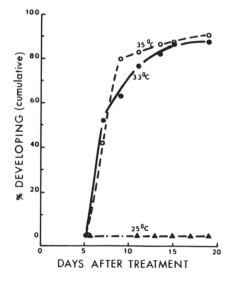

Fig. 1.   Influence of high temperatures on the rate of
diapause termination in the tobacco hornworm, <u>Manduca</u> <u>sexta</u>
(Control group was maintained at 25°C; N = 30–60 pupae in
each treatment.

cuticle of pharate adults.

Rates of oxygen uptake at different temperatures for
both diapausing and developing pupae are shown in Table 1.
Respiration in diapausing pupae was measured only at temperatures ranging from 17-35° C. When the temperature was
increased from 17 to 27° C, O_2 uptake in diapausing pupae
only increased 1.6X compared to a 2X increase in developing pupae. However, when the temperature increased from 27
to 35° C, there was a striking 3.8X increase in respiration
of diapausing as opposed to a 1.5X increase in developing
pupae. Within 6-7 days, some of the diapausing pupae held
at 33° resumed development and had undergone adult
differentiation. It was clear from the preceding
observations that high temperatures could stimulate
termination of diapause. Thus another experiment was
carried out in which 21 day old pupae were subjected to 33
and 35° C and observed daily for signs of development.
Control pupae were maintained at 25° C. Results (Fig.1)
showed that such high temperatures terminated diapause
within 1-2 weeks in most individuals. Since it was
subsequently found that 35°, but not 33° C, caused male
sterility, 33° C was routinely adopted as a simple and
practical method for rapid termination of diapause in this
insect (Rasul,1972).

The physiological role of high temperature in terminating diapause has received little study but experimental findings of other workers (Williams, 1952; Meola and Adkisson,
1977 and Denlinger, 1980) lead to the presumption that it
probably acts to precociously stimulate the secretion of the
brain prothoracicotropin (PTTH) neurohormone. PTTH then activates the prothoracic glands to secrete developmental promoting ecdysteroids. It is further suspected that such unusually high temperatures, when encountered by the insect
for prolonged periods, induces a rather sustained stress
response which can override the neural and/or neurohumoral
restraining mechanisms that actively maintain diapause by
preventing the secretion of PTTH.

ACKNOWLEDGEMENTS: This research was conducted at the Metabolism and Radiation Research Laboratory, USDA-ARS and in
cooperation with the Department of Entomology, North Dakota
State University, Fargo, ND. Data in Fig.1 was obtained by
Dr. C.G. Rasul while carrying out research for a PhD thesis
in the author's laboratory.

REFERENCES

Bell, R.A., Rasul, C.G. and Joachim, F.G. (1975) Photoperiodic induction of the pupal diapause in the tobacco hornworm, Manduca sexta J. Insect Physiol. 21, 1471-1480.

Bell, R.A. and Joachim, F.G. (1976) Techniques for rearing laboratory colonies of tobacco hornworms and pink bollworms. Ann. Entomol. Soc Am. 69, 365-373.

Denlinger, D.L., Campbell, J.J. and Bradfield, J.Y. (1980) Stimulatory effect of organic solvents on initiating development in diapausing pupae of the flesh fly, Sarcophaga crassipalpis, and the tobacco hornworm Manduca sexta Physiol. Entomol. 5: 7-15.

Meola, R.W. and Adkisson P.L. (1977) Release of prothoracicotropic hormone and potentiation of developmental ability during diapause in the bollworm, Heliothis zea. J. Insect Physiol. 23, 683-688.

Rasul, C.G. (1972) Factors affecting induction and termination of the pupal diapause in the tobacco hornworm Manduca sexta (L) PhD Thesis, North Dakota State University, Fargo

Williams, C.M. (1952) Physiology of Insect Diapause IV. The brain and prothoracic glands as an endocrine system in the Cecropia silkworm. Biol. Bull. 103, 120-138.

HOST-SEEKING IN A DIAPAUSING MOSQUITO

M. F. Bowen, D. A. Haggart, and E. E. Davis

SRI International, 333 Ravenswood Ave.,
Menlo Park, CA 94025   USA

Host-seeking behavior in the mosquito, Aedes aegypti
is absent at two distinct points in the life cycle.
Adult females do not host-seek until several days after
emergence (Christophers, 1960). Host-seeking is also
inhibited following a blood meal and resumes after
oviposition has occurred (Klowden and Lea, 1979a,b).
Host-seeking behavior is always accompanied by high
sensitivity in the antennal chemoreceptor neurons
responsive to the host attractant lactic acid (LA); the
absence of host-seeking is associated with low receptor
sensitivity (Davis, 1984a,b). We have developed a model
for the control of HSB in the mosquito which is based on
the state of the peripheral sensory system (Davis,
1984b). Our model suggests that control of HSB originates at the sensory level and that host-seeking is
modulated directly by changes in peripheral receptor
sensitivity. Such peripheral regulating mechanisms that
are based on variations in chemoreceptor sensitivity may
be more common in insects than previously thought (Blaney
et al., 1986). Available data on such phenomena are
limited; further study of chemoreception in different
species would contribute greatly to our understanding of
insect behavior, particularly mosquito host-seeking
behavior.

We therefore decided to extend our observations of
HSB to mosquito species that were chosen specifically for
their ability to overwinter as adults in the state of

diapause. These groups are interesting from an epidemio-
logical point of view because of their potential to act
as overwintering viral reservoirs (Reeves, 1974). They
are of interest from a behavioral point of view because
host-seeking can be absent or present during diapause
depending upon the species (Washino, 1977). We have
begun a study of host-seeking behavior in the diapausing
species Culex pipiens in order to test our model of HSB
and extend it to other hematophagous members of the
family Culicidae.

Many studies of mosquito blood-feeding during
diapause are not only contradictory (see review by
Washino, 1977), but are difficult to interpret, since the
experiments were not designed to distinguish between
host-seeking and the taking of a blood meal. Successful
blood-feeding by a mosquito is actually the end point of
a rather complex behavioral series. The performance of
each step in this series is probably cued by different
stimuli (e.g., Dethier, 1957). We circumvent the problem
of confounding these various stimulus-response patterns
by analyzing only HSB and its attendant physiological and
sensory correlates. The data presented here are concerned with our investigations into the daily behavior
patterns and sensory physiology of diapausing as well as
non-diapausing C. pipiens.

We assayed the host-seeking response to human breath
in C. pipiens using a behavioral protocol described
previously (Davis, 1984b). In addition to flight movement, we also noted whether the mosquitoes probed in
response to the host in each assay. Samples of adult
females to be assayed for HSB were taken from two mated
populations of the same age (1 to 3 weeks) reared under
identical conditions of temperature and humidity. One
population was reared under a long-day photoperiodic
regime which resulted in 0% diapause. The other group
was reared under short daylengths which resulted in 99.3%
diapause. Samples (n = 30) from each population were
removed at 3-hour intervals over a 24-hour period and
tested one time only for HSB in groups of 10 individuals.
The results showed that a daily pattern of HSB exists in
the non-diapausing population. Host-seeking anticipated
lights-off and rose to an average maximum of 50% in the
early part of the dark phase. It then decreased
gradually, reaching 0% at lights-on where it remained

during the rest of the day. Although minor peaks of
activity were observed at 0600 (15%) and 1500 hours (24%)
these did not represent true host-seeking individuals but
consisted solely of mosquitoes that did not exhibit
probing behavior. In contrast the diapausing population
failed to show HSB at any time during the 24-hour test
period. Although minor peaks of activity at 0600 (13%),
1600 (24%), and 2300 (10%) were observed these did not
consist of true host-seeking individuals since probing
was not observed at any time during any of the assays.
The peak of HSB observed in the first part of the dark
phase, so prominent in non-diapausing mosquitoes, is not
seen in the diapausing population. Diapause in C.
pipiens is thus characterized by the absence of HSB. HSB
in non-diapausing populations exhibits a strong diel
periodicity in its expression with a peak occurring early
in the dark phase.

Before we are able to assess possible changes in the
peripheral sensory system, we must determine the physiological response characteristics of the chemosensory
neurons associated with each morphologically distinct
type of antennal sensillum. Accordingly, we have begun
to look at the electrophysiologic response patterns in
non-diapausing C. pipiens elicited by potential airborne
host attractants. The location of our recording site is
corroborated by scanning electron microscopy. We have
found that the host-related odor, LA, is detected by
neurons in the grooved peg sensillum as described by
McIver (1974). We find both excited and inhibited
neurons sensitive to LA at stimulus intensities on the
order of 10^{-11} moles/sec. We have also observed excitatory responses to ammonia at 10^{-8} moles/sec. All of
these responses of the grooved peg neurons are relatively
specific. There are 2 receptor cells per sensillum in
Culex (Elizarov and Chaika, 1972). More work needs to be
done in order to clarify the stimulus-response characteristics of these cells and integrate this information with
that concerning the fine structure of the receptors.

Once the characterization of the grooved peg neurons
in C. pipiens is completed it will be possible to quantitatively describe chemoreceptor responsiveness and relate
it to patterns of HSB in non-diapausing as well as diapausing individuals. By doing so, we hope to investigate
the extent to which we can apply our model of peripheral

control of HSB to other species and to clarify the control of this behavior during mosquito diapause.

REFERENCES

Blaney, W. M., Schoonhoven, L. M., and Simmonds, M.S.J.
    (1986) Sensitivity variations in insect chemo-
    receptors: a review. Experientia 42, 13–19.
Christophers, R. (1960) The Yellow Fever Mosquito: Its
    Life History, Bionomics and Structure. Cambridge
    University Press, Cambridge.
Davis, E. E. (1984a) Regulation of sensitivity in the
    peripheral chemoreceptor systems for host-seeking
    behavior by a haemolymph-borne factor in Aedes
    aegypti. J. Insect Physiol., 30, 179–183.
Davis, E. E. (1984b) Development of lactic acid
    receptor sensitivity and host-seeking behavior in
    newly emerged female Aedes aegypti mosquitoes. J.
    Insect Physiol. 30, 211–215.
Dethier, V. G. (1957) The sensory physiology of blood-
    sucking arthropods. Exp. parasitol. 6, 68–122.
Elizarov, Yu.A. and Chaika, S.Yu. (1972)
    Ultrasturctural organization of olfactory sensilla of
    antennae and palpi of the mosquito Culex pipiens
    molestus (Diptera: Culicidae). Zoologich. Zhur. 51,
    1665–1674 (In Russian).
Klowden, M. J. and Lea, A. O. (1979a) Humoral
    inhibition of host-seeking in Aedes aegypti during
    oocyte maturation. J. Insect Physiol. 25, 231–235.
Klowden, M. J. and Lea, A. O. (1979b) Abdominal
    distention terminates subsequent host-seeking behavior
    of Aedes aegypti following a blood meal. J. Insect
    Physiol. 25, 583–585.
McIver, S. B. (1974) Fine structure of antennal grooved
    pegs of the mosquito, Aedes aegypti. Cell Tiss. Res.
    153, 327–337.
Reeves, W. C. (1974). Overwintering of arboviruses.
    Prog. Med. Virol. 17, 193–220.
Washino, R. K. (1977) The physiological ecology of
    gonotrophic dissociation and related phenomena in
    mosquitoes. J. Med. Entomol. 13, 381–388.

AN ACTH-LIKE DIURETIC HORMONE IN LOCUSTS

A. Rafaeli,   S.W.   Applebaum   and
P. Moshitzky

The Hebrew University,
Department of Entomology,
Rehovot 76-100, Israel.

For the past 10 years evidence has been accumulating from various sources (DeLoof, 1984) indicating the existence of vertebrate-type hormones in insect tissues. Much of the evidence is based on immuno-cytochemical techniques. The method provides information on the localization of the vertebrate-type hormone and suggests presumptive pathways in which processing, transport and release occur. However, the specificity of immuno-cytochemical identifications is limited, and related hormones, or fortuitous partial sequences are liable to be recognized. Further biochemical and physiological bioassays are therefore essential in order to attribute significance to such observations. This has been demonstrated in the case of an insulin-like peptide in Calliphora using a combination of immuno-cytochemical evidence, specific radio-immunoassays and physiological bioassays (Duve & Thorpe, 1984). Where vertebrate-like hormones are present in insects we must ask what function they serve and if this function is similar to that exhibited in the vertebrate systems.

Cyto-immunoreactivity to adrenocorticotrophic hormone (ACTH) has been previously demonstrated in insect tissues (DeLoof, 1984). In vertebrates ACTH is implicated in water balance where its humoral levels control the production of aldosterone.

We report here an ACTH-like peptide present in locusts which is responsible for fluid secretion by Malpighian tubules and, in fact, mimics the diuretic action of a native diuretic hormone (DH) found in locust corpora cardiaca (CC). The evidence we present is based on two sensitive bioassays using isolated Malpighian tubules and a specific radio-immunoassay for ACTH using both crude CC as well as HPLC purified extracts.

We tested the physiological significance of an ACTH-like hormone in insects using Malpighian tubules as target tissues. Synthetic ACTH (10^{-5}M) (obtained as a generous gift from Prof. J. Ramachandran, Genentech, CA. USA) successfully mimics the naturally occurring DH from CC extracts (Table I[A]). We found that even at physiological levels (10^{-8}M) ACTH produces a significant response by Malpighian tubules causing an increase in fluid secretion. Since the action of the native hormone is mediated by the production of c-AMP at such excess as to cause secretion of c-AMP into the lumen (Rafaeli et al., 1984) we tested this response using ACTH and indeed, a positive response was obtained (Table I[B]).

TABLE   I   DIURETIC   ACTION   AND   IMMUNOLOGICAL
          CROSS-REACTIVITY OF ACTH AND DH.

BIOASSAY	TREATMENT	RESPONSE
[A] Fluid secretion	0.5 gl CC	358.04 ± 151.6 (19)
(% rate increase)	10^{-5}M ACTH	317.30 ± 76.9 (13)
	10^{-8}M ACTH	109.25 ± 43.2 (6)
[B] c-AMP secretion	Basal level	10.43 ± 5.3 (7)
(fmole/tubule)	10^{-5}M ACTH	74.84 ± 13.9 (10)
(C) ACTH RIA (pg)	1.0 gl CC	900
	50 gl HPLC Purified DH	450

We then investigated a possible immuno cross-reactivity between ACTH and DH. Crude CC extracts were subjected to a sensitive, specific ACTH RIA using 3-[^{125}I] iodotyrosyl ^{23}ACTH (1-39) (Amersham, UK.) as the radioligand and sheep anti-ACTH (immunogen: synthetic 1-24 ACTH-BSA) as the anti-serum (Cambridge Medical Diagnostics Inc., USA). We found a response equivalent to 900 pg ACTH / locust CC (Table I[C]). This was further verified using RP-HPLC purified DH (Column: Merck, 10µm Lichrosorb RP-18; mobile phase: 2-propanol and 0.1% TFA; conditions: linear gradient 0-60% 2-propanol at 1%/min; flow: 1ml/min) where a response equivalent to 450 pg ACTH was detected by RIA (Table I[C]). Further evidence for cross-immunoreactivity was obtained when 67% diuretic activity (fluid secretion) was lost after incubations with ACTH anti-serum.

From these studies we suggest that a close similarity exists between vertebrate ACTH and insect diuretic hormone. However, DH differs in structure since it does not share identical retention times on reversed-phase HPLC with ACTH (Rafaeli et al., 1986).

Acknowledgements: This work was supported, in part, by a Hebrew University of Jerusalem Central Research Grant given to Dr. A. Rafaeli. The authors would like to thank Prof. J. Ramachandran for helpful advice.

REFERENCES:
DeLoof, A. (1984) Evolution of hormones controlling reproduction. In: Adv. Invert. Reproduction 3 (Edited Engels, W.) Elsevier Press pp.139-149.

Duve, H. & Thorpe A. (1984) Comparative aspects on insect-vertebrate neurohormones. In: Insect Neurochemistry and Neurophysiology (Edited Borkovec, A.B. & Kelly, T.J.) Plenum Press, N.Y. pp. 171-195.

Rafaeli, A., Pines, M., Stern, P. & Applebaum, S.W. (1984) Locust diuretic hormone-stimulated synthesis and excretion of cyclic-AMP: A novel Malpighian tubule bioassay. Gen. Comp. Endocr. 54, 35-42.

Rafaeli, A., Moshitzky, P. & Applebaum, S.W. (1986) Diuretic action and immunological cross-reactivity of corticotropin and locust diuretic hormone (in prep.)

THE BIOLOGY AND REGULATION OF POST-ECLOSION
DIURESIS IN ADULT HELIOTHIS ZEA

David W. Bushman, Ashok K. Raina, and Judd O. Nelson

Dept. of Entomology, University of Maryland

College Park, MD 20742

INTRODUCTION

Hormonal control of post-eclosion diuresis has been studied in several species of butterflies; Pieris brassicae (Nicolson 1976), Danaus plexippus (Dores, et al. 1979), Acraea horta, Danaus chrysippus, and Papilio demodocus (Nicolson 1980). We report here a post-eclosion diuresis in Heliothis zea (Lepidoptera: Noctuidae) that is responsible for a loss in weight equivalent to approximately 20% of the weight of a newly emerged adult during the first 6 hours after eclosion. Ligation between head and thorax immediately after eclosion reduced this diuretic weight loss to 7% of the emergence weight. In vivo assays of various nervous and neuroendocrine tissue homogenates suggest that a factor responsible for the post-eclosion diuresis may be located near the junction of the brain and the suboesophageal ganglion (SOG).

MAGNITUDE AND TIME COURSE OF THE DIURETIC RESPONSE

To determine the extent and timing of the diuretic response, insects were allowed to emerge normally and then handled briefly to ensure that all meconium and other fluids had been eliminated from the gut. The adults were then placed individually in screen cages for weighing and observation. At each weighing, the insects were gently squeezed to induce the elimination of fluids accumulated in

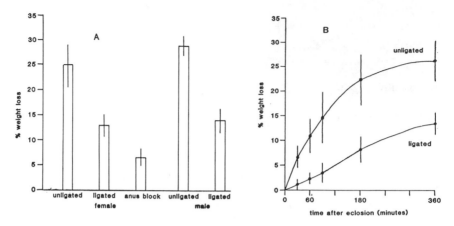

Fig. 1.   Total weight loss of unligated, ligated, and
           anus blocked H. zea (A) and the time course of
           total weight loss (B) during the first 6 hours
           following adult eclosion. Ligations and anus
           blocks were performed within 5 minutes of
           eclosion. Data presented as mean ± SD, n=20.

the hindgut. Weight loss stabilized by 6 hours post-eclosion
and this duration was used for all subsequent assays.
     Data from these observations (Fig. 1A) indicate a total
weight loss equivalent to 24.7% of emergence weight in
females and 29.0% in males. To account for weight loss due
to evaporation or metabolic activity, the anus of newly
emerged females was sealed with wax and subsequent weight
loss measured. These insects lost only 6.5% of their
emergence weight during the 6 hour assay period. When the
wax block was removed, clear fluid drained immediately from
the gut and the total weight loss then equaled that of
normal females. By subtracting the evaporative loss from the
total loss observed, an approximate weight loss due to
diuresis was calculated: 18.2% in females and 22.5% in
males, or an average of 19.7% for both sexes.

     When insects were ligated with cotton thread between
head and thorax immediately following emergence, total
weight loss was reduced to 13.0% of emergence weight in
females and 14.2% in males. The weight loss due to diuresis
alone averaged 7.0% in ligated insects of both sexes.

The time course of the total weight loss for males and females combined (Fig. 1B) indicated a rate of approximately 10% of emergence weight per hour for the first 90 minutes following eclosion. During the next 90 minutes, weight loss continued at a rate of 5.2% per hour. During the 3 hour interval from 180-360 minutes post-eclosion, the rate was only 1.3%/hour. This rate was comparable to the 1.1% per hour loss due to evaporation. Therefore, the elevated level of post-eclosion weight loss was essentially completed in the first 3 hours after emergence.

The time course of weight loss in insects ligated at emergence indicated a rate of 2.2% per hour for the first 90 minutes after eclosion, 3.3% per hour for the next 90 minutes, and 1.8%/hour for the subsequent 180 minutes of the observation period.

Despite the rapid rate at which fluid secretion occured in the first 90 minutes after eclosion, undisturbed insects generally retained the fluid in the hindgut until 120-180 minutes post-eclosion. It was at this time that the first attempts at flight occured and drops of clear fluid appeared from the anus. The purpose of this fluid retention may be to maintain maximum body volume during wing expansion and hardening, as suggested by Nicolson (1976).

LOCALIZATION OF THE DIURETIC FACTOR

Data from the ligation experiments and other evidence (Raina, unpublished) suggested hormonal control of the diuretic response. To examine this possibility, we tested homogenates of various nervous and neuroendocrine tissue in an in vivo weight loss assay modified from Dores, et al. (1979). Males and females were ligated at emergence as described previously and placed in a covered petri dish lined with filter paper. Ninety minutes following emergence, ligated insects were injected through the dorsal surface of the abdomen with 15 ul of saline only (Meyers and Miller, 1969) or 15 ul of a tissue homogenate. Final weights were measured 4.5 hours after injection (6 hours post-eclosion). The tissues tested were the frontal ganglion, brain + SOG + corpora cardiaca (CC)/corpora allata(CA) complex, thoracic ganglia, brain + SOG, brain alone, SOG alone, and CC-CA complex alone. Tissue donors were pharate adults within 0.5-4 hours of emergence. Injection of saline alone resulted in

a weight loss equivalent to 13.7% of the emergence weight.
Weight loss significantly higher than that from saline
controls occurred only following injection of brain alone
and SOG alone (19.5% and 19.8%, respectively). When the
brain and SOG from the same donor were assayed alone in
separate recipients, diuretic activity was found in only one
of the tissues, with the SOG showing the positive response
in 60% of the cases. These data suggest the putative H. zea
diuretic hormone to be produced and/or stored at the
junction of the brain and SOG. Fractionation of brain-SOG
homogenates by high performance liquid chromatography and
assays of the resulting fractions are being pursued to
isolate the hormonal factor.

ACKNOWLEDGEMENTS

    This material is based upon work supported under a
National Science Foundation Graduate Fellowship to DWB, and
Agricultural Research Service, USDA, Cooperative Agreement
No. 58-32U4-5-11. We acknowledge the assistance of Howard
Jaffe of Livestock Insects Laboratory, and Jeffrey Bray of
Insect Chemical Ecology Laboratory, both of USDA, ARS.
Scientific Article No. A-4488, Contribution No. 7481 of the
Maryland Agricultural Experiment Station.

REFERENCES

Dores R. M., Dallmann S. H. and Herman W. S. (1979) The
    regulation of post-eclosion and post-feeding diuresis
    in the Monarch butterfly, Danaus plexippus. J. Insect
    Physiol. 25, 895-901.
Meyers, T. and Miller, T. (1969) Starvation induced
    activity in cockroach Malpighian tubules. Ann. Ent.
    Soc. Am. 62, 725-729.
Nicolson, S. W. (1976) Diuresis in the cabbage white
    butterfly, Pieris brassicae: fluid secretion by the
    Malpighian tubules. J. Insect Physiol. 22, 1347-1356.
Nicolson, S. W. (1980) Diuresis and its hormonal control in
    butterflies. J. Insect Physiol. 26, 841-846.

ISOLATION AND PURIFICATION OF TWO AVP-LIKE FACTORS FROM

LOCUSTA MIGRATORIA

Jacques Proux[*], Christine A. Miller[†], Adrien
Girardie[*], Michel Delaage[§] and David A.
Schooley[†]

[*]Lab. Neuroendocrinologie, U.A.1138 CNRS, Univ.
Bordeaux I, 33405 Talence, FRANCE; [†]Sandoz Crop
Protection, Zoecon Research Institute, Palo
Alto, CA, USA; [§]Immunotech, Marseille, FRANCE

In 1979, Rémy et al. detected in two cells of the
suboesophageal ganglion (SOG) of an insect, the migratory
locust, a substance immunologically related to arginine
vasopressin (AVP), the antidiuretic hormone of mammals.
This substance was named "AVP-like hormone." It is transported in fibers and arborizations of two SOG cells
throughout the whole central nervous system (Rémy and
Girardie, 1980). These results, obtained with an immuno-
histological method, were confirmed with radioimmunoassay
(RIA) by Proux and Rougon-Rapuzzi in 1980. These authors
established also that the AVP-like hormone was released
into the hemolymph and transported to the insect diuretic
organ, the Malpighian tubules (MT). It plays a diuretic
role by increasing the excretion of primary urine (Proux
et al., 1981).

In an attempt to purify and characterize the AVP-like
hormone, 100,000 SOG and thoracic ganglia were dissected
and stored in acetone at 4°C. A part of this stock was
used to perform preliminary biological and biochemical
studies (Proux et al., 1981; Cupo and Proux, 1983). We
report here on the utilization of the rest to purify the
AVP-like hormone. Two purification procedures were used:
immunopurification and HPLC. The detection of the hormone
during the purification steps was done by RIA. At certain
important steps, in vitro or in vivo bioassays were

363

performed to confirm the RIA results and to ascertain the biological activity.

The RIA was generally performed using equilibrium dialysis with an AVP antiserum at 1/100,000 dilution and [^{125}I]AVP from New England Nuclear. However, to confirm certain results, we used another RIA with a different antiserum and second antibody precipitation.

The biological response of the MT to the diuretic effect of the AVP-like hormone was measured by means of two bioassays: the amaranth test based on the selective excretion of a dye, amaranth, by the MT in vivo and previously used in the preliminary studies on the AVP-like hormone (Proux et al., 1981), and a new in vitro test (Proux, unpublished) based on the excretion of urine by whole isolated MT.

The immunopurification was preceded by extraction of the ganglia in 0.2 M acetic acid, centrifugation, and elution of the supernatant through Sephadex G-50 gel with 0.2 M acetic acid as eluent. The AVP-like fractions were pooled and loaded on a gel prepared by coupling an AVP antiserum to Bio Rad Affi-Gel 10. With the AVP-like material bound to the antiserum, the nonimmunoreactive material was discarded by washing the support; then the AVP-like hormone was recovered by dissociating the antigen-antiserum complex. This AVP-like hormone had both biological and immunological activities, but dissociation of the antigen-antiserum complex by a pH shock caused a progressive alteration of the antiserum. Because of this degradation, the stock of antiserum was insufficient to permit the purification of all of the AVP-like hormone, and we switched to purification by reversed-phase LC.

For LC purification, the AVP-like hormone was extracted from ganglia using a strong acid buffer with protease inhibitors which gave slightly better recoveries than 0.2 M acetic acid. The extracts were centrifuged, and the supernatant was concentrated and then redissolved in 1% aqueous TFA. Salts, sugars, etc. were discarded by elution through a C_{18} Sep Pak cartridge with 1% TFA. After elution with 30% 1-propanol/0.1% TFA, the extract of predominantly peptides and proteins contained the AVP-like hormone as shown by RIA. This extract was then separated by reversed-phase LC using a Vydac C_4 semipreparative column with a CH_3CN/TFA gradient system. RIA detection revealed two separate AVP-like factors which we named Fl and F2, since it was now obvious that there was more than one AVP-like component. Bioassays revealed that only F2 had biological activity. Notwithstanding that, purifications of both Fl and F2 were carried out separately using different reversed-phase LC systems.

Fl was fully purified by two additional LC separations with a DuPont PEP RP/l C_8 analytical column and CH_3CN/HFBA followed by l-propanol/TFA gradients. The pure product we obtained was efficiently recognized by the AVP antiserum (the Fl and synthetic AVP displacement curves are parallel), but without any diuretic activity when assayed either in vivo or with MT in vitro.

F2 required four additional LC steps to be completely purified. A Vydac C_4 analytical column and different gradients (consecutively CH_3CN/HFBA, l-propanol/TFA, l-propanol/HFBA, and l-propanol/TFA) were used. The biological activity of pure F2 is fully equivalent to that elicited by crude extracts both in vivo and in vitro. The immunological activity is conserved, too, but a comparison of the slopes of the displacement curves of F2 and synthetic AVP are slightly different.

In conclusion, a diuretic neurohormone synthesized in the migratory locust SOG was brought to purity. Referred to as F2 during the purification procedure, we named this factor "AVP-like Insect Diuretic Hormone" because of its immunological relationship with mammalian AVP. A second AVP-like factor, Fl, was also purified to homogeneity, but was inactive in our biological tests. It could be a by-product of F2 formed during the long storage in acetone. Such storage might have affected its structure, as it is known to affect the structure of oxytocin, a peptide structurally related to AVP (Yamashiro et al., 1965).

REFERENCES

Cupo A. and Proux J. (1983) Biochemical characterization of a vasopressin-like neuropeptide in Locusta migratoria. Evidence of high molecular weight protein encoding vasopressin sequence. Neuropeptides 3, 309-318.

Proux J. and Rougon-Rapuzzi G. (1980) Evidence for vasopressin-like molecule in migratory locust. Radioimmunological measurements in different tissues: correlation with various states of hydration. Gen. Comp. Endoc. **42**, 378-383.

Proux J., Rougon G. and Cupo A. (1981) Enhancement of excretion across locust Malpighian tubules by a diuretic vasopressin-like hormone. Gen. Comp. Endocr. **47**, 449-457.

Rémy C., Girardie J. and Dubois M. P. (1979) Vertebrate neuropeptide-like substance in the suboesophageal ganglion of two insects: Locusta migratoria R. and F. (Orthoptera) and Bombyx mori L. (Lepidoptera). Immunocytological investigation. Gen. Comp. Endocr. **37**, 93-100.

Rémy C. and Girardie J. (1980) Anatomical organization of two vasopressin-neurophysin-like neurosecretory cells throughout the central nervous system of the migratory locust. Gen. Comp. Endocr. **40**, 27-35.

Yamashiro D., Aanning H. L. and Du Vigneaud V. (1965) Inactivation of oxytocin by acetone. Proc. Natn. Acad. Sci. U.S.A. **54**, 166-171.

FACTOR FROM BRAINS OF MALE HELIOTHIS VIRESCENS INDUCES ECDYSTEROID PRODUCTION BY TESTES

M.J. Loeb, C.W. Woods, E.P. Brandt
and A.B. Borkovec

Insect Reproduction Laboratory, A.R.S., U.S.
Dept. of Agriculture, Beltsville, MD. U.S.A.

Brain neuropeptides induce target glands to secrete ecdysteroid in two well-characterized insect systems: the prothoracicotropic hormone (PTTH) activates prothoracic glands to produce ecdysone, and the egg development neurosecretory hormone (EDNH) activates ovaries to produce ecdysteroid (reviewed by Wigglesworth, 1985). Bollenbacher et al.(1979) studied the PTTH system isolated in vitro by incubating brain extracts containing PTTH with prothoracic glands, detecting the resultant ecdysteroid by radioimmunoassay (RIA) (Borst and O'Connor 1972). Similarly, Hagedorn et al. (1979) elicited immunoreactive ecdysone production directly from mosquito ovaries by incubating them with head extracts containing EDNH. Testes of Heliothis virescens produce ecdysteroid in vitro at specific periods during development (Loeb et al. 1984). Incubation with a peptide-like factor from male brains can initiate ecdysteroid synthesis by testes in vitro.

Brains of male larvae and pupae of H. virescens were sonicated (Microprobe, Heat Systems Ultrasonics) in methanol:water: acetic acid (90:9:1), and centrifuged; supernatant was stored at $-20°C$ as crude brain extract.

367

Aliquots were dried (Savant Speed Vac) and
reconstituted in sterile Ringer with 0.5%
garamycin (Shering) (Loeb et al. (1984) for use
the same day. Testes from early-last-instar
larvae do not synthesize immunoreactive
ecdysteroid in vitro in Ringer or Ringer with
20-hydroxyecdysone (Loeb et al. 1984) and served
to bioassay ecdysiotropic factors. Bioassays
were run in duplicate with three testes in 200
µl of solution per Tek Plate (Falcon) well.
After 3 h at room temperature, samples were
sonicated in 67% methanol and centrifuged.
Supernatants were subjected to ecdysteroid RIA
(Loeb et al. 1984). Since crude tissue extracts
contained immunoreactive ecdysteroid, controls
without testes were assayed to calculate net
synthesis.

       Crude extracts of brains of last-instar
larvae and pupae, as well as extracts of
suboesophageal ganglia from   4th instar larvae,
prolegs from late-last-instar larvae, and testes
of late-last-instar larvae and 3-day pupae, were
bioassayed. Only brain extracts from late-last-
instar larvae and 3-6 day pupae (fig. 1, solid
line) induced RIA-detectable ecdysteroid production when bioassayed with young testes. However,
older testes produce ecdysteroid spontaneously
in vitro (dotted line, fig. 1) (Loeb et al.
1984). The coincidence of brain and spontaneous
testis activities suggests that testis ecdysiotropin (TE) plays a role in ecdysteroid production by testes in vivo. The TE-testis system may
be analogous to the PTTH-prothoracic gland and
EDNH-ovary ecdysiotropic systems. The dose-response of crude brain extract is linear (fig. 2)
to a maximum TE concentration; TE titer may
regulate ecdysteroid secretion by testes.

       Aqueous brain extracts lost activity when
stored at $4^{o}C$, or frozen and thawed. However,
extracts prepared in acid methanol were stable
at $-20^{o}C$ when stored in polyethylene tubes.
Boiling for 2 min reduced TE activity 45%.
Trypsin (Sigma, 1mg/ml, 1h) or nonspecific

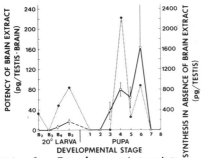

Fig.1: Brain extracts induce ecdysteroid in bioassay (solid line). Spontaneous secretion by older testes (dotted line).

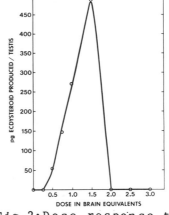

Fig.2: Dose response to pupal brain extract.

protease (Sigma, 1 Unit, 1h) destroyed TE activity. Thus, TE may be a peptide. High performance exclusion chromatography (HPEC) of crude extract with a Waters I-125 size exclusion column (trifluoroactic acid: acetonitrile: water 0.1:40:59.9) separated endogenous ecdysteroid from TE activity (Table 1). However, TE bioactivity was only detected when small amounts of 20-hydroxyecdysone (5-10 pg/μl) were returned to incubation medium. Activity co-chromatographed with insulin and cytochrome c markers, suggesting a molecular weight of 6-12 KD for TE. This is the molecular weight range of small PTTH (Bollenbacher et al. 1984) and EDNH (Hagedorn et al. 1979). Like PTTH (Bollenbacher et al. 1984) and EDNH (Hagedorn et al. 1979), TE is destroyed by proteases.

Table 1. HPEC of crude *H. virescens* brain extract (Waters I-125 size exclusion column): bioassay of fractions.*

Fraction #	Ecdyst content (pg)	Ecdysteroid produced in Ringer (pg)	20-HE** 5 pg/ul (pg)
0	0	0	0
1	0	0	0
2	20	0	0
3	18	18	1246
4	11	24	184
5	0	45	0
6	1538	0	30

* Mean of two experiments
** 20-hydroxyecdysone

However, PTTH and EDNH withstand boiling, aqueous and dry storage at 4° and −20°C, and freezing and thawing in aqueous solvents (Bollenbacher et al. 1984; Masler et al. 1983), and TE does not. Unlike PTTH (Bollenbacher et al.

1984) and EDNH (Masler et al. 1983), partial purification made TE ineffective to bioassay in Ringer; activity was restored with 20-hydroxy ecdysone. Therefore, TE may be structurally different from PTTH and EDNH and have different biochemical requirements for activity. TE may be a new insect ecdysiotropin.

REFERENCES

Bollenbacher W.E., Agui N., Granger N., and Gilbert, L.I.(1979) In vitro activation of insect prothoracic glands by the pro-thoracicotropic hormone. Proc. Nat. Acad. Sci. 76, 5148-5152.

Bollenbacher W.E., Katahira G.J., O'Brien M., Gilbert L.I., Thomas M.K., Agui N. and Baumhover A.H. (1984) Insect prothoracicotropic hormone: evidence for two molecular forms. Science 224, 1243-1245.

Borst D.W. and O'Connor J.D.(1974) Trace analysis of ecdysones by gas-liquid chro-matography, radioimmunoassay and bioass-ay. Steroids 24, 637-655.

Hagedorn H.H., Shapiro J.P. and Hanaoka K.(1979) Ovarian ecdysone secretion is controlled by a brain hormone in an adult mosquito. Nature 282,92-94.

Loeb M.J., Brandt E.P., and Birnbaum M.J. (1984) Ecdysteroid production by testes of the tobacco budworm, Heliothis virescens from last larval instar to adult. J. Insect Physiol. 30, 375-381.

Masler E.P., Hagedorn H.H., Petzel D.H., and Borkovec A.B. (1983) Partial purificat-ion of egg development neurosecretory ho-rmone with reverse-phase liquid chromato-graphic techniques. Life Sci. 33, 1925-1931.

Wigglesworth V.B. (1985) Historical Perspectiv-es. In Comprehensive Insect Physiology Biochemistry and Pharmacology vol 7 (Ed. Kerkut G.A. and Gilbert L.I.) pp 1-24. Pergamon Press, Oxford.

HEAD CRITICAL PERIODS FOR CONTROLLING APYRENE AND EUPYRENE SPERMIOGENESIS IN THE EUROPEAN CORN BORER, OSTRINIA NUBILALIS[1]

Dale B. Gelman and Alexej B. Borkovec

Insect Reproduction Laboratory
ARS USDA
Beltsville, MD   20705   USA

In the European corn borer as in all lepidopterans, two types of sperm are produced, eupyrene (nucleated) sperm that eventually will fertilize the egg, and apyrene (anucleated) sperm, the function of which is unknown. Early in larval life, individual germ cells in the apical region of each testis follicle undergo mitosis to form 2, 4, 8, 16, 32, and finally 64-celled eupyrene spermatocysts (primary spermatocytes). Meiosis results in the production of cysts containing 256 spermatids. Spermiogenesis (spermatid elongation and maturation) follows and soon is accompanied by the appearance of smaller apyrene cysts. These, too, will undergo meiosis and elongation forming bundles of sperm that are smaller than their eupyrene counterparts.

The progress or timing of spermatogenesis varies among insects. In the European corn borer, meiosis and spermiogenesis have been reported to occur early in the fifth (last) larval instar (Cloutier and Beck, 1963) although Chaudhury and Raun (1966) observed secondary spermatocytes in late penultimate instar testes. In the Ankeny strain of European corn borer reared under LD (light:dark) 16:8 and 30°C as described by Gelman and Hayes (1982), we have identified times for the onset of eupyrene and apyrene meiosis and elongation (Fig.1). The onset of the premolt ecdysteroid peak was correlated with the onset of apyrene meiosis and elongation.

---

[1]Mention of a commercial or a proprietary product does not constitute an endorsement of this product by the USDA.

371

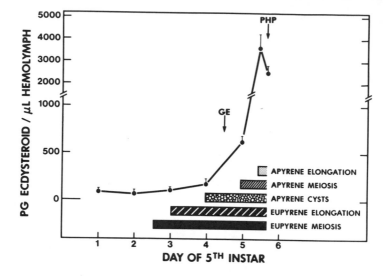

Fig. 1. Progress of spermatogenesis in 5th instars of the
European corn borer. Ecdysteroid titres are from Gelman and
Brents (1984). Spermatogenesis was monitored by placing
freshly dissected testes in insect Ringer's, teasing them
apart to release spermatocysts, and examining the preparation under Nomarski optics. Since "day 1" was used to denote
that period of time from ecdysis to lights out, day numbers
represent the early portion of any given day. Day 1 animals
were 6-12 h post-ecdysis, day 2, 30-36 h; day 3, 54-60 h,
etc. GE, gut purge. PHP, pharate pupa.

     Several factors have been implicated in the control of
insect spermatogenesis. Activating factors include a
hemolymph macromolecular factor and/or ecdysteroids (Schmidt
and Williams, 1969[2]; Kambysellis and Williams, 1971[2];
Chippendale and Alexander, 1973), factors of unknown origin
that initiate eupyrene and apyrene spermiogenesis
(Friedlander and Benz, 1981; 1982[2]), a brain factor that
stimulates eupyrene spermiogenesis (Loeb et al., 1985), and a
product of the testes that promotes spermatogenesis (Shimizu
and Yagi, 1978[2]; Giebultowicz et al., in press). Inhibitory
factors include juvenile hormone (Dumser and Davey, 1975[2];
Mitsui et al., 1976[2]), meiosis-blocking factors (Friedlander
and Benz,1982[2]) and a factor from the subesophageal ganglion
(Loeb et al. 1985). By means of ligation and injection

[2] See Loeb et al. (1985) for reference citations.

techniques we have been able to detect a head critical period for the initiation of apyrene spermiogenesis and have discovered that 20-hydroxyecdysone stimulates the elongation of apyrene spermatocysts.

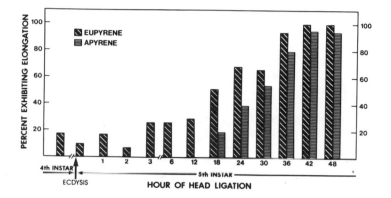

Fig. 2. Effect of head ligation on eupyrene and apyrene spermiogenesis. Insects were ligated behind the head at the times specified. After 7 days at LD 16:8 and 30°C, testes were dissected and examined as described for Fig. 1.

Fig. 2 depicts the progress of eupyrene and apyrene spermiogenesis in larvae ligated behind the head during the last day of the 4th instar and at specified hours after ecdysis to the fifth instar. While we could not detect a head critical period for eupyrene spermiogenesis, this process did not become completely independent of the head until 36-42 hours post-ecdysis to the last instar. Levels of eupyrene spermatocyst elongation remained low (less than 50 percent) until 18 h post-ecdysis, the time at which apyrene spermiogenesis was first observed. Thus, there did appear to be a head critical period between 12 and 18 h for the initiation of apyrene spermiogenesis; and for the initiation of both eupyrene and apyrene spermiogenesis, the testes appeared to receive signals from the head several days before these processes were observed in normally developing unligated larvae (Fig. 1).

To determine if 20-hydroxyecdysone could override the observed dependency on the head, larvae head-ligated at <16 h post-ecdysis were anaesthetized under dry ice and injected

with 2 ul of Grace's medium containing 250 ng of 20-
hydroxyecdysone.   Controls received 2 ul of Grace's medium.
The percentage of larvae exhibiting eupyrene spermiogenesis
increased from 40 (control) to 81 (experimental), and the
percentage of larvae exhibiting apyrene spermiogenesis
increased from 0 (control) to 22 (experimental).   Therefore,
20-hydroxyecdysone did promote both eupyrene and apyrene
spermiogenesis, and part of the observed dependency on the
head may have been due to a need for PTTH (prothoracicotropic hormone) to stimulate the prothoracic glands to
produce ecdysone.   However, since only partial stimulation
was observed, there may be regulatory factors in the head
other than PTTH that promote spermiogenesis.   Investigations
are now in progress to determine the existence of an apyrene-
spermiogenesis-promoting factor from the brain and/or other
structures in the head.

REFERENCES

Chaudhury M. F. B. and Raun E. S. (1966) Spermatogenesis
    and testicular development of the European corn borer
    Ostrinia nubilalis (Lepidoptera: Pyraustidae). Ann.
    ent. Soc. Amer. 59, 1157-1159.
Chippendale G. M. and Alexander B. R. (1973) Spermatogenesis
    of the Southwestern corn borer Diatraea grandiosella. 2.
    Resumption in diapause larvae. Ann. ent. Soc. Amer. 66,
    761-768.
Cloutier E. J. and Beck S. D. (1963) Spermatogenesis and
    diapause in the European corn borer, Ostrinia nubilalis.
    Ann. ent. Soc. Amer. 56, 253-255.
Gelman D. B. and Brents L. A. (1984) Hemolymph ecdysteroid
    levels in diapause- and nondiapause-bound fourth and fifth
    instars and in pupae of the European corn borer, Ostrinia
    nubilalis (Hubner). Comp. Biochem. Physiol. 78A, 319-325.
Gelman D. B. and Hayes D. K. (1982) Methods and markers for
    synchronizing the maturation of fifth stage diapause-bound
    and nondiapause-bound larvae, pupae and pharate adults of
    the European corn borer, Ostrinia nubilalis (Lepidoptera:
    Pyralidae). Ann. ent. Soc. Amer. 75, 485-493.
Giebultowicz J. M., Loeb M. J. and Borkovec A. B. (In
    press) In vitro spermatogenesis in lepidopteran larvae:
    role of the testis sheath. J. Invert. Reprod. Develop.
Loeb M. J., Brandt E. P. and Masler E. P. (1985) Modulation
    of the rate of spermatogenesis by the central nervous
    system of the tobacco budworm, Heliothis virescens. Int.
    J. Invert. Reprod. 8, 39-49.

ELECTRICAL PROPERTIES OF MEMBRANES OF CELLS OF THE CORPORA ALLATA OF THE COCKROACH DIPLOPTERA PUNCTATA: EVIDENCE FOR THE PRESENCE OF VOLTAGE-SENSITIVE CALCIUM CHANNELS

C.S. Thompson and S.S. Tobe

Dept. of Zoology
University of Toronto,
Toronto, Ontario, CANADA M5S 1A1

In adult cockroaches, as in many other insect species, juvenile hormone (JH) is necessary for the vitellogenic growth and maturation of eggs. In adult mated females of the cockroach, Diploptera punctata, rates of JH biosynthesis by the corpora allata (CA) undergo large cyclic changes in rates of JH biosynthesis and release in association with oocyte growth--high rates of JH biosynthesis are observed during vitellogenic growth whereas low rates are observed during previtellogenesis and during chorion formation (Tobe and Stay, 1977). The rate of JH biosynthesis by the CA appears to be regulated by both inhibitory and stimulatory signals reaching the CA by the nervous tracts and by humoral pathways (Tobe and Feyereisen, 1983; Tobe and Stay, 1985). Several second messengers, including cAMP, have been implicated in the transduction of these signals (Meller et al., 1985) and recent evidence suggests that calcium ions also play an important role in the regulation of JH biosynthesis and release (Kikukawa et al., 1986). JH production is almost totally inhibited in the absence of extracellular calcium and is maximal at concentrations of 3-5 mM. Also, pharmacological agents which modify intracellular calcium ion concentration can exert profound effects on JH production. The present studies were undertaken to examine the electrical characteristics of CA cell membranes and to determine whether or not calcium channels are present.

375

Each corpus allatum is composed of several thousand small parenchymal cells (Tobe et al., 1984) which are extensively intercoupled by gap junctions (Lococo et al., 1986). Thus it is possible to alter the membrane potential of all of the cells of a gland by injecting electrical current into one cell. The magnitude of the electrotonic potential recorded from a CA cell during current injection into another cell is determined by both the membrane resistance of the cells and by gap junctional conductance.

Current-voltage relations were linear in the hyperpolarizing direction but show a pronounced decrease in membrane resistance (and/or gap junctional conductance) in the depolarizing direction (Figure 1.A1). Following addition of pharmacological agents which block potassium ion conductance (tetraethylammonium chloride (TEA) or 4-aminopyridine (4AP)) to the saline bath, the apparent input resistance of the CA increased and active membrane responses were elicited by positive current injection (Figure 1.A1 and B2). These responses had a distinct threshold and appear to be the result of an influx of positive charge. They were followed by a large decrease in the apparent input resistance of the CA and by an after-hyperpolarization (AHP) which occurred at the conclusion of the current pulse. Thus, it is likely that such responses are the result of changes in membrane resistance rather than in gap junctional conductance.

Both calcium ions and sodium ions appear to be involved in the active membrane response to depolarization. Responses were greatly enhanced following replacement of calcium in the saline bath by barium and the AHP was eliminated (Figure 1.A3). Following addition of cobalt ions to the saline, the active responses and the AHP were reduced in amplitude. Micromolar concentrations of the sodium channel blocking agent tetrodotoxin (TTX) in the saline also reduced the active responses (Figure 1.B2), as did perfusion of glands with sodium-free saline. However, in the presence of TTX, or in sodium-free saline, large depolarizations still elicited an apparent reduction in input resistance and an AHP (Figure 1.B3).

A likely explanation for thse results is that the CA cell membranes have both voltage-sensitive calcium channels and voltage-sensitive sodium channels. Depolarizing current pulses activate these channels and result in an

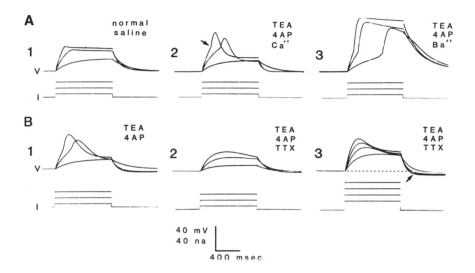

Figure 1: Membrane responses of CA cells to positive
current injection. A. 1. In normal saline, positive current
injection elicited apparently classic delayed
rectification. 2. When CA were perfused with 40 mM TEA and
1 mM 4AP, positive current pulses elcited an active
membrane response above a threshold value of about 20 mV
(arrow). Such responses were followed by a decrease in the
apparent input resistance of the CA. 3. Following
replacement of calcium ions in the saline with barium ions,
the active responses were greatly enhanced and the apparent
decrease in input resistance was reduced.
B. 1. Positive current injection also resulted in active
membrane responses in another CA preparation perfused with
TEA (40 mM) and 4AP (1 mM). 2. When CA were perfused with
6.2 μM TTX, the membrane response to positive current
injection was reduced. 3. Larger depolarizations in the
presence of TTX still elicited an apparent decrease in
input resistance. TTX did not eliminate the AHP (arrow).

inward current which is composed of both sodium ions and
calcium ions. The subsequent increase in intracellular
calcium then activates a third type of ion channel, the
calcium-activated potassium channel. Activation of the
latter type of channel would be expected to reduce the

apparent input resistance of the CA and to produce an
after–hyperpolarization.

REFERENCES

Kikukawa, S., Tobe, S.S., Solowiej, S., Rankin, S.M. and
Stay, B. (1986) Calcium as a regulator of juvenile hormone
biosynthesis and release in the cockroach, Diploptera
punctata. Insect Biochem., in press.

Lococo, D.J., Thompson, C.S. and Tobe, S.S. (1986)
Intracellular communication in an insect endocrine gland.
J. Exp. Biol. 121: 407–419.

Meller, V.H., Aucoin, R.R., Tobe, S.S. and Feyereisen, R.
(1985) Evidence for an inhibitory role of cyclic AMP in
the control of juvenile hormone biosynthesis by cockroach
corpora allata. Mol. Cell. Endocr. 43: 155–163.

Tobe, S.S. and Feyereisen, R. (1983) Juvenile hormone
biosynthesis: Regulation and assay. In Endocrinology of
Insects (ed. by R.G.H. Downer and H. Laufer), pp. 161–178.
Alan R. Liss, New York.

Tobe, S.S. and Stay, B. (1977) Corpus allatum activity in
vitro during the reproductive cycle of the viviparous
cockroach, Diploptera punctata. Gen. Comp. Endocr. 31: 138–
147.

Tobe, S.S. and Stay, B. (1985) Structure and regulation of
the corpus allatum. Adv. Insect Physiol. 18: 305–432.

Tobe, S.S., Clarke, N., Stay, B. and Ruegg, R.P. (1984)
Changes in cell number and activity of the corpora allata
of the cockroach, Diploptera punctata: A role for mating
and the ovary. Can. J. Zool. 62: 2178–2182.

PATCH CLAMP AND NOISE ANALYSIS STUDIES OF NEUROTRANSMITTER

RECEPTORS OF CULTURED INSECT NEURONES

David J. Beadle, Yves Pichon and Takeshi
Shimahara
School of Biological Sciences, Thames
Polytechnic, London SE18 6PF, U.K. and
Laboratoire de Neurobiologie Cellulaire du
CNRS, F-91190, Gif-sur-Yvette, France.

Studies of insect neuronal cultures prepared from the
brains of embryonic cockroaches have shown that the
in vitro neurones possess α-bungarotoxin binding sites and
are depolarised by acetylcholine (ACh) and approximately
60% of them are hyperpolarized by the application of gamma-
aminobutyric acid (GABA) (Lees et al., 1983; Lees et al.,
1984). These studies have now been extended using the
patch-clamp technique to obtain more information about the
ionic channels activated by these neurotransmitters. The
technique has been used in both the 'cell-attached' mode,
in which case the transmitters were added to the saline in
the pipette, or in the 'whole cell clamp' mode, in which
case current fluctuations resulting from pressure application of the transmitters were analysed.

The cultures were prepared from the brains of 23-26
day old embryos of Periplaneta americana as previously
described (Beadle et al., 1982). For electrophysiological
experiments the cells were allowed to equilibrate in a
saline consisting of 210 mM NaCl, 10 mM $CaCl_2$, 3.1 mM KCl
and 10 mM Hepes buffer at pH 7.2. The improved patch-clamp
technique of Hamill et al. (1981) was used and after the
establishment of a giga-seal, recordings were made either
in the cell-attached configuration or the membrane patch
was disrupted to obtain access to the cell interior for
whole cell voltage clamp recordings. In the latter case
the electrode contained 114 mM KCl, 5 mM EGTA, 1.6 mM
$MgCl_2$, 0.2 mM $CaCl_2$, 100 mM D-glucose and 10 mM Hepes buffer

379

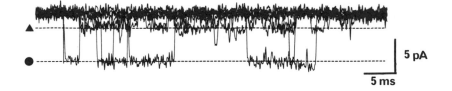

Fig. 1   Superimposed patch-clamp recordings of single-
channel activity induced by 5μM carbamylcholine applied
onto the membrane of a cultured cockroach neurone
illustrating the two conductance states of the channel.

at pH 7.2. Data were collected on analog tape (AMPEX PR
500) after low pass filtering at 1-3 kHz and single channel
data analysed on a PDP 11 computer and current fluctuations
on a spectrum analyser connected to an HP 9825T desk
computer and digital plotter.

Application of 50μM ACh or carbamylcholine (CCh) onto
cultured neurones held under whole cell voltage clamp
evoked an inward current that was accompanied by an
increase in current noise. The spectra of the current
fluctuations induced by ACh were best fitted by two
Lorentzian curves of half power frequencies around 11 Hz
and 130 Hz whereas those induced by CCh could be fitted by
a single Lorentzian (f_C = 30Hz). With 5μM CCh in the patch
electrode in the cell-attached configuration small, inward
unitary currents were observed of amplitude 2pA and 1-2
msec. duration at resting potential. These currents were
never observed in the absence of CCh. The duration of
these currents increased with membrane hyperpolarisation
and at -60mV two conductance states of the channel were
observed (18 pS and 48 pS) (Fig. 1). Analysis of these
channels at -60 mV gave amplitudes of 1.8pA and 6pA and
opening time constants of 0.7 msec. and 1.87 msec.
respectively.

When 50 μM GABA was applied to the neurones under
whole cell voltage clamp conditions with 114 mM potassium

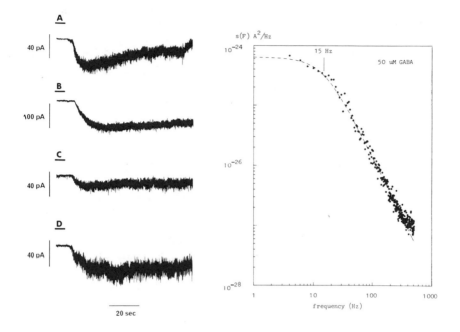

Fig. 2   Inward currents evoked by prolonged application
of agonist onto neurones held under whole cell voltage
clamp. (A) 50 μM GABA (B) 5μM muscimol (C) 50μM GABA (D)
50μM GABA in the presence of 2 x 10^{-6}M flunitrazepam.
(C) and (D) are neurones from the same culture. Also the
power spectrum of the noise induced by pressure application
of 50 μM GABA. The spectrum was fitted by a single
Lorentzian component of corner frequency 15Hz at a holding
potential of -50mV.

chloride in the intracellular electrode an inward current
was observed that was accompanied by an increase in
current noise (Fig. 2). The spectra of the current fluctuations induced by GABA were best fitted by a single
Lorentzian curve with a half power frequency of 13.5±2.53
Hz giving a mean channel opening time of 11.8 msec (Fig. 2).
When the values for the variance of the current noise were
taken with the mean current to calculate the single
channel conductance values of 18.6±6.16 pS were obtained
based on the assumption that chloride ions provided the
driving force of the response. A similar analysis of
currents evoked by the GABA agonist, muscimol, gave a

single channel conductance of 15±5.3pS and a mean channel
opening time of 6.5 msec. (f_c = 24.5±3.08 Hz).

Since the benzodiazepine, flunitrazepam, potentiates
GABA responses in locust neurones (Lees et al., in press),
GABA was applied to cultured cockroach neurones in the
presence of this compound. With 50 μM GABA applied to the
neuronal membranes in the presence of 2 x 10^{-6}M flunitrazepam the inward current evoked and the corresponding
noise increased when neurones within the same culture
were compared (Fig. 2C and D). When the single channel
conductance and half power frequency were calculated for
these responses values of 18.3 pS and 11.1±2.46 Hz were
obtained giving a mean channel opening time of 14.3 msec.
These values are very similar to those obtained with GABA
alone. However, when the variance of the noise was
compared values of 2.34±0.57 x 10^{-23} A^2 for GABA and
3.84±0.25 x 10^{-23} A^2 for GABA in the presence of flunitrazepam were obtained representing an increase of approximately 60%. This suggests that the benzodiazepine potentiates the GABA response by increasing the frequency of
channel events rather than by increasing the channel
opening time or conductance.

REFERENCES

Beadle, D.J., Hicks, D. and Middleton, C. (1982) Fine
    structure of neurones from embryonic Periplaneta
    americana growing in long term culture. J. Neurocytol.
    11, 611-626.
Hamil, O.P., Marty, A., Naher, E., Sakmann, B. and
    Sigworth, F.J. (1981) Improved patch-clamp techniques
    for high resolution current recording from cells and
    cell-free membrane patches. Pflugers Arch. 391, 85-
    100.
Lees, G., Beadle, D.J. and Botham, R.P. (1983) Cholinergic
    receptors on cultured neurones from the central
    nervous system of embryonic cockroaches. Brain Res.
    288, 49-59.
Lees, G., Beadle, D.J., Neumann, R. and Benson, J.A.
    (in press). Responses to GABA by isolated insect
    neuronal somata: pharmacology and modulation by a
    benzodiazepine and a barbiturate. Brain Res.
Lees, G., Benson, J.A. and Beadle, D.J. (1984) Putative
    insect neurotransmitters evoke electrophysiological
    responses from embryonic cockroach neurones in
    primary culture. Soc. Neurosci. Abst. 10, 688.

VOLTAGE DEPENDENT CONDUCTANCES IN CULTURED COCKROACH NEURONES

Y. Pichon[*], Y. Larmet [*], B.N. Christensen[+], T. Shimahara[*]
and D. Beadle[**]

[*]Laboratoire de Neurobiologie Cellulaire et Moléculaire du C.N.R.S., F-91190 Gif sur Yvette, France,[+]Department of Physiology and Biophysics, U.T.M.B., Galveston, Tx 77550, U.S.A. and [**]School of Biological Sciences, Thames Polytechnic,London SE18 6PF, U.K.

Neurones from the central nervous system of the american cockroach, <u>Periplaneta</u> <u>americana</u> can be kept in culture for several months. After a few days in culture, the neurones show growing cones which soon develop into fine neurites which connect the different cells and form a complicated network after a week or two (Beadle <u>et</u> <u>al.</u>, 1982). This morphological differentiation is accompanied with a physiological differentiation of the cell membrane. In this paper we shall summarize what is known so far of the voltage-dependent conductances.

The ionic currents which underlie excitability have been studied under voltage-clamp conditions using the two configurations of the patch-clamp techniques (Hamill <u>et</u> <u>al.</u>, 1981) : the whole cell clamp configuration ,which enables the recording of the currents which cross the entire surface of the cell and the associated current noise, and the patch-clamp proper (both cell-attached and cell-free) configuration which enables the recording of single channel events. The current and voltage traces were stored on video tape as 16 bit data and analysed off-line.
Square membrane depolarizations induced an increase in the membrane conductance to sodium, potassium and calcium ions. These changes have been analysed through a combination of three different techniques: a kinetic study of the total currents recorded in the whole cell clamp

383

configuration at various membrane potential levels, a statistical analysis of the single channel events and a spectrum analysis of the single channel noise induced by membrane depolarizations.

I. SODIUM CONDUCTANCE

Cockroach cultured neurones become excitable after about a week (Lees et al., 1985) and exhibit overshooting spikes which were blocked by the tetrodotoxin (TTX), a specific blocker of the sodium conductance. This activity is related to the existence of a fast inward sodium current. The complex geometry of the cells associated with the slow response time of the voltage-clamp amplifier made it impossible to fully analyse the kinetics of this current. Short unitary inward current which might correspond to the sodium current have been recorded and are presently being analysed.

II. POTASSIUM CONDUCTANCE

Delayed rectification appears very early during differenciation and is fully developed after two days in culture. There is some indication that it might decline after one or two weeks. The corresponding conductance has been analysed under various experimental conditions.

In the presence of TTX (1 uM), square membrane depolarizations in the whole cell configuration gave rise to a slow outward current which in some cases showed an initial hump, then declined to a plateau value. The intensity of this current increased with membrane depolarization and reached about 3 nA for a 150 mV depolarization, corresponding to about 3 mA cm^{-2}. This current reversed at EK, the equilibrium potential for potassium ions and was blocked by externally applied potassium channel blockers such as 4-aminopyridine (4-AP,1 mM) or tetraethylammonium ions (TEA, 10 to 50 mM) and is therefore analogous to the classical potassium current also seen in the giant axons.

Analysis of the single events associated with long term membrane depolarizations in the cell-attached configuration of the patch-clamp technique indicated that this

current corresponds to the opening of two distinct populations of potassium channels differing by their conductance (20 and 35 pS). Occasionally, channels with a much
larger single channel conductance were observed (about 80
pS). This last category of channel seemed to be related to
the existence of an inward calcium current.

The effect of membrane potential on these single
channel events has been analysed. The frequency of single
channel events increases with increasing membrane depolarization together with the duration of these events. For
large depolarizations, long bursts were observed which
were characterized by a large number of short closings.
These data suggest that an increase in membrane conductance reflects two different phenomena: an increase in the
probability of opening of any single channel associated
with an increase in channel lifetime. Statistical analysis
of a large number of channels in hyperpolarized inside-out
membrane patches revealed several conductance levels
which could correspond to several open states of the same
channel.

The noise spectra associated with the opening and
closing of the potassium channels could be fitted with a
combination of three Lorentzian with distinct corner frequencies of about 1 Hz, 30 Hz and 400 Hz. The low frequency component could correspond to a small number of
long openings, the middle frequency to a comparatively
large number of channels with a mean opening time of about
3 msec and the high frequency could reflect the combination of very short (<1 msec) openings and closings.

III. CALCIUM CONDUCTANCE

Besides fast sodium TTX sensitive spikes, cultured
neurones respond to long lasting depolarizations by graded
depolarizing responses which could be transformed into
spikes if TEA ions were are added to the external solution
(Lees et al., 1985). This response is associated with an
entry of calcium. It has thus been shown that, if the
external saline is replaced by an isotonic sodium-free
solution containing TEA and Barium ions, long duration
heart-like action potentials could be elicited by electrical stimulation and were often followed by spontaneous

fluctuations of the membrane potential.

Under voltage-clamp conditions, an inward calcium current was observed when the other conductances were blocked by TTX, TEA and 4-AP. This slow conductance inactivated very quickly if calcium was not replaced with barium, suggesting the existence under normal conditions of a calcium mediated calcium inactivation. In the presence of 20 mM Barium and in the absence of calcium, the current-voltage relationship exhibited a maximum of about 500 pA at 0 mV membrane potential. It has never been possible to obtain a reversal potential for this current The turning-off of the current was very slow (time constant of several hundreds of msec) and could be approximated with a single exponential. As for the sodium conductance, the unitary events associated with this conductance have not yet been fully characterized.

In many respects, the voltage dependent ionic conductances of cultured cockroach neurones resemble those of normal insect neurones. Experiments now in progress should provide interesting informations on the nature of the single channel events which are associated with these changes in conductance and thereby improve our understanding of the molecular events which underlie excitability.

REFERENCES

Beadle D.J., Hicks D. and Middleton C. (1982) Fine structure of Periplaneta americana neurones in long-term culture. J. Neurocytol. 11, 611-626.
Hamill O.P., Marty A., Neher E., Sakmann B. and Sigworth F.J. (1981) Improved patch-clamp techniques for high-resolution current recording from cells and cell-free membrane patches. Pflugers Arch. 391, 85-100.
Lees G., Beadle D.J., Botham R.P. and Kelly J.S. (1985) Excitable properties of insect neurones in culture: a developmental study. J. Insect Physiol. 31, 135-143.

VARIATION IN ACTIVITY OF A MAJOR MOLECULAR FORM OF ACHE IN BRAIN AND SMOOTH MUSCLE DURING TENEBRIO METAMORPHOSIS.

Jean-Jacques LENOIR-ROUSSEAUX

laboratoire de biologie animale ,UFR Sciences

Universite Paris 12,94010 Creteil cedex FRANCE

Acetycholine is the major transmitter of the central nervous system in insects whereas l-glutamic acid is restricted to the neuromuscular junction. Cholinergic transmission is rapidly terminated by the elimination of acetylchomine from the synaptic cleft due to hydrolysis by acetylcholiesterase [AChE;EC:3.1.1.7]. Changes in AChE activity are reported in insects during metamorphosis [1] and an increase in AChE activity was produced in a Drosophila cell line by the addition of molting hormone to the medium [2,3]. We report here the characterisation of an AChE activity due to a single molecular form found in several arthropods. We describe the variation of AChE activity during metamorphosis of a coleopteran Tenebrio molitor in the brain and in mesodermal organs: the male accessory glands. We also suggest that AChE activity during metamorphosis is correlated with ecdysterone levels.

Using the colorimetric method of Ellman [4], we found a "true" cholinesterase activity after extraction in a Tris buffer [pH 7] with 1% Triton X-100 and high salt [1 M NaCl]. This activity was inhibited by excess substrate [5 x 10^{-2} M acetylcholine iodide], by BW 284C51 [I_{50} = 5 x 10^{-7}M] but not by iso-OMPA [10^{-2} M] a specific inhibitor for pseudo-cholinesterase. AChE activity meeting these criteria was found in the brain of adult Tenebrio with a Km of 1.7 x 10^{-4}. Additional AChE was found in the male accessory reproductive glands with a Km of 2.9 x 10^{-4} and with an activity one hundred times lower than in the brain. AChE activity was restricted to the neuropile in the brain and to the muscular sheath of the glands as revealed with the histochemical method of Koelle [4].

1-AChE activity in brain and glands was due to a major hydrophobic molecular form with an S value of 5.6.

Brains were homogenized in buffers of high ionic strength [1 M NaCl] and 1 % Triton X-100 and centrifuged on a 5-20% linear sucrose gradient [Kontron TGA 50, TST 41 rotor]. 81% of the enzyme activity was recovered in a sharp single peak at 5.6 S. This peak was not altered by the addition of β-mercaptoethanol [10^{-3} M] or of dithiothreitol [10^{-3} M], both of which disrupt S-S bonds. This result suggests that the 5.6 S form of the enzyme is monomeric. At lower ionic strengths [0.1 M NaCl] with detergent, the major peak of AChE activity shifted from 5.6 S to 6.8 S, perhaps due to a conformational change in the protein or due to the fact that smaller molecules are ionically bound to the enzyme. Without detergent, there was AChE activity in fractions up to 16 S with no clear peaks. We believe that the aggregation seen without detergent at higher S-values may be due to an interaction between monomer and plasma membrane. The brain AChE seems to be a hydrophobic molecule linked to the membrane.

This major hydrophobic molecular form which was extracted with detergent and buffers of high ionic strength was also found in other Insects [6] such as : Periplaneta, Locusta, Gryllus, Leptinotarsa, Dysdercus, Pieris, Manduca, Musca, as well as in other arthropods : Astacus and Aranaria. We loaded the gradient with the same specific activity for each species. For Hemiptera and Lepidoptera, there was a slight peak asymmetry towards heavier forms whereas Diptera exhibited a minor shoulder towards lighter molecular forms.

After extraction of accessory glands of Tenebrio under the same conditions, the AChE activity was found in a single peak with an S value of 5.6, exhibiting the same properties as for the AChE extracted from the brain (fig.1).

2-During metamorphosis, AChE activity exhibited variations in both brain and glands.

The total AChE activity for the brain of the last larval instar was one half that of the adult [136 ηmol/h/brain]. The increase occured just after the pupal molt. The specific activity is similar in the newly ecdysed larva and in the adult [7000 μmol/h/g protein]. During prepupal and pupal stages, specific activity fluctuated with 3 maxima that coincide with the prepupal and pupal ecdysteroid peaks (fig.2a).

In the glands, there was an increase in total AChE activity from the newly-ecdysed pupa [3 ηmol/h] to the mature adult [30 ηmol/h]. The AChE activity per wet weight [5 μmol/h/g wet weight] and specific activity [70 μmol/h/g protein] was identical in the very young pupa and in the adult. There was a transitory increase to 170 μmol/h/g protein (fig. 2b) in the mid-pupa [Stage 3].

-3- **Variations in AChE specific activity presented peaks that correlated with the peaks of ecdysterone in the hemolymph.** (fig. 2)

During the prepupal stage, the AChE specific activity of the brain showed 2 peaks corresponding to the ecdysterone peak (fig.2) at stage 7 and stage 12. The highest AChE specific activity in the pupa was seen at stage 4 during the major peak in molting hormone [7].

The peak of AChE specific activity in the glands occurred during the pupal ecdysterone peak [7]. At the time of the highest ecdysteroid peak, AChE specific activity reached its maximum [193 μmol/h/g proteins].

The in vitro culture of glands for 2 weeks revealed that ecdysterone influenced AChE activity. When 0 day accessory glands were cultured in the inorganic fraction of Landureau's medium [8], a pulse of ecdysterone [10^{-5} M, 1.5 days] produced a secretion of AChE into the medium but no increase in activity within the glands. When the hormone pulse was applied to glands cultured in Landureau's complete medium, there was an increase in total and specific AChE activity in the glands. This result strongly suggests that AChE accumulation and release during the instar is sensitive to ecdysterone.

The next question to ask is to know how hormone levels affect AChE activity in the accessory glands. For example, it is possible that storage of AChE occurs during the rise of hormone titer and that secretion is triggered by the decline of the titer. In addition, it is important to know whether the increase in AChE activity in the brain is likewise affected by the molting hormone.

References

1.Smallman B.N. & Mansingh A.**Ann. Rev. Entomol.**,1969,14,
     387-408.
   Dewhurst S.A.,McCaman R.E. & Kaplan W.D. **Biochem.
     Genet.**,1970,4,499-508
   Chaudharay K.D. , Srivastava U.& Lemonde
     A.,**Arch.Inter.Physiol. biochem.**,1966,74,416-428
2.Cherbas P.,Cherbas L. & Williams C.M., **Science,**
     1977,197,275-277.
3.Best-Belpomme M.,Courgeon A.M. & Echalier G., **Progress in
   Ecdysone Research**, Hoffman J. Ed.Elsevier, 1980,379-392.
4.Ellman G.,Courtney K.,Andres V. & Featherstone R.,
       biochem. Pharmacol.,1961,7,88-95.
5.Koelle G.B. & Friedenwald J.S.,**Proc.Soc. Exp. Biol.
     Med.**,1949,70,617-622.
6.Lenoir-Rousseaux J.J.,**C.R.Soc.Biol.**,1985,179,741-747.
7.Lenoir-Rousseaux J.J.& Gautron J.,**C.R.Soc.Biol.**,1986,180,
     22-28.
8.Landureau J.C.,**Invertebrate tissue culture.** Kurstak E.&
     Maramorosh K.Ed.Acad.Press,1976.

Figures
1-Sedimentation profiles with 5-20% sucrose gradient of the
single molecular form of AChE in Triton X-100 and 1M NaCl
extracted from accessory glands (AG) and brain (SN).
β GAL : β -galactosidase; CAT : catalase; ADH : alcohol
dehydrogenase (horse =7.4 S)
2-AChE Specific activity in Tenebrio brain (a) and accessory
glands (b) compared to ecdysteroid level in the hemolymph
[Delbecque J.P.,Thesis,Dijon,1983].
Specific stages of ecdysteroids peaks : T ;7 ;12 ; 4m ;
mitotic period :M.Ordinates:development of Tenebrio from the
last larval instar(L), including the prepupal stages (SO),
the pupal stages (N) and the adult (J) in days.Open circle
:maximum value in AChE activity for glands correlated with
the ecdysteroid peak.

Acknowledgments

   We wish to thank J.Gautron for his advice and G.M. Happ for
contructive comments about this manuscript.

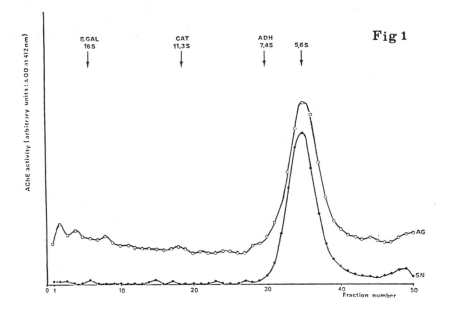

Fig 1

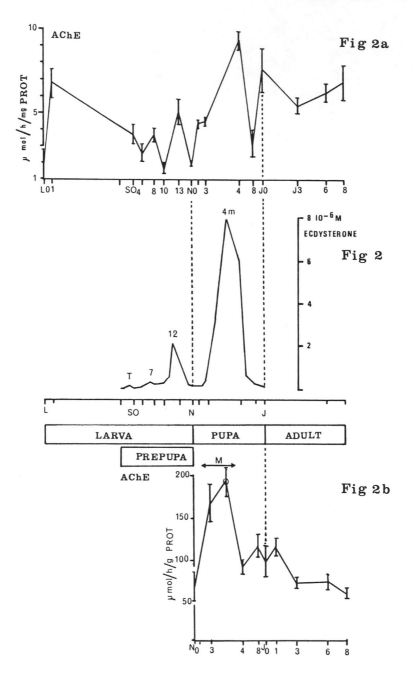

Fig 2a

Fig 2

Fig 2b

SODIUM CHANNELS OF THE COCKROACH CENTRAL NERVOUS SYSTEM SYNAPTOSOMES

Ash K. Dwivedy

A.F.R.C. Unit of Insect Neurophysiology and Pharmacology, Dept. of Zoology, University of Cambridge, Downing Street, Cambridge, CB2 3EJ, U.K.

INTRODUCTION

The inward movement of sodium ions during the propagation of action potential in most nerve membranes occurs through voltage-sensitive sodium channels. The molecular properties of sodium channels are defined predominantly by the interaction of specific neurotoxins with the discrete receptor sites of the voltage-sensitive sodium channels, such as sodium influx inhibitors (tetrodotoxin, saxitoxin) sodium channel activators (batrachotoxin, veratridine, aconitine), and inactivation inhibitors (sea anemone toxin, scorpion toxin) (Lazdunski and Renaud, 1982; Catterall, 1984). Because the neurotoxins bind to their receptor sites with a high affinity and specificity, they have been used as pharmacological probes to study the structural and functional properties of voltage-sensitive sodium channels in variety of excitable membranes. However, the responses of specific toxins for sodium channels differ considerably between the rat brain synaptosomes, membrane vesicles (Buonanno and Villegas, 1983), mouse brain synaptosomes (Ghiasuddin and Soderlund, 1984), neuroblastoma cells (Jacques et al., 1978), and rat brain cultured neurons (Couraud et al., 1986). The present investigation describes appropriate conditions for measuring sodium channel-dependent $^{22}Na^+$ uptake in the synaptosomal fraction isolated from the central nervous system (CNS) of cockroach, Periplaneta americana, and the responses of

393

this preparation to various neurotoxins have been examined for their effect on the voltage sensitive sodium channels. ^{22}NaCl (carrier free, 100mCi/mg) was purchased from Amersham International, U.K. Batrachotoxin (BTX) was a generous gift of Dr John N. Daly, National Institute of Health, Bethesda, Md., U.S.A. Anemonia sulcata II (ATX) was obtained from Calbiochem, veratridine (VTD), aconitine (ACN), and other chemicals were purchased from Sigma Chemicals, U.K.

The central nervous system (CNS) of the cockroach, Periplaneta americana, was used for the preparation of CNS-homogenate, and the synaptosomal fraction was obtained by using the discontinuous sucrose-Ficoll gradient centrifugation method (Dwivedy, 1985, 1986). The synaptosomal preparation was suspended in a sodium free buffer (5mM Glucose, 3mM MgCl$_2$, 8mM KC1, 2mM ouabain, 2mM KCN, 0.5 mM CaCl$_2$, 0.5mM EGTA, 20 mM Hepes-Tris, pH 7.4). Aliquots of synaptosomal suspension (10-20μg) were incubated at 21°C for 30s with various toxins in a sodium incubation buffer (suspension buffer plus 20mM choline chloride, 1mg/ml BSA, 10mM ^{22}Na$^+$). The uptake of ^{22}Na$^+$ was measured by rapid filteration technique (Whatman C filter, 0.45 μm pore). Each incubation tube and filter were washed twice with 2ml ice-cold stopping solution (5mM Glucose, 3mM MgCl$_2$, 20mM choline chloride, 20mM Hepes-Tris, pH 7.4). The filters were dried, dissolved in 20% methyl cellosolve and counted in ACS (Amersham). The protein content were measured by the method of Lowry (Dwivedy, 1985). The synaptosomal preparation was routinely incubated with the ouabain, KCN; and ^{22}Na$^+$ flux measured in presence of EGTA and BSA in order to eliminate the possibility of the factors which could have contributed to Na$^+$ uptake other than toxins in the experiments. The ^{22}Na$^+$-channel activity was measured at room temperature (20-22°C) because the VTD-stimulated ^{22}Na$^+$ flux was not observed at 0°C. The assay times shorter than 10-15 seconds gave extremely variable results, therefore ^{22}Na$^+$ flux were studied upto 30 seconds. The composition of loading and incubation buffers were adjusted as necessary using otherwise similar conditions those described for rat and mouse brain synaptosomes (Buonanno and Villegas, 1983; Ghiasuddin and Soderlund, 1984).

A comparison of the pharmacological properties of insect synaptosomal sodium channels by BTX, VTD and ACN to those previously reported in vertebrate brain synaptosomes and neurons are given in Table 1. It shows that BTX is a

Table 1. Comparison of apparent affinities, $K_{0.5}$, of neurotoxin effects on $^{22}Na^+$ uptake. Data are fitted to a modified form of Michaelis-Menten equation (Catterall, 1980). n.d. not determined.

Preparations	Activators $K_{0.5}$ (μm)		
	Batra-chotoxin	Veratr-dine	Aconi-tine
Mouse brain synaptosomes[α]	0.5	34.5	19.6
Rat brain synaptosomes[α]	0.5	13.0	14.0
Rat brain (10 days old) membrane vesicles [§]	n.d.	17.0	n.d
Mouse neuroblastoma cells[α]	0.7	29.0	3.6
Cultured mouse neuronal cells*			
(a) 2 day fetal brain	6.0	90.0	n.d.
(b) 9-12 day fetal brain	0.3	8.6	"
Insect CNS (present study)			
(a) CNS homogenate	0.2	9.5	15.0
(b) synaptosomes#	0.2	3.0¶	7.6

$K_{0.5}$ Concentration (μM) giving a half-maximal activation.
[α]   Data from Ghiasuddin & Soderlund, 1984.
*   Data from Couraud et al., 1986.
#   The initial rate of $^{22}Na^+$ uptake in absence of activator was 57 nmole/30sec/mg of protein.
¶   Veratridine activation of $^{22}Na^+$ influx was inhibited by tetradotoxin (I_{50},23.5nM) and potentiated by anemonia sulcata-II ($K_{0.5}$, 2.0nM).
§   Data from Buonanno and Villegas (1983).

full agonist in its potency similar to other vertebrate preparation except two days fetal brain cultured neuron cells. However, the responses of partial agonists VTD and ACN appear to be more effective in insect synaptosomes than rat and mouse brain synaptosomes. The identification of sodium channels relies on the specificity of tetrodotoxin (TTX) to block VTD-induced $^{22}Na^+$ flux. The $K_{0.5}$ value for TTX in presence of VTD (50μM) was approx. 23.5nM, a slightly higher value than reported for vertebrate synaptsomes (Buonanno and Villegas, 1983; Ghiasuddin and Soderlund, 1984). An increase in $^{22}Na^+$ flux produced by ATX was smaller than that of VTD.

However, a synergistic effect between VTD and ATX was observed ($K_{0.5}$, 2nM at 50μM VTD) which was abolished by the TTX. Results suggest the presence of three specific receptor sites associated with $^{22}Na^+$ channels in the cockroach CNS synaptosomes. The data show that voltage sensitive sodium channels as defined by their specific activation by the alkaloids and polypeptides toxin are qualitatively distinct in insect synaptosomes than that reported in vertebrate studies. Also, the potency and efficacy of various neurotoxins in their response to the sodium channels of insect CNS synaptosomes reflect a quantitative differences between vertebrate brain synaptosomes, membrane vesicles and cultured neuron cell.

REFERENCES

Buonanno A. and Villegas R. (1983) Sodium channel activity in brain membrane fractions isolated from rats of different ages. Biochim. biophys. Acta. 730, 161-172.

Catterall W.A. (1984) The molecular basis of neuronal excitability. Science., 223, 653-661.

Couraud F., Martin-Moutot N., Koulakoff A. and Berwald-Netter Y. (1986) Neurotoxin-sensitive sodium channels in neurons developing in vivo and vitro. J. Neurosc. 6, 192-198.

Dwivedy A.K. (1986) Synaptic nerve-endings (synaptosomes) with cholinergic properties from a simple nervous system. Biochem. Soc. Trans. 14, 359-360.

Dwivedy A.K. (1985) Cholinergic properties of pinched off synaptic nerve endings from the central nervous system of the cockroach Periplaneta americana. Pestic. Sci. 16, 615-626.

Ghiasuddin S.M. and Soderlund D.M. (1984) Mouse brain synaptosomal sodium channels. Comp. Biochem. Physiol. 77, 267-271.

Jacques Y., Fosset M. and Lazdunski M. (1978) Molecular properties of the action potential Na^+ ionophore in neuroblastoma cells: Interactions with neurotoxins. J. biol. Chem. 253, 7383-7392.

Lazdunski M. and Renaud J.F. (1982) The action of cardiotoxins on cardiac plasma membranes. Ann. Rev. Physiol. 44, 463-473.

ISOLATION AND BIOLOGICAL ACTIVITY OF SYNAPTIC TOXINS FROM THE VENOM OF THE FUNNEL WEB SPIDER, AGELENOPSIS APERTA

Michael E. Adams*, F. E. Enderlin, R. I. Cone and D. A. Schooley

*Dept. of Entomology, University of California, Riverside, CA 92521 and Zoecon Corp., 975 California Ave., Palo Alto, CA 94304

The structure elucidation of novel transmitter and hormone substances, especially peptides, has proceeded rapidly in recent years. Several reports at this conference indicate that the identification of insect neuropeptides will continue with even greater intensity in the near future. In contrast, our knowledge of corresponding receptor systems has lagged because of the absence of potent and selective antagonists. This has been particularly true for the insect neuromuscular system, where non-cholinergic excitatory transmission involves amino acids (e.g., glutamate, aspartate) and peptides (e.g., proctolin). The discovery of antagonists specific for synaptic receptors would offer opportunities for their isolation and characterization, and would permit incisive experiments designed to chemically ablate individual transmitter systems in cellular preparations and in intact insects. This would be useful in assessing the relative importance of different ligand-receptor systems to the survival of pest insects and may assist in defining those systems most promising for design of new insect control agents.

Many predators and parasites have evolved venoms to immobilize their insect prey. It is becoming evident that toxins evolved against insect receptors are present in venoms of certain spider groups which could prove useful in studies of synaptic pharmacology and as leads for the creation of receptor antagonists. Previous work by Piek et al., (1985); Kawai et al., (1982); and Usherwood et al., (1985), have indicated the presence of low molecular weight synaptic toxins in wasp and spider venoms.

We report here on the isolation and structure elucidation of multiple synaptic toxins from the venom of the funnel web spider, Agelenopsis aperta. These toxins can be classified into two groups on the basis of chemical structure and by their different modes of action on synaptic receptors.

Our search for synaptic antagonists began by screening crude venoms from a variety of spider species for paralytic activity. Crude venom was obtained by an electrical milking technique developed by Mr. C. Kristensen of Spider Pharm, Black Canyon City, AZ, and was assayed by intra-abdominal injection into cockroaches, Periplaneta americana and/or flies, Musca domestica. Since we were interested in finding low molecular weight toxins, we separated active venoms on a Sephadex G-25 column (100 x 1.6 cm., useful for molecular weight determination in the 400-5000 dalton range). Many of the venoms contained exclusively large molecular weight toxins, and these were usually slower acting but more toxic (Phidippus johnsonii, Loxosceles reclusa, Filistata spp.). Other venoms also contained low molecular weight toxins (orb weaver, funnel web species). A. aperta venom was further fractionated on reversed phase LC, using a Vydac TP-214 C4 analytical column and either acetonitrile or 1-propanol as the organic eluant. Either trifluoroacetic acid or heptafluorobutyric acid were used as ion-pairing reagents. Venom was usually prepurified by passage through Sep-Pak C18 cartridges to remove particulates, salts and lipids.

The most potent toxins from A. aperta are approximately 4800 daltons and tentatively are called T1, T1', T2 and T3. A second class of toxins are lower molecular weight (<1000) and showed minimal retention on C4 columns, The most active of these is referred to as TH or "hydrophilic toxin." "Big Agelenopsis toxins" (>5000 daltons) eluted at high organic concentrations.

At the present time, we know most about the T1-T3 toxins. They are unaffected by boiling and all except for T1 are resistant to trypsin and chymotrypsin. However, preliminary analyses of amino acid composition showed that they

were peptides containing multiple cysteine residues. The
toxins were reduced and alkylated to form carboxymethyl-
cysteine prior to sequencing. Samples of 0.5-1.0 nanomole
were sequenced using an Applied Biosystems 470A gas phase
sequencer, with a 120A PTH analyzer.

T1, T1', T2 and T3 have unblocked amino-termini and contain eight cysteine residues, indicating the presence of 4
disulfide bridges. We are currently verifying the iden-
tities of the C-terminal residues and testing for possible
C-terminal amidation.

T1, T1' and T2 show strong sequence homology. The positions of the eight cysteine residues are identical
(allowing for a shift due to an additional N-terminal
amino acid in T1), which suggests a similar, if not identical disulfide bridging pattern. Twenty of the 37 residues found in T1, T1' and T2 occur in identical positions.
For T3, the first two cysteines have spacing identical to
those of the first three toxins, but very little additional similarity is evident.

We have tested the actions of the toxins on neuromuscular
preparations from flies (Musca domestica, Drosophila melanogaster) and cockroaches (Periplaneta americana). The
most obvious action of T1 was an abrupt appearance of
repetitive EPSP activity. This effect occurred intermit-
tently in response to single shocks applied to the nerve
or spontaneously in the absence of stimuli. Continued
washing for hours did not reverse the repetitive firing.
The structure of the T1-T3 toxins and their induction of
repetitive firing indicates a similar mechanism of action
to that of certain scorpion toxins described by Zlotkin et
al. (1985), that is, a presynaptic action on the sodium
channel.

In contrast, the TH toxins appear to act postsynaptically.
Application of TH to the fly synaptic preparation caused a
gradual diminution of the EPSP and eventual block. We
observed no repetitive firing as was the case with T1.

In summary, the venom of A. aperta contains multiple peptide toxins of low molecular weight which antagonize
synaptic transmission in insects. These toxins are
distinguished on the basis of structure and sites of

action at the neuromuscular junction. T1, T1', T2 and T3
contain 36-37 amino acids and four disulfide bridges.
Preliminary physiological studies indicate a presynaptic
action on the sodium channel. Several hydrophilic toxins,
of which TH is the most active, are much smaller (<1000
daltons) and probably affect postsynaptic receptors on the
muscle. A. aperta appears to use multiple synaptic toxins
synergistically to cause paralysis in prey insects.
Studies on the structures and physiological actions of low
molecular weight arachnid toxins may eventually lead to
new pharmacological tools and possibly to fundamentally
new leads for chemicals affecting synaptic transmission in
pest insects.

REFERENCES

Kawai, N., A. Niwa and T. Abe (1982) Spider venom contains
    specific receptor blocker of glutaminergic synapses.
    Brain Res. 247: 169-171.

Piek, T., K.S. Kits, W. Spanjer, J. Van Marle and H. Van
    Wilgenburg (1985) Neurotoxic effects of wasp venoms -
    synaptic and behavioral aspects.   Neurotoxicology 6:
    251-260.

Usherwood, P.N.R. (1985) The action of spider toxins on
    the insect nerve-muscle system. In: Approaches to New
    Leads for Insecticides, (H.C. von Keyserlingk, A. Jager
    and Ch. von Szczepanski, Ed.) Springer Verlag, Berlin.

Zlotkin, E., D. Kadouri, D. Gordon, M. Pelhate, M.F.
    Martin and H. Rochat (1985) An excitatory and a
    depressant insect toxin from scorpion venom both affect
    sodium conductance and possess a common binding site.
    Arch. Biochem. Biophys. 240: 877-887.

PEPTIDE TOXINS FROM ARTHROPOD VENOMS DISRUPT FEEDING AND UTILIZATION OF DIET IN THE COTTON BOLLWORM

David C. Ross[1,2], Gary A. Herzog[3], and
Joe W. Crim[1]

Departments of Zoology[1] & Entomology[2], University of Georgia, Athens, GA, & Department of Entomology[3], University of Georgia Coastal Plain Experiment Station, Tifton, GA

The broad spectrum insecticides in current use cause indiscriminate mortality among pest and beneficial species alike. Resistance to synthetic insecticides continues to be a major problem. At least twenty-three insect species have developed resistance to the newest class of insecticides, the pyrethroids (Georghiou, 1986). As an alternative to these synthetic insecticides, we propose the development of highly selective insect-specific control systems based on peptides with new target sites and/or novel modes of action. Peptide toxins with activity specific to insects are of great interest in this respect. Candidate peptides would conceivably disrupt feeding or utilization of food in addition to any acute toxic effect they have. The goal of this study was to identify model peptides which reduce overall gastroenteric activity or efficiency in the corn earworm, *Heliothis zea*.

Acute toxicity and effects on diet consumption and utilization were tested by injecting peptides into fourth instar *H. zea* larvae. Standardized indicies and growth rates (Slansky and Scriber, 1985) were used to compare the effects of peptide toxins with arthropod venoms and the insecticides chlorpyriphos and cyromazine (an inhibitor of chitin synthesis). Automated data acquisition and analysis were performed via an electronic balance-computer interface.

401

We found that melittin, from honey bee venom (*Apis mellifera*), and mastoparan, from *Vespula* wasp venom, were responsible in large part for the effects of their parent venoms. These toxins significantly decreased amount of food ingested, larval weight, and fecal production in *H. zea* at doses of 12.5 µg/larva and greater. Mellitin severely inhibited assimilation of nutrients and overall conversion of ingested food to body tissue. Mastoparan reduced the efficiency with which larvae converted digested and assimilated food to biomass. Slansky and Scriber (1985) attribute such effects to increased catabolism of digested nutrients. These effects may be related to the mastocytolytic activity of mastoparan (Nakajima, 1984).

All wasp kinins contain the peptide sequence of bradykinin (Nakajima, 1984). Bradykinin decreased food assimilation, but this effect was countered by increased utilization of assimilated nutrients. Thus, bradykinin left larval weight unchanged.

Venoms of the scorpions *Androctonus australis* and *Centruroides sculpteratus* decreased larval weight and amount of food consumed at 12.5 µg/larva, and they reduced the efficiency of conversion of food to biomass (Ross *et al.*, submitted).

Larvae displayed a spastic-contractile paralysis (SCP) when injected with melittin, mastoparan, and venoms. The response was similar to that previously reported for scorpion venoms injected into *Sarcophaga* larvae (Lazarovici *et al.*, 1982). Bradykinin and the insecticides did not ellicit an SCP response.

The LD_{50}s of melittin and mastoparan were 27 and 41 µg/larva, respectively, at 24 hrs post-injection, while for venoms they ranged from 40 to 51 µg/larva. Although chlorpyriphos was comparatively more toxic (LD_{50}), at sublethal doses it showed no effects on feeding or digestion. In contrast, cyromazine caused no mortality after 48 hrs at doses tested and only slightly diminished growth rate and assimilation efficiency.

Sublethal doses of venoms and peptides generally acted to reduce gastroenteric activity and efficiency. Notably, these actions were distinct from acute toxic effects, suggesting a novel physiological basis for insect control.

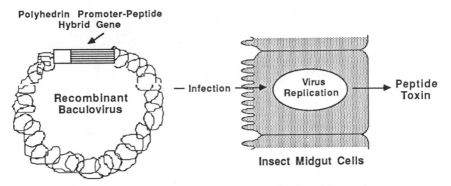

Figure 1. Delivery system for peptide toxins using genetically engineered baculovirus.

Accordingly, our system for determining disruption of in-
gestion, digestion, and utilization of food appears particularly suitable for testing additional compounds as well.

Toxins such as melittin, mastoparan, and those found
in scorpion venoms serve as model peptides for new insect
control agents. However, appropriate methods of delivery
must first be devised. For example, genetically engineered,
species-specific insect baculoviruses infecting the midgut
could act to synthesize and secrete toxins from the insect's own cells (Fig. 1). Smith *et al.* (1983) have inserted the gene for human beta interferon into *Autographa*
californica and obtained interferon secretion in *Spodoptera*
frugiperda cells *in vitro*. An alternate approach would involve the development of non-labile peptides that would act
directly as stomach poisons (Primor and Zlotkin, 1980),
perhaps by disruption of digestive and metabolic activities.
Peptides thus exhibit high potential for development of
very selective insect control systems.

ACKNOWLEDGEMENTS

We thank Drs. A.O. Lea, M.R. Brown, and U.E. Brady for
critical discussions. Supported by the University of Georgia Program in Biological Resources and Biotechnology.

REFERENCES

GEORGHIOU G.P.   (1986)   The magnitude of the resistance
        problem.   In Pesticide Resistance: Strategies and
        Tactics for Management (Edited by Glass E.H. *et al.*)
        pp. 14-43. National Academy Press, Washington, DC.
LAZAROVICI P., YANAI P., PELHATE M., and ZLOTKIN E.   (1982)
        Insect toxic components from the venom of a chactoid
        scorpion, *Scorpio Maurus palmatus* (Scorpionidae).
        J. Biol. Chem. 257, 8397-8403.
NAKAJIMA T.   (1984)   Biochemistry of vespid venoms.   In
        Handbook of Natural Toxins, Vol. 2, Insect Poisons,
        Allergens, and Other Invertebrate Venoms (Edited by
        Tu A.T.) pp. 109-133. Marcel Dekker, Inc., New York.
PRIMOR N. and ZLOTKIN E.   (1980)   Penetrability of proteins
        through the digestive system of *Sarcophaga falculata*
        blowfly.   Biochem. Biophys. Acta 627, 82-90.
ROSS D.C., CRIM J.W., and HERZOG G.A.   Comparative toxicity
        of the venoms of two scorpions (*Androctonus aus-
        tralis* and *Centruroides sculpturatus*) and a sea
        anemone *(Condylactis gigantea)*: Acute and antifeed-
        ing actions in the cotton bollworm (*Heliothis zea*).
        (Submitted).
ROSS D.C., CRIM J.W., BROWN M.R., HERZOG G.A., and LEA A.O.
        Toxic and antifeeding actions of melittin on *Helio-
        this zea* (Boddie): Comparisons to bee venom, chlor-
        pyriphos and cyromazine. (Submitted).
SLANSKY F. and SCRIBER J.M.   (1985)   Food consumption and
        utilization.   In Comprehensive Insect Physiology,
        Biochemistry, and Pharmacology, Vol. 4 (Edited by
        Kerkut G.A. and Gilbert L.I.) pp. 87-163.   Pergamon,
        New York.
SMITH G.E., SUMMERS M.D., and FRASER J.   (1983)   Production
        of human beta interferon in insect cells infected
        with a baculovirus expression vector.   Mol. Cell.
        Biol. 3, 2156-2165.

WHOLE-CELL AND SINGLE-CHANNEL RECORDING FROM INSECT MYOSACS.

Isabel Bermudez, George Lees*, Roger P. Botham*
and David J. Beadle
School of Biological Sciences, Thames Polytechnic, London SE18 6PF, U.K. and *Wellcome
Research Laboratories, Berkhamsted, Herts, U.K.

Insect muscle cultures have been successfully used to study several aspects of insect myogenesis (Bermudez et. al. 1986). Although the unrestricted access to the surface of the cultured cells has facilitated intracellular recording studies (Bermudez et, al., 1986), the tubular and branched structure of the myotubes has impeded the use of other electrophysiological techniques such as voltage clamp to study many basic questions related to the pharmacology and electrophysiology of the cultured insect myotube. In this communication we describe a technique to produce spherical insect muscle cells (myosacs) by treating insect muscle cultures with colchicine and show that they are suitable for whole-cell and single-channel recording.

Myogenic monolayer cultures were grown in Leibovitz L-15/Yunker's modified Grace's medium plus 5% horse serum (Bermudez et. al., 1986) following cell dissociation from thoracic and abdominal muscle somites of 11-day old embryos of Periplaneta americana. Colchicine was added, at a final concentration of 10^{-8} M (Fukuda et. al.,1976), to the culture medium after 7-9 days of growth in vitro. After 4 days of growth in the colchicine medium the cells were transferred to normal culture medium. Cells for electron microscopy were fixed in situ with glutaraldehyde, post-fixed in osmium, dehydrated in an ethanol series and embed-ded in Epon. Thin sections were stained with uranyl acetate and counterstained with lead citrate.

Myosac cultures were equilibrated in a bathing saline
(200 mM NaCl, 3.1 mM KCl, 9 mM $CaCl_2$, 5 mM Hepes; pH 7.2)
containing 1μM concanavalin A for 15 min before being
mounted on the stage of an inverted, phase contrast microscope. For whole-cell recording under current clamp or
voltage clamp conditions, patch-clamp micropipettes filled
with a saline consisting of 140 mM KCl, 0.1 mM $CaCl_2$, 2 mM
$MgCl_2$, 1 mM EGTA, 100 mM D-glucose, 10 mM Hepes; pH 7.2,
were used. After establishment of a giga-seal the patch
membrane was disrupted by suction to get access to the cell
interior (Hamill et. al., 1981). Outside-out patches were
obtained by withdrawing the pipette from the cell. 10^{-4} M
L-glutamate was applied to the surface of whole-cell clamped
myosacs, or to the extracellular face of outside-out patches,
by pressure ejection. Data were collected on a Racal FM tape
recorder after low pass filtering at 1-3 kHz.

After 3-4 days of colchicine treatment, most myotubes
became smooth-surfaced spheres and ellipsoids that ranged
from 20 μM to 200 μM in diameter. Electron microscopy
revealed the presence of thin and thick myofilaments in the
cytoplasm of the myosacs. These filaments were arranged
into parallel arrays of thin and thick filaments but they
were not organised into orderly sarcomeres.

The resting potentials of the myosacs ranged between
-35 mV and -70 mV. The time constant of these cells was
64 ± 15 msec (mean ± s.e.m; range: 60-90 msec). The input
resistances measured using hyperpolarising test pulses
ranged between 70 and 200 M . Appropriate current pulses
revealed slowly rising and broad, non-overshooting action
potentials.

Application of 10^{-4} M L-glutamate onto the surface of
whole-cell clamped myosacs induced fast depolarising responses in all cell tested (N=10). In current clamped myosacs, a large, transient decrease in resistance was observed
which was always accompanied by a marked depolarisation.
Under voltage clamp conditions, at resting membrane potentials, the underlying current was always inwardly directed.
The reversal potential was in the range -5 to +10 mV.

L-glutamate, applied to the extracellular face of
outside-out patches elicited inward currents by activation
of receptor associated ion channels. The frequency of
channel opening was apparently related to agonist pulse

duration. Thus, 50-100 msec pulses evoked discrete, well-resolved openings which were randomly distributed whereas prolonged L-glutamate pulses (500-900 msec) elicited high frequency bursts of channel opening within which several open conductance states were evident (Fig. 1). The approximate conductance of these glutamate activated channels was 100 pS.

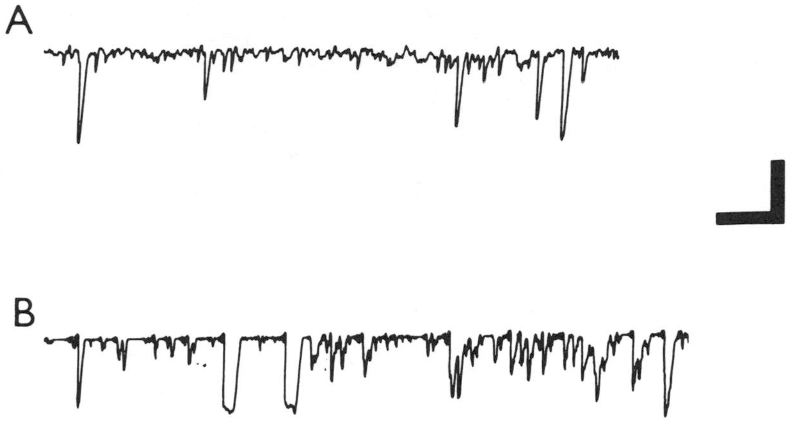

Fig. 1. L-glutamate activated channels recorded from an outside-out membrane patch of an insect myosac. In trace A the current was obtained after application of a 200 msec L-glu pulse. The current in trace B was observed after a 600 msec pulse of L-glu. These two recordings were obtained at a holding potential of -40 mV. Inward current shown in downward direction. Calibration: 80 msec, 4 pA.

Successful sealing of the patch-electrode on the myosac surface often revealed presumptive voltage dependent potassium channels. At least two classes of such channels were observed and often several discrete conductance states could be discerned within the same membrane patch.

Most of the passive and active electrophysiological properties of the plasma membrane of myosacs were similar to those of cultured myotubes (Bermudez et. al., 1986). Myosacs were highly sensitive to L-glutamate and the reversal potential for the current activated by this agonist ranged between -5 to +10 mV. The approximate conductance of the glutamate gated channels was 110 pS. A reversal potential of about O mV and a channel conductance of 90-120 pS were reported for glutamate gated channels in cultured locust muscle by Duce and Usherwood (1986). Thus, cultured myosacs would appear to be suitable for studies of insect muscle pharmacology. The smooth surface and the spherical shape of myosacs make these cells especially suitable for whole-cell and single-channel recording studies.

I. Bermudez was supported by the Science Research Council and Wellcome Research Laboratories as a CASE student.

REFERENCES

Bermudez, I., Lees, G. Botham, R.P. and Beadle, D.J. (1986) Myogenesis and neuromuscular junction formation in cultures of Periplaneta americana myoblasts and neurones. Dev. Biol. In Press.

Duce, J.A. and Usherwood, P.N.R. (1986) Primary cultures of muscle from embryonic locusts (Locusta migratoria, Schistocerca gregaria): developmental, electrophysiological and patch-clamp studies. J.exp.Biol. In press.

Fukuda, J., Henkart, M.P., Fischbach, G.D. and Smith, T.G. (1976) Physiological and structural properties of colchicine treated chick skeletal muscle cells grown in tissue culture. Dev. Biol. 49, 395-411.

Hamill, P., Marty, A., Neher, E., Sakman, B. and Sigworth, F.I. (1981) Improved patch-clamp techniques for high-resolution current recording from cells and cell-free membrane patches. Pflugers Arch. 391, 85-100.

ENHANCED INCREASE OF TYRAMINE AND OCTOPAMINE THROUGH THE INHIBITION OF MONOAMINE OXIDASE AND THE SYNERGISTIC EFFECT OF CHLORDIMEFORM TOWARDS DELTAMETHRIN

Joseph T. Chang, H. L. Tang and Y. S. Ni

Department of Biology, Peking Univ.

Beijing, Peoples' Republic of China

Early in 1949, Bot reported the presence of a certain toxic substance in the haemolymph of DDT-prostrate American cockroaches; when 20 ul of the haemolymph were injected into Calliphora erythrocephala, the flies were killed. Sternburg and Kearns (1952) confirmed the result, and found that the toxin is neither DDT nor its metabolites. It was tentatively identified as an aromatic amine or ester (Sternburg, 1960; Hawkins and Sternburg, 1964). Tashiro et al (1972) re-identified the toxin as L-leucine, using however a different insect, the common silkworm. We identified the neurotoxin in American cockroaches as tyramine, a phenolic amine, using the same chromatographic systems as Sternburg and confirmed with HPLC. When the insects are treated with DDT or deltamethrin, (Chang et al, 1984a, 1984b), tyramine increases excessively, but small amount of phenyl ethyl amine can not be excluded, for the neurotoxin may be a composite. Tyramine is produced by the action of tyrosine decarboxylase on tyrosine, the activity of the enzyme is greatly enhanced by treatment with DDT or deltamethrin (Lou et al 1985).

Evans (1978), Davenport and Evans (personal communication) found that octopamine increases

409

also under physical and chemical stress in Locusta, so they suggested that octopamine may be
a part of the neurotoxin. Lake et al (1970)
reported that tyramine β-hydroxylase is present
in the haemolymph of the American cockroaches,
the enzyme is responsible for the conversion of
tyramine into octopamine. Preliminary study in
our laboratory showed that about 50% of the
tyramine produced under the treatment of deltamethrin is converted into octopamine within half
an hour, so tyramine and octopamine increase
almost simultaneously. The auto-intoxication
of the insect is probably due to the excessive
increase of tyramine and/or octopamine, since
tyramine has been shown to cause excitation at
low concentration and blockage of nerve conduction at high concentration when applied to the
nerve cord of the cockroaches (Chang et al, 1984
a).

The neurotoxin is very unstable, it loses
its activity completely when left at room temperature for 1 - 2 hours, and DDT prostrate
cockroaches can recover from the effect of the
neurotoxin if not killed. So it was assumed
that the neurotoxin is degraded by some enzyme
in the haemolymph, and one of the possible
candidate for the degrading enzyme is monoamine
oxidase (MAO), which oxidatively deaminates
tyramine and octopamine into non-toxic aldehydes,
eventually into ρ-hydroxymandelic acid.

Plapp (1975,1979) reported that chlordimeform is an effective synergist for many insecticides, notably the pyrethroids. Chlordimeform
is a known inhibitor of MAO (Azia and Knowles,
1973; Beeman and Matsumura, 1974), so a possible
mechanism for the synergistic action of chlordimeform towards pyrethroids might be the increase of tyramine and octopamine through the inhibition of MAO.

The synergistic action of chlordimeform
towards deltamethrin was thus first ascertained.
The LC_{50} of deltamethrin applied topically to
American cockroaches is 0.011%, that of chlordimeform is 0.012%, but the LC_{50} of deltamethrin

plus chlordimeform is only 1.33×10^{-3}%. The
co-toxicity index calculated according to Sun
and Johnson (1960) is as high as 890.3, showing a very high synergistic effect.

MAO activity was then determined in the haemolymph and other organs of American cockroach,
using the same method of Blaschko et al. (1961),
which measures spectrophotometrically the
amount of benzylaldehyde produced when benzyl
amine is incubated with insect tissue. MAO
is found generally present in all tissues, but
particularly high in malpighian tubules and
muscles. Treatment with chlordimeform significantly inhibits the activity of MAO, the
residual activity of MAO after treatment with
chlordimeform (100 ppm, 2 ul/ insect) is only
19.0 to 36.6%.

A parallel study was made to measure the
amount of tyramine and octopamine produced in
insects treated with deltamethrin alone and in
insects treated with deltamethrin plus chlordimeform. The amount of the two biogenic
amines was determined by TLC scanning. The
results showed that when deltamethrin was used
alone, tyramine increased 255%, octopamine
208%, but when de-ltamethrin was used together
with chlordimeform, the amount of tyramine
increased was 925% and that of octopamine 500%.

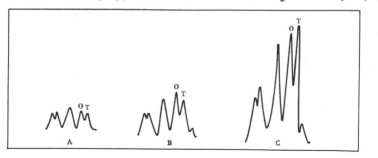

Fig. I Scanning graph showing the relative
    amount of tyramine and octopamine in
    (A) control, (B) treatment with delta-
    methrin alone, (C) treatment with delta-
    methrin plus chlordimeform.

   Apparently, these two biogenic amines increase greatly when chlordimeform is added to
deltamethrin. The increase amount of tyramine
and/or octopamine will eventually result in
increased toxicity, so the inhibition of MAO by
chlordimeform, resulting in the increase of
tyramine and octopamine might be the possible
cause of synergism of chlordimeform towards
deltamethrin and other neurotoxic insecticides.

Literature cited

Azia, S. A. & Knowles, C. O. (1973) Inhibition
   of MAO by pesticide chlordimeform and related
   compounds. Nature 242:417.
Beeman, R. W. & Matsumura, F. (1974) Studies
   on the action of chlordimeform in cockroaches.
   Pestic. Biochem. Physiol. 4:325.
Blaschko, H. Colhoun, E. H. & Frontall, N.
   (1961) Occurrence of amine oxidase in an
   insect, Periplaneta americana. J. Physiol.
   156:28.
Bot, J. A. (1949) Ph. D. Thesis Leiden Univ.
Chang, J. T., Wu, S. H. & Chin, H. L. (1984a)
   Studies on insect neurotoxin: tyramine as the
   neurotoxin released in the haemolymph of DDT-
   prostrate cockroaches. Acta Ent. Sinica 27:15.
Chang, J. T., Wu, S. H., Chen, N. S. & Yao, Y. H.
   (1984b) Studies on insect neurotoxin: the release of neurotoxin under treatment of different neurotoxic insecticides. Acta Ent.
   Sinica 27:165.
Evans, P. D. (1978) Octopamine, a highly affinity uptake mechanism in the nervous system
   of the cockroach. J. Neurochem. 30:1015.
Hawkins, W. B. & Sternburg, J. (1964) Some chemical characteristics of a DDT-induced neuro-
   active substance from cockroach and crayfish.
   J. Econ. Ent. 57:241.
Lake, C. R., Mills, R. R. & Brunet, P. C. J.
   (1970) β-hydroxylation of tyramine by cockroach haemolymph. Biochim. Biophys. Acta
   215:226.

Lou, Y.,Ni, Y. S. & Chang, J. T. (1985) Studies
	on insect neurotoxin: DDT induction of tyro-
	sine decarboxylase in the American cockroach.
	Acta Ent. Sinica 28:241
Plapp, jr. F. W. (1975) Chlordimeform as a
	synergist for insecticides against the toba-
	cco budworm. J. Econ. Ent. 69:91
Plapp, jr. F. W. (1979) Synergism of pyre-
	throids insecticides by formamidines against
	Heliothis pests of cotton. J. Econ. Ent. 72:
	667
Sternburg, J. (1960) Effects of insecticides
	on neurophysiological activity in insects.
	Agr. Food Chem. 8:257
Sternburg, J. & Kearns, C. W. (1952) The pre-
	sence of toxin other than DDT in the blood of
	DDT-poisoned roach. Science 116:144
Tashiro, S., Taniguchi, E. & Eto, M. (1972)
	L-leucine, neuroactive substance in insects.
	Science 175:448
Tashiro, S., Taniguchi, E., Eto, M. & Mackawa,
	K. (1975) Isoamylamine as the possible
	neuroactive metabolites of L-leucine. Agr.
	Biol. Chem. 39:569
Sun, Y. P. & Johnson, E. R. (1960) Analysis
	of joint action of insecticides against house
	flies. J. Econ. Ent. 53:889

ECDYSTEROID LEVELS IN EGGS AND LARVAE OF <u>TOXORHYNCHITES</u> <u>AMBOINENSIS</u>

R. J. Russo and A. L. Westbrook

Indiana Univ.-Purdue Univ. at Indianapolis

Indianapolis, IN 46223

<u>Toxorhynchites</u> are mosquitoes that prey upon larvae of other mosquito species or small aquatic arthropods (Steffan and Evenhuis, 1981). As adults they do not bloodfeed. During the last 4 years, research in this laboratory has focused on various aspects of the predatory behaviour of <u>Toxorhynchites</u> larvae. Surplus killing, a behaviour exhibited during the last larval instar, has been of particular interest. The surplus killing behaviour involves the attack and killing of a prey larva by the predator, but the prey is not consumed.

The purpose of this investigation is to elucidate the hormonal basis of surplus killing in <u>Toxorhynchites</u> <u>amboinensis</u>. We suspected that surplus killing was analogous to behavioural changes in other insects that are associated with a rise in ecdysteroid levels. Ecdysteroid-associated behavioural changes include cessation of feeding (Bollenbacher <u>et al</u>., 1975)and gut purging (Fujishita <u>et al</u>., 1982). These behaviours occur during the last larval stadium and are associated with feeding, as is surplus killing in <u>Toxorhynchites</u>. Because of these similarities, we began measuring the levels of ecdysteroids via radioimmunoassay (Borst and O'Connor, 1974) at daily intervals during the course of pre-pupal development. While these data describe the pattern of ecdysteroid fluctuations during development, they also demonstrate the necessity for improved staging procedures. We sought morphological markers to decrease the heterogeneity of the sample and to obtain individuals of similar physiological age. Ligation experiments were performed to determine the head critical periods during the last larval instar.

The results of ecdysteroid studies to date show the general pattern of ecdysteroid fluctuations throughout the egg and larval stages. The ecdysteroid content of eggs was approximately 550 pg/egg during the first 20 h after oviposition (Fig. 1). The ecdysteroid level then rose to 1673 pg/egg at 32 h after oviposition before diminishing during the next 12 h. These data suggest that Toxorhynchites embryos do not undergo embryonic moulting.

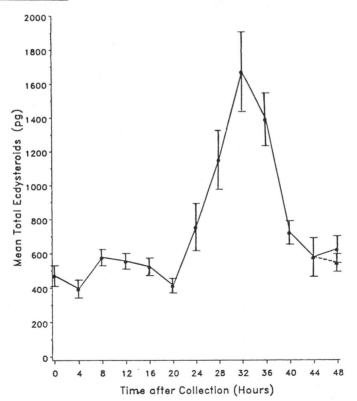

Fig. 1. Ecdysteroid Levels in T. amboinensis Eggs. Each point represents the mean ecdysteroid content (pg/egg) + SEM of 17-20 individual eggs. The dashed line indicates the ecdysteroid level in first instars hatched between 44 and 48 hours after oviposition.

Although embronic moulting occurs in several insect species (Hagedorn, 1983), its apparent absence in Toxorhynchites agrees with Oelhafen's (1961, as cited by Hagedorn, 1983) finding with Culicine mosquitoes.

The ecdysteroid titre fluctuated greatly in larvae of the lower weight classes (early instars) (Fig. 2). During most of the fourth stadium, ecdysteroids remained at a basal level of 25 pg/mg until 42.5 mg when the hormone level increased prior to pupation. This large pulse is similar to that observed in Manduca, Drosophila, and Calliphora (review: Richards, 1981) and is probably responsible for the deposition of pupal cuticle. Surplus

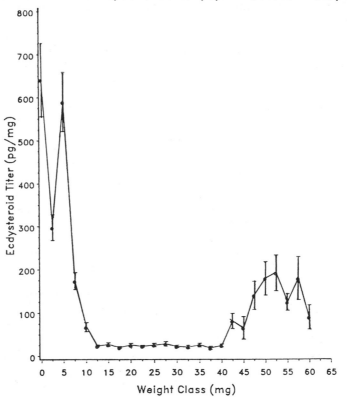

Fig. 2. Ecdysteroid Titres in T. amboinensis Larvae in 2.5 mg Weight Classes. Each point represents the mean ecdysteroid titre (pg/mg) + SEM of 2 to 63 individual larvae. All weight classes less than 55 mg include 15 or more larvae.

killing began at 22.5 mg, significantly before the large pulse of ecdysteroids in this instar. Our inability to detect a small pulse of ecdysteroids early in the stadium, as is found in other insects, suggested that our staging methods needed greater accuracy.

We examined larvae at hourly intervals during the fourth stadium to select characteristics which could be used to identify individuals of the same physiological age. Three characteristics were chosen which, when used with a synchronization procedure, could subdivide the 200 h fourth stadium into significantly shorter intervals (Fig. 3). These characteristics were the developing compound eye, pupal trumpets, and the presence of certain abdominal setae.

The results of the ligation experiments identified two periods during the fourth instar that required the head be attached to the body for continued normal development (Fig. 3). These head critical periods (Truman, 1972) suggest that some factor from the head plays a necessary role at that point in development. If the factor is not supplied, development ceases. Bollenbacher and Gilbert (1981) have shown that during the two head critical periods of the fifth instar of Manduca sexta, prothoracicotropic hormone (PTTH) is released from the brain and stimulates a short pulse of ecdysteroid production by the prothoracic gland. This pulse of ecdysteroids initiates a set of preparatory behaviours for the pupal stage.

In summary, our results did not show as close a correlation between ecdysteroid fluctuations and surplus killing as we had hoped. The intensity of surplus killing was greatest at the time when the ecdysteroid level was at higher levels during the fourth instar. The poor correlation between the onset of surplus killing and ecdysteroid increase may be attributable to the coarseness of our staging methods used during the early ecdysteroid assays. We have since identified morphological characters which should refine our staging procedure to achieve more homogeneous groups. If the onset of surplus killing by fourth instar Toxorhynchites is evoked by ecdysteroids, these improved methods should allow us to detect a small pulse occurring prior to the behavioural change and closely associated with the first HCP.

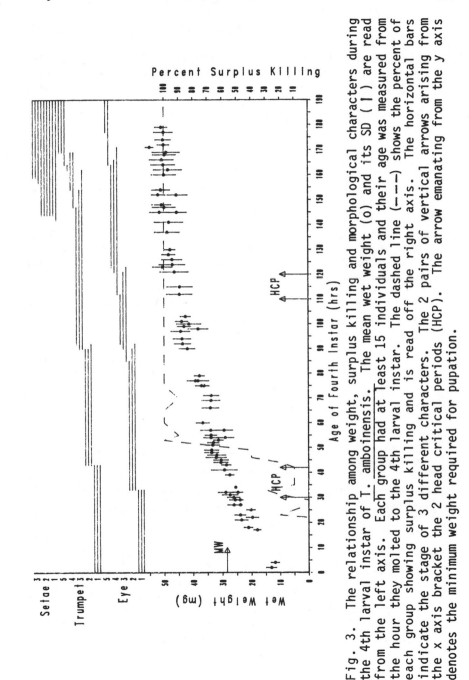

Fig. 3. The relationship among weight, surplus killing and morphological characters during the 4th larval instar of T. amboinensis. The mean wet weight (o) and its SD (|) are read from the left axis. Each group had at least 15 individuals and their age was measured from the hour they molted to the 4th larval instar. The dashed line (— —) shows the percent of each group showing surplus killing and is read off the right axis. The horizontal bars indicate the stage of 3 different characters. The 2 pairs of vertical arrows arising from the x axis bracket the 2 head critical periods (HCP). The arrow emanating from the y axis denotes the minimum weight required for pupation.

REFERENCES CITED

Bollenbacher W. E. and Gilbert L. I. (1981)
Neuroendocrine control of post embryonic development:
the prothoracicotropic hormone. Proc. VIII Int. Symp.
on Neurosecretion: Molecules, Cells, Systems (Ed. by
Farner D. S. and Lederis K.), pp. 361-370. P l e n u m
Press, New York.

Bollenbacher W. E., Vedeckis W. V., Gilbert L. I., and
O'Connor J. D. (1975) Ecdysone titers and prothoracic
gland activity during the larval-pupal   development
of Manduca sexta. Devl. Biol. 44, 46-53.

Borst D. W. and O'Connor J. D. (1974) Trace analysis of
ecdysone   by   gas-liquid   chromatography,
radioimmunoassay, and bioassay. Steroids 24, 637-656.

Fujishita M., Ohnishi E., and Ishizaki H. (1982) The role
of ecdysteroids in the determination of gut-purge
timing in the saturniid Samia cynthia ricini. J.
Insect Physiol. 28, 961-967.

Hagedorn H. H. (1983) The role of ecdysteroids in the
adult insect. In Endocrinology of Insects (Ed. by
Downer R. G. H. and Laufer H.), pp. 271-304. Alan R .
Liss, Inc., New York.

Oelhafen F. (1961) Zur Embryogenese von Culex pipiens:
Markierungen und Extirpation mit UV-Strahlenstick.
Wilhelm Roux Arch. Entomol Organ. 153, 120-157.

Richards G. (1981) The radioimmune assay of ecdysteroid
titres in Drosophila melanogaster.   Molec. Cell.
Endocrinol. 21, 181-197.

Steffan W. A. and Evenhuis N. L. (1981) Biology of
Toxorhynchites. Ann. Rev. Entomol. 26, 159-181.

Truman J. W. (1972) Physiology of insect rhythms. I.
Circadian organization of the endocrine events
underlying the moulting cycle of larval tobacco
hornworms. J. exp. Biol. 57, 805-820.

EFFECT OF BIOGENIC AMINES ON MOULTING IN SOME

LEPIDOPTEROUS LARVAE.

T. Thangaraj,   C.A. Vasuki,   R. Jeyaraj
and   M. Aruchami
Department of Zoology
Kongunadu Arts and Science College
Coimbatore 641 029,   INDIA.

Moulting of insects is a complex process which starts with apolysis followed by the synthesis and hardening of new cuticle and ends with ecdysis. This process is manifested by the involvement of neurohormones and hormones. The neurosecretory hormone, Prothoracico trophic hormone is known to exert a trophic influence on the prothoracic glands to release the moulting hormone,   ecdysone which is converted to active state as 20 hydroxyecdysone. This steroid hormone inducts the apolysis and secretion of a new cuticle by the epidermis.

Biogenic amines are considered to be the important neurosecretory products of the neurosecretory cells of pars intercerebralis in many insects. These amines include Dopamine, epinephrine,nor-epinephrine and octopamine. Recently it has been observed that the biogenic amines are playing important role on the moulting of some insects. (Idriss et al., 1984). The present study was undertaken to test the role of certain biogenic amines (Dopa, dopamine, epinephrine and tyramine) on the moulting of decapitated larvae of silkworm Bombyx mori and Spodoptera litura. Fifth instar Bombyx mori and Sixth instar, Spodoptera litura larval forms were used before the critical period of the action of ecdysone. The decapitated larval forms pupated 16 days (2 days normal) in Bombyx mori and Spodoptera litura

421

pupated after 10 days (one day normal). The effects of
tyramine, epinephrine, Dopa and dopamine have been observed
to induce apolysis of decapitated Spodoptera litura and the
results suggest the possible role of catacholamines on the
induction of insect development.

    The neurosecretory cells of anterior pars intercerebralis
of several insects are known to contain various biogenic amines
and the amine containing cell bodies are considered to be
different from peptidergic neurosecretory cells. (Orchard,
1984). It has been suggested that the biogenic amines
released from the aminergic neurous perform the function of
neurohormones or they may have controlling effect on the
release of peptidergic hormone. Recent reports by
Ruegg et al. (1981, 82) have indicated that dopamine mimics
the function of ecdysone in initiating the release of
peptidergic hormone derived from neurosecretory cells in
the brain. The present study provides evidences that the
biogenic amines induce apolysis as like that of the action
of ecdysone which is under the control of the peptidergic
neurohormone, prothoracicotropic hormone, released from the
lateral neurosecretory cells.

                         REFERENCES

IDRISS M., SHERBY S., MORSHEDY M and MANSOUR N.A. (1984)
   Prothoracicotropic Hormone - like effects of Biogenic
   amine in lepidopterous larvae. In Insect Neurochemistry
   and Neurophysiology (Ed. by Borkovec A.B and Kelly T.J)
   pp.385-387, Plenum Press, New York.

ORCHARD I. (1984) The role of Biogenic amines in the
   regulation of peptidergic neurosecretory cells. In
   Insect Neurochemistry and Neurophysiology (Ed. by
   Borkovec A.B. and Kelly T.J) pp. 115-134 Plenum Press,
   New York.

RUEGG R.P., KRIGER F.L., DAVEY K.G. and STEEL C.G.H. (1981)
Ovarian ecdysone elicits release of a myotropic ovulation
  hormone in Rhodnius (Insecta Hemiptera). Int. J. Invert.
  Reprod. 35, 357-361.

RUEGG R.D., ORCHARD I. and DAVEY K.G. (1982)
20-Hydroxy ecdysone as a modulator of electrical activity
  in neurosecretory cells of Rhodnius prolixus J. Insect.
  Physiol. 28, 243-248.

Neuroanatomy

OVERVIEW: PROGRESS IN INSECT NEUROANATOMY

Grant M. Carrow

Department of Biochemistry
Brandeis University
Waltham, MA 02254 USA

Over the last several years, there have been
fundamental changes in approaches to insect neuroscience
that have blurred the distinctions between neuroanatomy,
neurochemistry, and neurophysiology. As is made evident by
the work presented throughout this volume, many recent
advances in insect neuroscience have stemmed from the
convergence of objectives in these three fields: the
acquisition of comprehensive anatomical, biochemical, and
physiological profiles of individual neurons as well as
small systems of functionally related neurons. Recent
progress toward this end has been fueled by extensive
application of immunochemistry to the study of insect
nervous systems. The papers presented herein reveal
aspects of the anatomy and biochemistry of both new and
previously identified neurons and neurosecretory sites by
means of immunocytochemistry. Moreover, the physiology of
several of the neurosecretory sites was probed through
immunochemical or biochemical detection of neurosecretion.
Finally, as indicated by the work presented in this volume,
recent progress in insect neuroanatomy has also provided
insights into nervous system organization, clues to the
development and differentiation of nerve cells, information
on comparative aspects of neuropeptide distribution, and
indications of the anatomical and neurochemical complexity
of insect central and peripheral nervous systems. These
principal trends in insect neuroanatomy are elaborated
here.

Probing the Nervous System with Immunocytochemistry

Consideration of this volume as a barometer of progress in insect neuroscience leads to the recognition of immunocytochemistry as the predominant tool for revealing the distribution of new or previously identified neuropeptides and nonpeptidergic neurotransmitters. Several years ago, the earliest efforts in immunocytochemistry uncovered the distribution of immunoreactive homologues of vertebrate neuropeptides in insect nervous systems. Lately, this approach has been expanded to encompass study of the distribution of immunoreactive products homologous to immunogens originally isolated in insects or other invertebrates. Finally, study of the biochemical differentiation of cells as well as comparative analysis of the distribution of neuropeptides has been greatly facilitated by applications of immunocytochemistry. A general finding from this work is the apparently broad distribution of neuropeptides with great degrees of sequence homology, leading to the realization that the distinction between vertebrate and invertebrate neuropeptides may no longer be appropriate.

Focusing on Identified Neurons

One of the approaches evident throughout this volume is the study of uniquely identifiable neurons by the combination of immunocytochemistry and cellular marking (intracellular dye injection or cobalt backfilling). Anatomically identified neurons may be accessible to further biochemical and neurophysiological analysis. For example, demonstration of neuropeptide or neurotransmitter release from immunocytochemically identified release sites frequently marks the work presented here. This work shows that release of neurochemicals by high potassium depolarization is a powerful probe for neurohemal or synaptic release sites. These accomplishments are significant; yet, much work remains to be done before we obtain a complete anatomical, biochemical, and physiological profile of any single neuron.

Uncovering New Neurohemal Sites

An important theme that issues from the work presented here is recognition of the extent of extracerebral neurohemal areas in insects. Justifiably, workers have focused attention on the brain's neurohemal sites. However, as immunocytochemical and ultrastructural analysis indicates, there are diverse release sites for neurochemicals outside of the cerebral neuroendocrine system. In addition, peripheral neurons sometimes associated with these sites offer unique opportunities for the study of neurons in a simple environment due to their isolation from the complex central neuropil.

Deciphering Nervous System Organization

Beyond the localization of neuron somata, rapid progress is also being made in revealing neuropil organization in the central nervous system through intracellular staining and immunocytochemistry. Obtaining more detailed knowledge of the anatomy and neurochemistry of simple systems of neurons dramatically increases our understanding of how these cells may interact to coordinate physiology and behavior.

Prospects

Immunochemistry has become an important first step in probing insect nervous systems for new neuropeptides and neurotransmitters. It should also serve as the final step in defining the distribution and functions of previously isolated neurochemicals.

One of the problems with using immunochemistry to search for new neuropeptides and neurotransmitters is the difficulty in unraveling the functions of the new neurochemicals. The target organs are usually undefined and many of the effects of neuropeptides and neurotransmitters are difficult to detect. The proliferation of sensitive in vitro assays for demonstrating the release and binding of neurochemicals may help alleviate this problem. Indeed, in vitro bioassays were instrumental in isolating a number of the new insect neuropeptides reported herein.

Application of immunochemistry as a final step in characterizing newly isolated and sequenced neuropeptides is an approach less likely to require a quest for physiological activity. Since isolation of the neurochemicals is often based upon their action in a bioassay, biochemical and physiological analysis of the neuropeptides would precede neuroanatomical investigation. Undoubtedly, this may prove to be an expedient approach to characterizing the source cells for neurosecretions. However, nonsecreted neuropeptides are missed by this approach and their importance to cellular function should not be ignored.

Already in progress is work designed to reveal the ultrastructural distribution of new neurochemicals in insects by combining immunocytochemistry and electron microscopy. The combination of multiple immunolabeling with ultrastructural analysis also promises to reveal much about the interrelationship of identified neurons, particularly in neuropilar regions.

One practical application of advances in insect neuroanatomy is the increased possibility of targeting small systems of biochemically unique neurons. Identification of neurons on the basis of their neurochemistry raises the prospect of discovery of neurochemicals essential for normal function of a critical process in selected species. Thus, neuroanatomical information might lead to biorational approaches to narrow spectrum insect control. Moreover, as a research tool, the ability to target for disruption the biochemical apparatus of subsets of identified neurons would greatly facilitate study of the functions of those neurons.

Conclusions

The work presented throughout this volume is indicative of the progress toward integration of insect neuroanatomy with neurochemistry and neurophysiology. Mapping the distribution of new and previously identified neurochemicals aides in clarifying the roles of the neurochemicals and their source cells in insects. A prevalent objective appears to be that of obtaining comprehensive structural, biochemical, and functional profiles of small systems of related neurons. Combinations of approaches should facilitate advance toward this goal.

LOCALIZATION AND RELEASE OF FMRFAMIDE-LIKE IMMUNO-

REACTIVITY IN AN INSECT CEREBRAL NEUROENDOCRINE SYSTEM

Leslie S. Carroll, Grant M. Carrow, and
Ronald L. Calabrese

Department of Cellular and Developmental
    Biology
Harvard University
Cambridge, MA 02138 USA

Neuropeptides secreted by insect cerebral neuro-secretory cells (NSCs) are primary endocrine regulators of physiology, development, and behavior. Recently, many workers have shown that insect NSCs and neurohemal organs are immunoreactive to antisera directed against vertebrate and invertebrate neuropeptides. Thus, the insect homologues of these neuropeptides may serve a neurohormonal function. A possible neuroendocrine role in insects for the tetrapeptide Phe–Met–Arg–Phe–NH_2 (FMRFamide) (Price and Greenberg, 1977) or analogues thereof was suggested by immunocytochemical staining of neurohemal corpora cardiaca (Boer et al., 1980). Because cerebral NSCs of the moth, Manduca sexta, have been identified (Carrow et al., 1984), we used this model system to further clarify the role of FLI in insects. We briefly describe here the distribution of FMRFamide-like immunoreactivity (FLI) in this cerebral neuroendocrine system and provide evidence for storage and calcium dependent release of FLI from the neurohemal organs (Carroll et al., 1986).

The protocerebral neuroendocrine system of Manduca comprises two bilaterally symmetrical clusters of identified NSCs, one in the pars intercerebralis and the other in the pars lateralis. In pupae, the axons of the NSCs exit each brain hemisphere via the fused nervi corporis cardiaci I and II (NCC I + II) and terminate in

431

either of the two cerebral neurohemal organs, the corpus
cardiacum (CC) or the corpus allatum (CA) (Nijhout, 1975;
Carrow et al., 1984). The cells in each cluster have been
classified according to the size and location of their
somata as well as the architecture of their neurites,
axons, and terminal ramifications in the neurohemal organs
(Nijhout, 1975; Buys & Gibbs, 1981; Carrow et al., 1984;
Copenhaver & Truman, 1986).

FLI was detected in approximately 25 bilaterally
symmetrical pairs of somata in whole mounts of the pupal
protocerebrum by indirect immunofluorescence (Li and
Calabrese, 1985); the rabbit polyclonal antiserum (No.
232, O'Donohue et al., 1984) was a gift from Thomas
O'Donohue. In addition, FLI was evident in neurites in
the brain, axons in the nervi corporis cardiaci and nervi
corporis allati, and terminals in the neurohemal organs.
Similar immunolabelling was observed in premetamorphic
larvae. All immunolabelling was specific since it was
completely blocked by preabsorption of the anti-FMRFamide
antiserum with synthetic FMRFamide.

FLI was localized to a subset of the protocerebral
NSCs of pupae by combining intracellular injection of the
dye Lucifer Yellow with indirect immunofluorescence.
Among the NSCs in each hemisphere, FLI was observed in
both group IIa cells (lateral cluster), in most group IIb
(lateral) cells, and in two NSCs of group Ib (medial
cluster). None of the group Ia (medial) NSCs showed FLI.
Immunoreactive neurons in the protocerebrum outside of
these two clusters are likely to be non-endocrine since
they do not stain with paraldehyde fuchsin (Nijhout,
1975).

Confirmation of the presence of FLI in the brain and
neurohemal organs was obtained by radioimmunoassay of
extracts using the same anti-FMRFamide antiserum as used
for immunocytochemistry. Organs were isolated from premetamorphic (wandering) larvae and were homogenized in
acidified methanol. FLI could be quantified since brain
extracts and synthetic FMRFamide showed complete cross-
reactivity with the antiserum. The following proportions
of FLI were obtained, here expressed as a percentage of
the total FLI in the cerebral neuroendocrine system:
brain, 81%; CC pair, 14%; CA pair, 5% ($n \geq 6$). When

normalized for organ wet weight, the amounts in brain and CC were equivalent; however, the CC still yielded about 2.5 times the amount of FLI extracted from the CA.

In order to further determine whether FLI is a neurosecretory product, we attempted to evoke release of FLI in vitro by high potassium depolarization of neurohemal organs (combined CC and CA isolated from premetamorphic larvae) (Carrow et al., 1981). Released FLI was quantified by radioimmunoassay of the bathing medium; about 30 CC-CA were pooled for each sample. Neurohemal organs released small amounts of FLI when preincubated for 2 hr in a calcium-free, 100 mM potassium saline. By contrast, there was more than a three-fold increase in FLI release upon subsequent incubation of the same organs in a 5 mM calcium, 100 mM potassium saline (n=5). When CC-CA were preincubated in a 5 mM calcium, 7 mM potassium saline, there was a basal level of FLI release significantly greater than that in calcium-free saline (p<0.01). Nevertheless, the same organs released twice the basal level during subsequent incubation in 5 mM calcium, 100 mM potassium saline (n=4). Thus, release of FLI from the neurohemal organs was calcium dependent and the amount of FLI released was about 16% of that extracted from the organs.

The data presented here indicate that FLI, as with other neurosecretory products, is stored in cerebral NSCs and can be released by depolarizing terminals in the neurohemal organs. Therefore, the FLI detected here may have a neurohormonal or neuromodulatory function and may be colocalized with other neurosecretory products. In addition, FLI might act as a neurotransmitter in the brain since neurites and non-endocrine neurons were also immunoreactive. The accessible system of identified FLI neurons described here should facilitate clarification of the functions of this class of neuropeptides in insects.

We thank T. O'Donohue for his generous gift of anti-FMRFamide antiserum. The research was supported by grants from NSF and NIH.

REFERENCES

Boer, H.H., Schot, L.P.C., Veenstra, J.A., and Reichelt, D. (1980) Immunocytochemical identification of neural elements in the central nervous systems of a snail, some insects, a fish, and a mammal with an antiserum to the molluscan cardio-excitatory tetrapeptide FMRF-amide. Cell Tissue Res. 213, 21-27.

Buys, C.M. and Gibbs, D. (1981) The anatomy of neurons projecting to the corpus cardiacum from the larval brain of the tobacco hornworm, Manduca sexta (L.). Cell Tissue Res. 215, 505-513.

Carroll, L.S., Carrow, G.M., and Calabrese, R.L. (1986) Localization and release of FMRFamide-like immuno-reactivity in the cerebral neuroendocrine system of Manduca sexta. J. exp. Biol. In press.

Carrow, G.M., Calabrese, R.L., and Williams, C.M. (1981) Spontaneous and evoked release of prothoracicotropin from multiple neurohemal organs of the tobacco hornworm. Proc. Natl. Acad. Sci. USA 78, 5866-5870.

Carrow, G.M., Calabrese, R.L., and Williams, C.M. (1984) Architecture and physiology of insect cerebral neurosecretory cells. J. Neurosci. 4, 1034-1044.

Copenhaver, P.F. and Truman, J.W. (1986) Metamorphosis of the cerebral neuroendocrine system in the moth Manduca sexta. J. Comp. Neurol. 249, 186-204.

Li, C. and Calabrese, R.L. (1985) Evidence for proctolin-like substances in the central nervous system of the leech Hirudo medicinalis. J. Comp. Neurol. 232, 414-424.

Nijhout, H.F. (1975) Axonal pathways in the brain-retrocerebral neuroendocrine complex of Manduca sexta (L.) (Lepidoptera: Sphingidae). Int. J. Insect Morphol. & Embryol. 4, 529-538.

O'Donohue, T.L., Bishop, J.F., Chronwall, B.M., Groome, J., and Watson, W.H., III (1984). Characterization and distribution of FMRFamide immunoreactivity in the rat central nervous system. Peptides 5, 563-568.

Price, D.A. and Greenberg, M.J. (1977) Structure of a molluscan cardioexcitatory neuropeptide. Science 197, 670-671.

ASSOCIATION OF THE NEUROPEPTIDE PROCTOLIN WITH OVIDUCT

VISCERAL MUSCLE

Angela B. Lange and Ian Orchard

Department of Zoology, University of Toronto,

Toronto, Ontario, Canada, M5S 1A1

Although first discovered in the cockroach, Periplaneta americana, and proposed as a neurotransmitter associated with hindgut visceral muscle (Brown, 1975), the pentapeptide proctolin clearly has functional importance throughout the arthropods. Proctolin has been shown to be extremely active upon a variety of arthropod muscle preparations (see Lange et al., 1986) and to be widely distributed in insects (Brown, 1977). Proctolinergic neurons have been described in insects and crustacea, and some of these are uniquely identifiable skeletal muscle motoneurons (Adams and O'Shea, 1983; Witten and O'Shea, 1985; Bishop et al., 1984). Physiological studies subsequently revealed proctolin to be a co-transmitter at skeletal neuromuscular junctions (Adams and O'Shea, 1983; Bishop et al., 1985). It is ironic that, in spite of being first described as an insect visceral muscle transmitter, more is known about proctolinergic innervation of insect skeletal muscle. Indeed, the literature on proctolinergic innervation of insect visceral muscle is rather scanty, with the exception of cockroach and locust hindgut (Penzlin et al., 1981; Keshishian and O'Shea, 1985).

Locust oviducts are proving to be a useful preparation with which to examine the octopaminergic and proctolinergic regulation of insect visceral muscle. These oviducts are innervated by a relatively small group of neurons located in the VIIth abdominal ganglion (Lange and Orchard, 1984). Two of these neurons are octopaminergic and exert potent

435

modulatory actions upon the oviducts (Orchard and Lange, 1985). There are also three bilaterally paired oviducal motoneurons located within the VIIth abdominal ganglion. Action potentials in each of these three motor axons results in EJPs capable of summation and facilitation (Orchard and Lange, 1986). Although proctolin causes elevated basal tension, increased amplitude of neurally-evoked contractions and an overall stimulation of myogenic activity of the oviducts, these changes are not accompanied by any major change in membrane potential or EJP. It seems likely that the release of a more conventional transmitter such as glutamate is responsible for the EJP (see Orchard and Lange, 1986). In the present paper we shall provide evidence for the presence of a substance indistinguishable from proctolin in the VIIth abdominal ganglion, oviducal nerves and oviducts of <u>Locusta</u> <u>migratoria</u>. Furthermore, we shall demonstrate the release of this substance induced by high potassium saline and oviducal nerve stimulation.

Proctolin produces a dose-dependent tonic contraction of locust oviducts, easily measured with a force transducer. Threshold for this contraction lies close to 2 fmol (Lange et al., 1986). By use of this dose response curve we have been able to quantify the amount of material showing proctolin-like bioactivity (PLB) in samples following RP-HPLC. Tissues were extracted into methanol: water : acetic acid (100:10:1) and subsequently run through Sep-Pak C18 cartridges followed by chromatography on RP-HPLC. Most of the PLB extractable from these tissues co-migrated with ^{3}H - proctolin. Quantification of PLB expressed as proctolin equivalents in the various tissues associated with the oviducts was obtained following separation on RP-HPLC (Lange et al., 1986). Substantial levels of proctolin were present in the oviduct muscle which receives extensive innervation (172.7 ± 58.8 fmol or 1.85 ± 1.1 pmol/mg protein). Muscle taken from more anterior regions of oviducts (areas which receive little of no innervation) contained about ten fold less proctolin. Measurable levels of proctolin were also found to be associated with the oviducal nerves (1.19 ± 0.38 pmol/mg protein) which contain the motor axons to the oviducts. In addition, the VIIth abdominal ganglion which contains the oviduct motoneurons also contained substantial levels of proctolin (20.4 ± 10.3 pmol/mg protein).

If proctolin performs a physiological function within

the oviduct system, then proctolin should be released
following stimuli which depolarise nerve terminals on the
oviducts. Depolarisation with the use of high potassium
saline (100mM) was capable of releasing measurable amounts
of proctolin (unpublished observation). About 0.4% of the
total store of proctolin was released during a 5 min
incubation in high potassium. In addition, stimulation of
the oviducal nerves also resulted in the release of
proctolin (unpublished observation). This release was
frequency dependent with maximum release at about 30 Hz. At
this frequency 6.3 % of the total proctolin store was
released by 5 min of stimulation. The neurally-evoked
release of proctolin was calcium-dependent, with
calcium-free saline capable of inhibiting the
neurally-evoked release by 87%.

Finally, to locate the source of proctolin we turned to
immunohistochemistry (Lange et al., 1986). Proctolin-like
immunoreactivity was observed in a number of neurons in the
VIIth abdominal ganglion. In particular three positively
stained bilaterally-paired neurons were located in a similar
position to that of the three oviduct motoneurons revealed
by cobalt backfilling. Furthermore, immunoreactive axons
were observed in the sternal nerve and tracing of these
immunoreactive axons revealed that some of them passed along
the branches to the oviducts.

Taken together, the presence of a substance
indistinguishable from proctolin using RP-HPLC and the
presence of proctolin-like immunoreactivity within the
oviducal nerves indicates that proctolin is indeed
associated with this visceral muscle. The calcium-dependent
release of proctolin implies that proctolin performs a
physiological role. However we have previously shown that
the EJP is probably produced by glutamate and that proctolin
has a minimal influence upon EJP amplitude and membrane
potential. We are left with the logical conclusion that
proctolin must be a co-transmitter within this visceral
muscle preparation. Thus, locust oviducts present a model
system for not only the study of peptidergic regulation of
visceral muscle, but also peptides as co-transmitters.

References

Adams M.E. and O'Shea M. (1983) Peptide cotransmitter at a
neuromuscular junction. Science 221, 286-289.

Bishop C.A., Wine J.J. and O'Shea M. (1984) Neuropeptide proctolin in postural motoneurons of the crayfish. J. Neurosci. 4, 2001-2009.

Bishop C.A., Wine J.J. and O'Shea M. (1985) Neural release of a peptide co-transmitter greatly enhances tension generation in a crayfish tonic muscle. Neurosci. Abst. 15, 327.

Brown B.E. (1975) Proctolin: A peptide transmitter candidate in insects. Life Science 17, 1241-1252.

Brown B.E. (1977) Occurrence of proctolin in six orders of insects. J. Insect Physiol. 23, 861-864.

Keshishian H. and O'Shea M. (1985) The acquisition and expression of a peptidergic phenotype in the grasshopper embryo. J. Neurosci. 5, 1005-1015.

Lange A.B. and Orchard I. (1984) Dorsal unpaired median neurons, and ventral bilaterally paired neurons, project to a visceral muscle in an insect. J. Neurobiol. 15: 441-453.

Lange A.B., Orchard I. and Adams M.E. (1986) Peptidergic innervation of insect reproductive tissue: The association of proctolin with oviduct visceral musculature. J. Comp. Neurol. (in press).

Orchard I. and Lange A.B. (1985) Evidence for octopaminergic modulation of an insect visceral muscle. J. Neurobiol. 16, 171-181.

Orchard I. and Lange A.B. (1986) Neuromuscular transmission in an insect visceral muscle. J. Neurobiol. (in press).

Penzlin H., Agricola H., Eckert M. and Kusch T. (1981) Distribution of proctolin in the sixth abdominal ganglion of Periplaneta americana L. and the effect of proctolin on the ileum of mammals. Adv. Physiol. Sci. 22, 525-535.

Witten J.L. and O'Shea M. (1985) Peptidergic innervation of insect skeletal muscle: Immunochemical observations. J. Comp. Neurol. 242, 93-101.

HOMOLOGIES BETWEEN PERINEURONAL SEROTONIN FIBER SYSTEMS IN RHODNIUS PROLIXUS AND IN MANDUCA SEXTA: EVIDENCE FOR NEUROHEMAL ORGANS FOR SEROTONIN RELEASE

Thomas Flanagan, Allan Berlind[*] and
Walter E. Bollenbacher

Department of Biology, University of North
Carolina, Chapel Hill, NC 27514 and [*]Biology
Department, Wesleyan University, Middletown,
CT 11724

Serotonin is a monoamine neurotransmitter present in a broad distribution of metazoans. In insects, serotonin-containing fibers project throughout the central nervous system, toward peripheral targets, and into neurohemal (endocrine) organs (NHO) (Flanagan and Berlind, 1984; Orchard et al., 1986). The presence of serotonin within an NHO could mean the transmitter is released in a paracrine manner regulating neurosecretion and/or that it is released into the hemolymph as a neurohormone. In this report, we review evidence which demonstrates serotonin is released from fibers in an NHO of Rhodnius prolixus, and present evidence for the existence of a serotonergic NHO in Manduca sexta.

Five paired abdominal nerves in Rhodnius contain a perineuronal fiber system (an NHO) derived primarily from neurosecretory cells lying at the posterior margin of the terminal thoracic ganglion. One neurohormone released from this NHO is the diuretic hormone (Maddrell and Gee, 1974). Treatment of this ganglion with glyoxylic acid yields a weak, rapidly-fading yellow histofluorescence indicative of serotonin in nerves containing the NHO. Immunocytological studies have also shown serotonin-containing fibers projecting throughout the NHO and have allowed identification of serotonergic neurons within the ganglion. This serotonin immunoactivity is blocked by preabsorbing the antisera with BSA-conjugated serotonin.

The serotonin-containing fibers in the abdominal NHO

439

possess a high-affinity serotonin-uptake system. [^{3}H]-
Serotonin is selectively taken up into nerve trunks which
carry the serotonin-containing fibers. Release of this
material from the fibers provides a means of measuring
their secretory activity. Depolarization of the fibers
with K^+ evokes a Ca^{+2}-dependent release of [^{3}H] from the
abdominal nerves. The kinetics of this release are
characteristic of the induced release of diuretic hormone
from the abdominal NHO (Flanagan and Berlind, 1984;
Maddrell and Gee, 1974). Thus the serotonin fibers in the
NHO have functional properties similar to those of the
peptidergic fibers.

In Rhodnius, feeding stimulates neurohormone release,
e.g. diuretic hormone, from the abdominal NHO. To see if
serotonin is released, release of ^{3}H-serotonin from the
NHO was compared before and after feeding. A pronounced
Ca^{+2}-dependent release of [^{3}H] occurred after feeding,
indicating feeding activates the serotonin-containing
fiber system. This release of serotonin may be either
paracrine or endocrine. Endocrine effects might activate
fluid translocation by malphigian tubules (Maddrell and
Gee, 1974) and/or increase gut and malphigian tubule
contractions (unpublished). Paracrine effects might
facilitate depolarization-evoked release of secretory
products from the NHO (Flanagan and Berlind, 1984).

Serotonin-containing fibers which are not distributed
close to paracrine targets probably have purely endocrine
functions. The mandibular nerves of the tobacco hornworm,
Manduca sexta, carry a perineuronal system of serotonin-
containing fibers similar to those in the abdominal NHO of
Rhodnius. This fiber system originates from two pair of
large neurons in the anterior margin of the suboesophageal
ganglion. Projections from these neurons enter the
mandibular nerve and extend into the perineuronal matrix.
These fibers are distributed as if their only function
were endocrine, unlike in Rhodnius, where the fibers are
within a well established peptidergic NHO.

To differentiate between NHO-associated and non-NHO-
associated perineuronal serotonin systems in Rhodnius and
Manduca, these fiber systems have been compared
ultrastructurally. Based on secretory vesicle size and
density (20–40nm, hollow core vesicles), Manduca posess a
single fiber type projecting within the extracellular

matrix of the mandibular nerve, presumably the serotonin-containing fibers. Peptidergic secretory fibers were not evident in the mandibular nerves. In the abdominal NHO of Rhodnius, however, there exist, in addition to a corresponding population of fibers, classical peptidergic neurosecretory fibers containing 150-200nm, electron-dense granules. Thus serotonin and neuropeptide secretory fibers are present in the abdominal NHO.

The organization of these serotonin-containing fiber systems described in Rhodnius and Manduca suggests they are neurosecretory. Such a system appears to exist in the pericardial NHO of crustaceans from which both neuropeptide and serotonin are believed to be secreted (Beltz et al., 1984). Thus it is conceivable that serotonergic NHOs exist in a variety of arthropods and that the transmitter functions as a neurohormone. These conclusions are supported by the demonstrations of serotonin-containing perineuronal fibers in the mandibular nerve of Periplaneta americana (Davis, 1985) and along nerve trunks of Calliphora erythrocephala (Nassel and Elekes, 1985).

Acknowledgements—The authors thank Ms. Kathy Luchok for preparing this manuscript. This research was supported by grants NS-18791 and AM-31642 to W.E.B.

REFERENCES

Beltz B., Eisen J.S., Flamm R., Harris-Warrick R.M., Hooper S.L. and Marder E. (1984) Serotonergic innervation and modulation of the stomatogastric ganglion of three decapod crustaceans (Panulirus interruptus, Hormarus americanus and Cancer irroratus). J. Exp. Biol. 109, 35-54.

Davis N.T. (1985) Serotonin-immunoreactive nerves and neurohemal systems in the cockroach Periplaneta americana (L.). Cell Tiss. Res. 240, 593-600.

Flanagan T.R.J. and Berlind A. (1984) Serotonin modulation of the release of sequestered [^{3}H]Serotonin from nerve terminals in an insect neurohemal organ In Vitro. Brain Res. 306, 243-250.

Maddrell S.H.P. and Gee J.D. (1974) Potassium-induced
release of the diuretic hormones of Rhodnius prolixus
and Glossina austeni: Ca dependence, time course and
localization of neurohemal areas. J. Exp. Biol. 61,
155-171.

Nassel D.R. and Elekes K. (1985) Serotonergic terminals
in the neural sheath of the blowfly nervous system:
Electron microscopical immunocytochemistry and 5,7-
dihydroxytryptamine labelling. Neurosci. 15, 283-307.

Orchard I., Martin R.J., Sloley B.D. and Downer R.G.H.
(1986) The association of 5-hydroxytryptamine,
octopamine, and dopamine with the intrinsic lobe of the
corpus cardiacum of Locusta migratoria. Can. J. Zool.
64, 271-274.

LOCALIZATION OF AKH/RPCH-RELATED PEPTIDES IN INSECTS AND OTHER INVERTEBRATES

H. Schooneveld

Department of Entomology, Agricultural University, Wageningen, the Netherlands

It is becoming clear that adipokinetic hormone (AKH) is only one representative of a rapidly expanding family of neuropeptides extracted from different insect species. A similar peptide, the red-pigment concentrating hormone (RPCH), has been isolated from shrimps and the question arises whether these AKH/RPCH-related peptides are evolutionary related and perhaps are derived from some ancestor molecule. If these peptides are important for invertebrates, it seems logical to expect that the peptides are more widespread among related arthropods and perhaps also in common ancestor species.

We explored this possibility by screening the CNS of several invertebrates by immunocytochemical methods and utilizing antisera to different groups of the AKH/RPCH molecule for categorizing the types of immunoreactive (IR) substances thus detected. The following questions have been considered.

A. Are antisera against the C- and N-terminal regions of the AKH/RPCH-like peptides useful for cell and peptide differentiation?

Rabbits were immunized with synthetic peptide fragments representing the N- and C-terminal regions:

Antigens	Serum code	Specificity
$AKH_{1-4}-R$	433	N-terminus of several peptides
$R-[Tyr^1]-AKH$	241	C-terminus of AKH-I

The antisera were applied on paraffin-embedded tissues according to the unlabelled antibody technique (Schooneveld et al., 1983). In locusts, glandular corpus cardiacum cells (CC-GC) are readily and specifically stained by these anti-

sera. It is assumed that serum 241 is specific for AKH
whereas serum 433 recognizes AKH and AKHII-L as well as
their precursors. In general, substances stained with only
one of the antisera must be different from AKH and we found
that these are rather common (see below); their nature
should be revealed by chemical methods. We found that some
CC-GC in Locusta and some other species were left
unstained. Also, substances stained with different antisera
are not always colocalized. This suggests a CC-GC diversity
which is perhaps related to the production of different
peptides in the CC, only some of which are revealed here.

B. Do all insect species contain AKH/RPCH-like peptides?

The CC of a variety of species belonging to the orders
Collembola, Thysanura, Dermaptera, Dictyoptera, Phasmida,
Orthoptera, Coleoptera, Hymenoptera, Diptera, and Lepidoptera were studied with both antisera. Cells reacting
strongly with serum 241 were present only in the acridid
locusts and in the collembolan species Folsomia (Schooneveld et al., 1985). Antiserum code 433 gave strong reactions with the majority of the species, but in flies no
reactions we observed (Schooneveld et al., 1986 b,c).

It thus seems that the distribution of AKH is restricted to locusts whereas related peptides sharing with AKH
some or all of the 4 N-terminal amino acids are fairly
common.

C. Do AKH/RPCH-like peptides occur also outside the CC?

The screening of tissues other than CC reveals immunoreactive cells elsewhere: (a) in the mid-gut epithelia of
several species, probably representing endocrine cells (unpublished); (b) neurons in the brain and several other
ganglia of the central nervous system. These neurons are
present in all species investigated (Schooneveld et al.,
1983, 1985) but they react only with antiserum code 241,
and contain a peptide different from AKH. The neurons
greatly vary in number and position in different species.
They are most common in the lateral part of the protocerebral lobes and in the most ventral part of the ventral
ganglia in head, but may be present in all ganglia of the
CNS, rarely in the frontal ganglion. Considering the course
of their axons and axon collaterals, most of the neurons
are interneurons. Neurosecretory neurons were found in a
few species and these released their product through the
storage compartment of the CC or -in some species- through
the neurohemal nerves leaving the ventral ganglia (Raabe

1985). It is still difficult to distinguish a general pattern among species as the immunoreactive product can be revealed only in a certain state of accumulation. Many neurons may therefore escape detection.

D. Can we distinguish evolutionary trends in morphology or staining of the AKH/RPCH immunoreactive CC–GC?

A variety of a–, hemi–, and holometabolous species were screened to detect possible evolutionary patters in CC organization. The shape, number, and grouping of CC–GC was quite variable. There appears to be a gradual increase in association between glandular and storage parts of the CC. For instance, in Collembola these compartments are widely separated whereas the CC–GC are completely enveloped by neural CC tissue in Coleoptera and Lepidoptera. In many species, including locusts, the glandula and storage zones are adjacent. Extremely long cell processes are present in some species, for instance Thermobia and Calleria and no developmental pattern in the morphology of individual CC–GC could be discerned.

The antisera now available would be suitable to study problems of ontogeny and innervation of these cells.

E. Do AKH/RPCH–related peptides occur in other invertebrates?

Antiserum code 433 should be able to recognize the N-terminal tetrapeptide of RPCH. Immunoreactive material is indeed present in certain groups of neurons, i.e. in the x-organ of the medulla terminalis in the crayfish Astacus leptodactylus (Schooneveld et al. 1986a) and in the shrimp Palaemon serratus (Bellon–Humbert et al. 1986). The product is carried to the sinus glands, presumably for release into the hemolymph. Other neurons stain with antiserum code 241 and although no axon tracts could yet be identified, some axons terminate also in the sinus gland.

As the eyestalk is considered to be homologous to the protocerebral lobes of insects, the neurons in both arthropods may have gone through a homologous development.

We also extended our routine assays to species belonging to classes and phyla which developed before insects: a myriapod (Lithobius forficatus), an annelid (Allobophora cupulifera), and a snail (Lymnaea stagnalis). In Lithobius, neurons were detected with antiserum code 241 which occupied positions in the brain similar to insects. No neurohemal organs were found. Allobophora contains numerous neurons, not only in the brain but particularly in the ventral segmental ganglia; no neurohemal organs were identified. Lymnaea contains numerous and very large neurons in most of the ganglia. Those in the pedal and visceral ganglia appeared to release their product via the periphery of certain neurons leaving these ganglia (Schooneveld et al. 1986). In another mollusc, Hirudo medicinalis, numerous neurons were reported in the chain of ganglia (O'Shea and Schaffer, 1985). In the coelenterate Hydra attenuata no positive neurons could be detected (Grimmelikhuijzen, personal communication).

In neither of these species could neurons be identified with antiserum code 433, indicating that the immunoreactive materials demonstrated with antiserum code 241 are all different from the known AKH/RPCH-related peptides.

Therefore, if we may assume that the antiserum 241-immunoreactive substances are related, the results indicate that these substances originated early in evolution, but some time after the coelenterates, whereas the antiserum 433-immunoreactive materials are of a much later date and occur only in the higher arthopods. Chemical analysis of all immunoreactive materials is clearly needed.

REFERENCES

Bellon-Humbert C, van Herp F. and Schooneveld H. (1986) Biol. Bull. Wood's Hole (in press).
O'Shea M. and Schaffer M. (1985) Ann. Rev. Neurosci. 8, 171-198.
Raabe M. (1985) Compt. Rend. Acad. Fr. 301, 407-412.
Schooneveld H, van Herp F. and van Minnen J. (1986a) Brain Res. (in press).
Schooneveld H, Romberg-Privee H.M. and Veenstra J.A. (1985) Gen. comp. Endocrinol. 57, 184-194.
Schooneveld H, Romberg-Privee H.M. and Veenstra J.A.(1986b) Cell Tissue Res. 243, 9-14.
Schooneveld H, Romberg-Privee H.M. and Veenstra J.A.(1986c) J. Insect Physiol. (in press).
Schooneveld H, Tesser G.I., Veenstra J.A. and Romberg-Privee H.M. (1983) Cell Tissue Res. 230, 67-76.

THE PERICARDIAL SINUS, A NEUROHEMAL ORGAN IN

CYCLORRHAPHAN FLIES

S. M. Meola, B. J. Cook, and P. A. Langley

USDA, ARS, VTERL

P.O. Drawer GE, College Station, TX 77841

An ultrastructural study of the heart of the stable fly, Stomoxys calcitrans (Meola and Cook, 1986) revealed that the pericardial sinus (PS) associated with this heart is a neurohemal site for the release of products of the segmental nerves. This neurohemal area extends from the metathoracic apodeme to that of the 5th abdominal segment and is bounded dorsally by the heart (Ht) and ventrally by the pericardial septum (dorsal diaphram) (Figs. 1, 2). A study was initiated to determine whether this neurohemal area was present in other cyclorrhaphans. Eight species consisting of four families were prepared for ultrastructural analysis. In addition to S. calcitrans, four other species of muscids were studied: Haematobia irritans, Musca domestica, Glossina morsitans and Orthellia caesarion. Other families included Sarcophagidae (Ravinia derelicta), Calliphoridae (Lucilia cuprina), and Drosophilidae (Drosophila melanogaster).

The structure of the pericardial septum of all eight species consisted of a central band of longitudinal muscle (LM) upon which the alary muscles (AM) attach (Fig. 1). The fibers of the LM and AM are fenestrated (loosely connected by dense stroma). Based on the variety of species used in this study, this type of septum is unique to the Cyclorrhapha. In all other insects studied thus far the pericardial septum is composed solely of alary muscles.

447

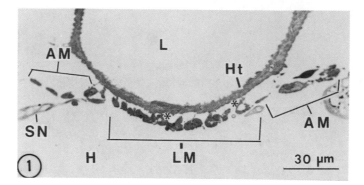

Fig. 1. Transverse section through the heart (Ht) and pericardial septum bounding the pericardial sinus (*) of a
muscid, S. calcitrans. The septum is composed of a longitudinal muscle (LM) and paired alary muscle (AM). L, lumen
of heart; SN, segmental nerve; H, hemocoel.

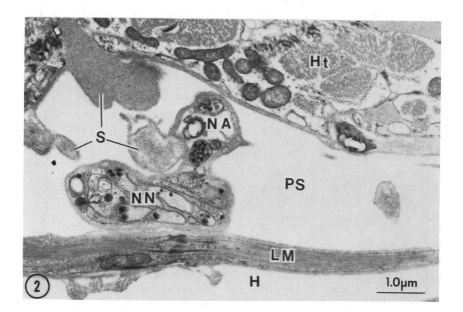

Fig. 2. Sagittal section through pericardial sinus (PS) of
a stable fly. A large neurosecretory nerve (NN) and axon
terminals (NA) lie between the heart (Ht) and the longitudinal muscle (LM). H, hemocoel.

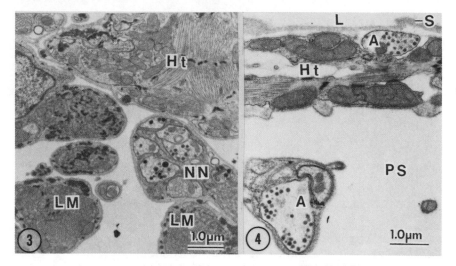

Fig. 3. Axons of a large nerve (NN) terminating in the
pericardial sinus (PS) of the tsetse fly. LM, longitudinal
muscle; Ht, heart.
Fig. 4. Transverse section through the PS of R. derelicta
shows neurosecretory axons (A) terminating beneth the
stroma (S) on the luminal surface of the heart (HT) as well
as in the pericardial sinus (PS). L, lumen of heart.

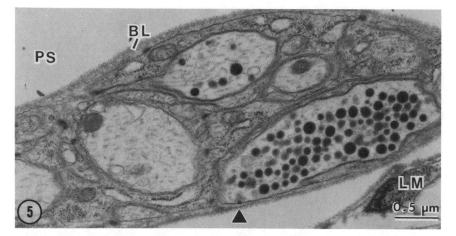

Fig. 5. A large neurosecretory nerve forming terminals in
the pericardial sinus (PS) of a calliphorid, L. cuprina.
Triangle, release site containing synaptoid vescicles; BL,
basal lamina; LM, Longitudinal muscle.

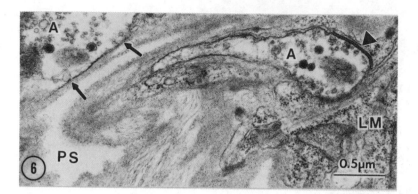

Fig. 6. Neurosecretory axons (A) in the pericardial sinus
of the horn fly H. irritans form neurosecretomotor
junctions (triangle) with the longitudinal muscle (LM) as
well as release products into the hemolymph of the sinus
(arrows).

The segmental nerves of all eight species formed
abundant neurohemal terminals in the pericardial sinus
(Figs. 2, 3, 4, 5, 6). The extent of the development of
the neurohemal area varied between the species, depending
upon the number of axons carried by the SN and arborization
of the axons prior to termination. In R. derelicta and G.
morsitans, axon terminals were not only found in the PS but
extended between fibers of the heart, terminating in the
stroma (S) lining the lumen of the vessel (Fig. 4). In
addition to forming neurohemal terminals, the axons of the
SN also formed neurosecretomotor junctions on the
longitudinal muscle of all eight species (Fig. 6).

The neurohormones released in this area may not only
modulate the activity of the heart, but also that of other
tissues in these insects due to the dual contractions of
the heart and longitudinal muscle and the fenestrated
structure of these muscles.

REFERENCES

Meola S. M. and Cook B. J. (1986) Neuroendocrine plexus in
the pericardial septum of the stable fly, Stomoxys
calcitrans (L.) (Diptera: Muscidae). Int. J. Insect
Morphol. Embryol. (In Press)

PARTICIPANTS

IBRAHIM M. ABALIS
Walter Reed Army
    Institute of Res.
Bldg. 40, Rm. B087
Washington, DC   20307
USA

MICHAEL E. ADAMS
Dept. of Entomology
Univ. of California
Riverside, CA   92521
USA

SHALOM W. APPLEBAUM
Faculty of Agriculture
The Hebrew University
P.O. Box 12
Rehovot 76-100
Israel

MARTINE ARPAGAUS
Labortatoire de
    Zoologie
Institut National de
    la Recherche
    Agronomique
Ecole Normale Superieure
46 rue d'Ulm
75 230 Paris Cedex 05
France

ROBERT A. BELL
Inect Reprod. Lab.,
    USDA
Rm. 206A, Bldg. 306,
    ARC-E
Beltsville, MD   20705
USA

ISABEL BERMUDEZ
School of Biological
    Sciences
Thames Polytechnic
Wellington Street
London SE18 6PF
United Kingdom

MARK J. BIRNBAUM
Dept. of Biology
Univ. of North Carolina
Wilson Hall 046A
Chapel Hill, NC   27514
USA

STEPHEN T. BISHOFF
Dept. of Biology
Univ. of North Carolina
403A Smith Avenue
Chapel Hill, NC   27514
USA

ALEXEJ B. BORKOVEC
Insect Reprod. Lab., USDA
Rm. 323, Bldg. 306, ARC-E
Beltsville, MD   20705   USA

M. F. BOWEN
SRI International
333 Ravenswood Avenue
Menlo Park, CA   94025
USA

HEINZ BREER
Dept. of Zoophysiology
Univ. of Osnabruck
4500 Osnabruck
Federal Republic of
    Germany

GARY L. BROOKHART
Dept. of Entomology
Purdue Univ.
West Lafayette, IN   47907
USA

MARK R. BROWN
Dept. of Entomology
Univ. of Georgia
Athens, GA   30602   USA

DAVID W. BUSHMAN
Dept. of Entomology
Univ. of Maryland
College Park, MD   20742
USA

PIERRE CARLE
Procida-Roussel UCLAF
CRBA Procida St. Marcel
13360 Cedex 11
Marseille
France

EDWARD L. CARMINES
Hoecht-Roussel Agri-Vet
    Company
Rt. 202-206 North
Somerville, NJ   08502
USA

GRANT M. CARROW
Dept. of Biochemistry
Brandeis Univ.
Waltham, MA   02254   USA

ALISON CHALMERS
Agricultural Res. Div.
American Cyanamid Co.
P.O. Box 400
Princeton, NJ   08540
USA

WENDELL L. COMBEST
Dept. of Biology
Univ. of North Carolina
Chapel Hill, NC   27514
USA

BENJAMIN J. COOK
VTERL, USDA
P.O. Box GE
College Station, TX
    77841   USA

JOSEPH W. CRIM
Dept. of Zoology
Univ. of Georgia
Athens, GA   30602   USA

KENNETH G. DAVEY
York Univ.
4700 Keele Street
North York, Ontario
    M3J 1P3
Canada

NORMAN T. DAVIS
Dept. of Physiology and
    Neurobiology
Univ. of Connecticut
U-42, TLS 416
75 North Eagleville Road
Storrs, CT   06268   USA

ALBERT B. DEMILO
Insect Reprod. Lab., USDA
Rm. 323, Bldg. 306, ARC-E
Beltsville, MD   20705   USA

ELVIRA DOMAN
National Science
    Foundation
1800 G Street, N.W.
Washington, DC   20550
USA

MICHAEL J. DUGGAN
Biochemistry Dept. (4W)
Univ. of Bath
Claverton Down, Bath
Avon, BA2 7AY
United Kingdom

ASH K. DWIVEDY
Dept. of Zoology
Univ. of Cambridge
Downing Street
Cambridge CB2 3EJ
United Kingdom

PETER EVANS
Dept. of Zoology
Univ. of Cambridge
Downing Street
Cambridge CB2 3EJ
United Kingdom

MARK FELDLAUFER
Insect & Nematode Horm.
   Lab., USDA
Bldg. 467, ARC-E
Beltsville, MD   20705
USA

HOWARD W. FESCEMYER
Univ. of Maryland
Baltimore Campus
5401 Wilkens Avenue
Catonsville, MD   21228
USA

THOMAS R. FLANAGAN
Dept. of Biology
Univ. of North Carolina
Coker Hall 010A
Chapel Hill, NC   27514
USA

GERD GÄDE
Institut fur Zoologie IV
Universitat Dusseldorf
Gebaude 26.12/100
UniversitatsstraBe 1
D-4000 Dusseldorf
Federal Republic of
   Germany

DALE B. GELMAN
Insect Reprod. Lab., USDA
Rm. 323, Bldg. 306, ARC-E
Beltsville, MD   20705   USA

JADWIGA GIEBULTOWICZ
Dept. of Zoology
Univ. of Maryland
College Park, MD   20742
USA

LAWRENCE I. GILBERT
Dept. of Biology
Univ. of North Carolina
Wilson Hall 046A
Chapel Hill, NC   27514
USA

GRAHAM J. GOLDSWORTHY
Dept. of Zoology
Univ. of Hull
North Humberside HU6 7RX
United Kingdom

NOELLE A. GRANGER
Dept. of Anatomy
Univ. of North Carolina
Rm. 111 Swing Building
Chapel Hill, NC   27514
USA

WILLIAM G. HAAG
Stauffer Chemical Co.
1200 S. 47th Street
Richmond, CA   94804   USA

HERBERT W. HAINES
Merck, Sharp & Dohma
    Research Lab.
P.O. Box 2000
80T, Rm. 132
Rahway, NJ   07065   USA

FRANK E. HANSON
Univ. of Maryland
Baltimore Campus
5401 Wilkens Avenue
Catonsville, MD   21228
USA

DORA K. HAYES
Livestock Insects Lab.,
    USDA
Rm. 120, Bldg. 307,
    ARC-E
Beltsville, MD   20705
USA

TIMOTHY K. HAYES
Dept. of Entomology
Texas A&M Univ.
College Station, TX
    77843   USA

WILLIAM S. HERMAN
Genetics & Cell Biology
Univ. of Minnesota
St. Paul, MN   55108
USA

JOHN G. HILDEBRANT
Arizona Research
    Laboratories
Div. of Neurobiology
Univ. of Arizona
60 Gould-Simpson
    Science Building
Tucson, AZ   85721   USA

G. MARK HOLMAN
VTERL, USDA
College Station, TX
    77803   USA

CALEB W. HOLYOKE
Agricultural Products
    Department
E. I. DuPont de
    Nemours & Co., Inc.
Bldg. 402/3101
Wilmington, DE   19898

IVAN HUBER
Dept. of Biology
Fairleigh Dickinson Univ.
Madison, WI   07940   USA

S. N. IRVING
Wellcome Res. Laboratories
Ravens Lane, Berkhamsted
    HP4 2DY
United Kingdom

R. ELWYN ISAAC
Dept. of Pure & Applied
    Zoology
Univ. of Leeds
Leeds, LS2 9JT
United Kingdom

HOWARD JAFFE
Livestock Insects Lab.,
    USDA
Rm. 120, Bldg. 307,
    ARC-E
Beltsville, MD   20705
USA

KENT R. JENNINGS
American Cyanamid Co.
P.O. Box 400
Princeton, NJ   08540
USA

DAVY JONES
Dept. of Entomology
Univ. of Kentucky
Lexington, KY   40546
USA

HIROSHI KATAOKA
Zoecon Corporation
975 California Avenue
Palo Alto, CA   94304
USA

LARRY L. KEELEY
Laboratories for
    Invertebrate
    Neuroendocrine
    Research
Dept. of Entomology
Texas A&M Univ.
College Station, TX
    77843
USA

THOMAS J. KELLY
Insect Reprod. Lab.,
    USDA
Rm. 323, Bldg. 306,
    ARC-E
Beltsville, MD   20705
USA

WILLIAM B. KEZER
FMC Corporation
Rt. 1, Box 8
Princeton, NJ   08540
USA

WALDEMAR KLASSEN
USDA, ARS
Rm. 227, Bldg. 003,
    ARC-W
Beltsville, MD   20705
USA

ROBERT M. KRAL
FMC Corporation
Rt. 1, Box 8
Princeton, NJ   08540
USA

CHRISTINE KUKEL
American Cyanamid Co.
P.O. Box 400
Princeton, NJ   08853
USA

CRAIG LAMISON
DuPont Experimental
    Station
E402/5220
Wilmington, DE   19898
USA

ANGELA B. LANGE
Dept. of Zoology
Univ. of Toronto
25 Harbord Street
Toronto, Ontario
    M5S 1A1
Canada

ARDEN O. LEA
Dept. of Entomology
Univ. of Georgia
Athens, GA   30602   USA

JEAN-JACQUES LENOIR-
   ROUSSEAUX
Lab. de Biologie
   Animale
UFR Sciences
Universite Paris 12
94010 Creteil Cedex
France

YAACOV LENSKY
Triwaks Bee Res.
   Center
Hebrew Univ.
76 100 Rehovot
P.O. Box 12
Israel

MARCIA J. LOEB
Insect Reprod. Lab.,
   USDA
Rm. 323, Bldg. 306,
   ARC-E
Beltsville, MD   20705
USA

MICHAEL MA
Dept. of Entomology
Univ. of Maryland
College Park, MD   20742
USA

S. H. P. MADDRELL
Dept. of Zoology
Cambridge Univ.
Downing Street
Cambridge CB2 3EJ
United Kingdom

R. J. MARRESE
Hoechst-Roussel
   Agri-Vet
Route 202-206 North
Somerville, NJ   08876
USA

E. PETER MASLER
Insect Reprod. Lab.,
   USDA
Rm. 323, Bldg. 306,
   ARC-E
Beltsville, MD   20705
USA

SHOGO MATSUMOTO
Dept. of Entomology
Univ. of Georgia
Athens, GA   30602   USA

JULIUS J. MENN
National Program Staff
USDA, ARS
Bldg. 005, ARC-W
Beltsville, MD   20705
USA

ROBERT E. MENZER
Dept. of Entomology
Univ. of Maryland
College Park, MD   20742
USA

SHIRLEE MEOLA
VTERL, USDA
P.O. Box GE
College Station, TX
   77840
USA

NANCY S. MILBURN
Dept. of Biology
Tufts Univ.
Medford, MA   02155
USA

JAMES R. MILLER
Dept. of Entomology
Michigan State Univ.
East Lansing, MI   48824
USA

STEPHEN MORRIS
National Science
    Foundation
1800 G Street, N.W.
Washington, DC   20550
USA

THOMAS M. MOWRY
Dept. of Entomology
Michigan State Univ.
203 Pesticide Res.
    Center
East Lansing, MI   48824
USA

D. MURALEEDHARAN
Dept. of Zoology
Univ. of Kerala
Kariavattom 695 581
India

ISAMU NAKAYAMA
Sumitomo Chemical
    Company, Ltd.
2-1 4-chome
    Takatsukasa
    Takarazuka
Hyogo, 665
Japan

JAMES A. NATHANSON
Dept. of Neurology
Harvard Medican School
Massachusetts General
    Hospital
Boston, MA   02114   USA

KLAUS NAUMANN
Bayer/Pflanzen-
    schutzzentrum
509 Leverkusen-
    Bayerwerk
Federal Republic of
    Germany

JUDD O. NELSON
Dept. of Entomology
Univ. of Maryland
College Park, MD   20742
USA

DONALD E. NYE
FMC Corporation
Rt. 1, Box 8
Princeton, NJ   08540
USA

MARTHA O'BRIEN
Dept. of Biology
Univ. of North Carolina
Coker Hall 010A
Chapel Hill, NC   27514
USA

IAN ORCHARD
Dept. of Zoology
Univ. of Toronto
25 Harbord Street
Toronto, Ontario
    M5S 1A1
Canada

MICHAEL O'SHEA
Laboratoire de
    Neurobiologie
Universite de
    Geneve
20 Boulevard D'Yvoy
CH-1211 Geneve 4
Switzerland

MICHAEL D. PAK
Dept. of Biology
Univ. of North Carolina
Wilson Hall 046A
Chapel Hill, NC   27514
USA

THOMAS PANNABECKER
Dept. of Zoology
Univ. of Toronto
25 Harbord Street
Toronto, Ontario
    M5S 1A1
Canada

VASUKI PARMASIVAN
Dept. of Zoology
Kongunadu Arts &
    Science College
Coimbatore 641029
Tamilnadu
India

YVES PICHON
Dept. of Biophysique
C.N.R.S.
F-91190 Gif Sur Yvette
France

JACQUES P. PROUX
Laboratoire de
    Neuroendocrinologie
ERA du CNRS 850
Universite de Bordeaux I
33405 Talence Cedex
France

MARIE RAABE
P. M. Curie Univ.
    Paris & CNRS
12 Rue Clivier
Paris 75005
France

ASHOK K. RAINA
Dept. of Entomology
Univ. of Maryland
College Park, MD   20742
USA

SUSAN M. RANKIN
Dept. of Biology
Univ. of Iowa
Iowa City, IA   52242
USA

BALAKKRISHNA R. RAO
East Stroudsburg Univ.
East Stroudsburg, PA
    18301
USA

STUART E. REYNOLDS
School of Biological
    Sciences
Claverton Down
Bath BA2 7AY
United Kingdom

DAVID C. ROSS
Dept. of Entomology
Univ. of Georgia
Athens, GA   30602   USA

RAYMOND J. RUSSO
Dept. of Biology
Indiana Univ.-Purdue
    Univ. of Indianapolis
1125 East 38th Street
Indianapolis, IN   46223
USA

VINCENT L. SALGADO
Rohm & Haas Company
727 Norristown Road
Spring House, PA   19477
USA

DAVID S. SAUNDERS
  Dept. of Zoology
Univ. of Edinburgh
West Mains Road
Edinburgh EH9 3JT
United Kingdom

MICHAEL E. SCHNEE
Agricultural Products
    Dept.
E. I. DuPont de
    Nemours Experimental
    Station
Rm. 5233, Bldg. 402
Wilmington, DE   19898
USA

DAVID A. SCHOOLEY
Zoecon Corporation
P.O. Box 10975
Palo Alto, CA   94303
USA

HUGO SCHOONEVELD
Dept. of Entomology
Agricultural Univ.
Binnenhaven 7
6709 PD Wageningen
Netherlands

DANIEL SEGAL
Dept. of Neurobiology
Weizmann Institute of
    Science
Rehovot 76-100
Israel

WENDY SMITH
Dept. of Biology
Northeastern Univ.
414 Mugar
Boston, MA   02115   USA

THOMAS SMYTH, JR.
Pennsylvania State Univ.
2 Patterson Bldg.
University Park, PA
    16802
USA

TERRY B. STONE
Monsanto Company
700 Chesterfield
    Village Parkway
Chesterfield, MO   63198
USA

AKINORI SUZUKI
The Univ. of Tokyo
Bunkyo-ku
Tokyo 113
Japan

PAUL H. TAGHERT
Dept. of Anatomy &
    Neurobiology
Washington Univ.
    School of Medicine
Box 8108
660 S. Euclid Avenue
St. Louis, MO   63110
USA

T. THANGARAJ
Dept. of Zoology
Kongunadu Arts and
    Science College
Coimbatore, 641029
India

CHARLES S. THOMPSON
Dept. of Zoology
Univ. of Toronto
25 Harbord Street
Toronto, Ontario
    M5S 1A1
Canada

B. S. THYAGARAJA
RSRS, Central Silk
    Board
Paramahamsa Road
Kollegal, Karnataka
India

STEPHEN S. TOBE
Dept. of Zoology
Univ. of Toronto
25 Harbord Street
Toronto, Ontario
    M5S 1A1
Canada

T. H. TOLBERT
Monsanto Company
800 N. Lindberg
St. Louis, MO   63166
USA

JEAN-PIERRE TOUTANT
Institut de le Recherche
    Agronomique
ENSA-INRA
Station de Physiol.
    Animale
9 Place Viala
34060 Montpellier Cedex
France

PETER VERHAERT
Catholic Univ of
    Leuven
Zoological Institute
Naamsestraat 59
B-3000 Leuven
Belgium

R. D. WATSON
Dept. of Biology
Univ. of North Carolina
Wilson Hall 046A
Chapel Hill, NC   27514
USA

COLIN H. WHEELER
Dept. of Zoology
The Univ. of Hull
North Humberside
    HU6 7RX
United Kingdom

LAVERN R. WHISENTON
Dept. of Biology
Univ. of North Carolina
Coker Hall 010A
Chapel Hill, NC   27514
USA

CARROLL WILLIAMS
The Biological
    Laboratories
Harvard Univ.
16 Divinity Avenue
Cambridge, MA   02138
USA

KEITH D. WING
Research Laboratories
Rohm and Haas Company
727 Norristown Road
Spring House, PA   19477
USA

CHARLES W. WOODS
Insect Reprod. Lab.,
    USDA
Rm. 323, Bldg. 306,
    ARC-E
Beltsville, MD   20705
USA

CHIH-MING YIN
Dept. of Entomology
Univ. of Massachusetts
Amherst, MA   01003   USA

AUTHOR INDEX